Local bifurcation and symmetry

A Vanderbauwhede

Institute for Theoretical Mechanics, State University of Ghent

Local bifurcation and symmetry

Pitman Advanced Publishing Program

BOSTON · LONDON · MELBOURNE

PITMAN BOOKS LIMITED
128 Long Acre, London WC2E 9AN

PITMAN PUBLISHING INC
1020 Plain Street, Marshfield, Massachusetts 02050

Associated Companies
Pitman Publishing Pty Ltd, Melbourne
Pitman Publishing New Zealand Ltd, Wellington
Copp Clark Pitman, Toronto

First published 1982

AMS Subject Classifications: (main) 58E07, 47H15
(subsidiary) 34C25, 35B32, 73H05

Library of Congress Cataloging in Publication Data

Vanderbauwhede, A.
Local bifurcation and symmetry.

(Research notes in mathematics; 75)
"Grew out of a Habilitation thesis . . . presented at the State University of Gent in 1980"—Introd.
Bibliography: p.
Includes index.
1. Differential equations, Nonlinear—Numerical solutions. 2. Differential equations, Partial—Numerical solutions. 3. Bifurcation theory. I. Title. II. Series.
QA372.V34 1982 515.3′5 82-13249
ISBN 0-273-08569-7

British Library Cataloguing in Publication Data

Vanderbauwhede, A.
Local bifurcation and symmetry.—(Research notes in mathematics; 75)
1. Bifurcation theory
I. Title II. Series
515.3′5 QA371

ISBN 0-273-08569-7

ISBN 0 273 08569 7

Reproduced and printed by photolithography
in Great Britain by Biddles Ltd, Guildford

To ANITA, BERT and LEEN

Preface

This book grew out of a Habilitation thesis (in Dutch) which was presented at the State University of Gent in 1980. An English version of this thesis has circulated among a small group of mathematicians working in differential equations and bifurcation theory. It is through the encouragement of several of these colleagues that I have reworked the original text into book form. I would like to express here my sincere gratitude to all those who by their direct or indirect help and support greatly contributed to the accomplishment of a major project like this.

I am especially indebted to Professor R. Mertens, who was the promotor of my thesis, and whose patience, interest and practical advise have been a constant source of encouragement during the past years. Of no less importance were my regular contacts with Professor J. Mawhin of the Institut de Mathématique in Louvain-la-Neuve; I am particularly grateful for his friendship, his stimulating enthusiasm and most of all for his great mathematical skill which I could witness during the numerous seminars he organizes.

My personal interest in bifurcation problems started during the academic year 1976-1977, while I was staying at the Lefschetz Center for Dynamical Systems of Brown University, Providence. I am especially grateful to Professor J.K. Hale who offered me this unique opportunity, for his many advices and for his friendship. His work on bifurcation theory has been very stimulating for my own research, and was an ever present guide while writing this book. I also owe much to all those who provided me during my stay at Brown University with a solid basis for my further work : J. Mallet-Paret, who introduced me to bifurcation theory, D. Henry, T. Banks, F. Kappel, J. Infante, W. Strauss and C. Dafermos for their lectures and seminars on differential equations, P. Taboas for many discussions, and H.M. Rodrigues for the pretty collaboration. Finally I acknowledge the Scientific Committee of NATO for providing the financial support for this stay.

I want to thank N. Chafee, O. Diekmann, S.-N. Chow, D. Chillingworth and M. Golubitsky for a number of discussions which started part of the work contained in this book or which influenced the presentation. I am also grateful to N. Dancer for pointing out an error in the earlier draft of the book.

Finally, I thank Professor K. Kirchgässner and Pitman Publishing for the interest which they took in the publication of this book; my colleagues at the Institute of Theoretical Mechanics, for their moral support; my family for enduring the side-effects; V. Ross for taking care of the figures; and H. Vermis for the extreme care with which he prepared the manuscript.

A. Vanderbauwhede

Gent, May 1982.

Contents

1 Introduction

Many problems in applied mathematics reduce, after the introduction of an appropriate model, to that of solving a set of equations, which in an abstract form can be written as

$$M(x,\lambda) = 0 \ . \tag{1}$$

In this equation x is the unknown of the problem, λ stands for the relevant parameters of the model, and M is a mapping which is in general nonlinear. The basic problem of bifurcation theory is to study how the solutions of (1) change as the parameter λ is changed.

In order to make (1) a mathematically well posed and treatable problem we have to give precise definitions of the spaces X and Λ to which x and λ belong, and of the mapping M. The elements of X will usually be functions of some time and (or) space variables, which have to satisfy certain smoothness, initial value, boundary value or range conditions. Taking these side-conditions into the definition of X and introducing an appropriate topology will generally result in giving X the structure of a manifold, usually an infinite-dimensional submanifold of some function space. Also the parameter λ will belong to a manifold Λ; for practical applications it is usually sufficient to consider finite-dimensional Λ, but since some mathematically interesting problems require an infinite-dimensional parameter space, we will not impose any restrictions on the dimension of Λ. Finally, the mapping M in (1) is defined for $(x,\lambda) \in X \times \Lambda$, and takes its values in a space (or manifold) Z, which in many cases will be an ambient space of X.

The ultimate goal of the theory is of course to give a description, as complete as possible, of the solution set of (1); i.e. of the set

$$S = \{(x,\lambda) \in X \times \Lambda \mid M(x,\lambda) = 0\} \ . \tag{2}$$

More precisely, since we have made the distinction between the unknown x and the parameter λ, we are in fact interested in the λ-sections of S :

$$S_\lambda = \{x \in X \mid (x,\lambda) \in S\} \tag{3}$$

The problem of bifurcation theory is to study how S_λ varies with λ.

The structure of the solution set can be rather wild and chaotic, as follows from the fact that for each closed subset S of $X \times \Lambda$ one can find a continuous M such that the solution set of M coincides with the given subset S. This remark shows that we will have to restrict the class of mappings M in order to come to an efficient treatment. Usually we will impose on M certain smoothness conditions. Another restriction which we impose onto ourselves is that we will make only a local study of S and S_λ, i.e. in a neighbourhood of a given point $(x_0,\lambda_0) \in S$. This is what we mean by "local bifurcation theory". The restriction to a local study has a few important consequences. First, we can use local charts in the manifolds X, Z and Λ, and consequently we may assume that X, Z and Λ are Banach spaces. Second, it will appear that the linearization of M at the point (x_0,λ_0) around which we work will play an important role in the theory.

Before we give some definitions, let us look at a simple but instructive example, namely the eigenvalue problem for a linear operator $A \in L(\mathbb{R}^n)$. We take $X = \mathbb{R}^n$, $\Lambda = \mathbb{R}$ and define $M : \mathbb{R}^n \times \mathbb{R} \to \mathbb{R}^n$ by $M(x,\lambda) = Ax-\lambda x$. We know from elementary algebra that for most values of λ the equation $M(x,\lambda) = 0$ has only the zero solution. It is only when λ equals a real eigenvalue of A that there will be a nontrivial subspace of solutions; at such eigenvalues there is a branch of nontrivial solutions bifurcating from the line $\{0\} \times \mathbb{R} \subset \mathbb{R}^n \times \mathbb{R}$ of zero solutions.

This example shows that the most interesting parameter values are those at which the structure of S_λ changes; such critical parameter values will be called *bifurcation points* (we give a precise definition further on). In the linear example above the bifurcation points are the real eigenvalues of A; in general bifurcation points are parameter values at which certain solutions appear, disappear, coalesce or split up into several branches. In the literature one can find several definitions of a bifurcation point. Classically (see e.g. Krasnoselskii [134], Crandall and Rabinowitz [50], Sattinger [193]) one considers a one-parameter family of equations (i.e. $\Lambda = \mathbb{R}$), and assumes that for each $\lambda \in \mathbb{R}$ a solution $x(\lambda)$ of (1) is given, where $x(\lambda)$ depends continuously on λ. Then bifurcation is defined with respect ot this given curve of solutions.

Here we will use the more general definition given by Chow, Hale and Mallet-Paret ([39],[84]). The idea is to consider all equations near to a given equation in a certain sense, and to study all solutions of these perturbed equations near a given solution of the unperturbed equation. The precise framework in which we will work is as follows.

Let X and Z be real Banach spaces, and $\Omega \subset X$ an open subset. For each integer r, let $C^r(\Omega;Z)$ denote the Banach space of all r-times continuously Fréchet-differentiable mappings $m : \Omega \to Z$ such that

$$\|m\|_r = \sup_{x\in\Omega} \|m(x)\| + \|Dm(x)\| + \ldots + \|D^r m(x)\| \} < \infty . \tag{4}$$

Here D is differentiation, and $\|.\|$ denotes the norm in the appropriate space. Usually we will assume $r \geqslant 1$.

Definition. Let $r \in \mathbb{N}$ and $A \subset C^r(\Omega;Z)$. Let $x_0 \in \Omega$ and $m_0 \in A$ be such that

$$m_0(x_0) = 0 . \tag{5}$$

Then m_0 is a *bifurcation point at* x_0 *with respect to* A if for each neighbourhood U of (m_0,x_0) in $C^r(\Omega;Z)\times X$ we can find (m,x_1) and (m,x_2), both belonging to $U \cap (A\times\Omega)$ and with $x_1 \neq x_2$, such that $m(x_1) = m(x_2) = 0$.

Roughly speaking, m_0 is a bifurcation point at x_0 when there is nonuniqueness of the solutions of $m(x) = 0$ near x_0, for some $m \in A$ near m_0. With this definition in mind one can pose the following problems.

Problem 1. Given $A \subset C^r(\Omega;Z)$ and $(m_0,x_0) \in A\times\Omega$, satisfying (5), show that m_0 is a bifurcation point at x_0 with respect to A.

Problem 2. Given $A \subset C^r(\Omega;Z)$, $(m_0,x_0) \in A\times\Omega$ satisfying (5) and a neighbourhood U of (m_0,x_0) in $C^r(\Omega;Z)\times X$, describe the local solution set $\{(m,x)\in U \mid m\in A,\ m(x) = 0\}$.

It is clear that when we can solve problem 2, then also problem 1 can be solved. In this book we will mainly be interested in problem 2; in particular we will want to find the number of solutions of $m(x) = 0$ with $(m,x)\in U$, for each $m \in A$ close to m_0. When A is an open subset of $C^r(\Omega;Z)$ we call this problem the *generic bifurcation problem*; in the opposite case we speak about a *restricted bifurcation problem* (see [84]).

As an example of a restricted bifurcation problem, let ω be an open subset of a Banach space Λ, $\eta : \omega \to C^r(\Omega;Z)$ a continuous mapping, and $A = \{\eta(\lambda) \mid \lambda \in \omega\}$. Then the corresponding restricted bifurcation problem amounts to the study of S_λ for $\lambda \in \omega$, where S_λ is defined by (3), with $M : \Omega \times \omega \to Z$ defined by $M(x,\lambda) = \eta(\lambda)(x)$. A point $\lambda_0 \in \omega$ will be a bifurcation point for (1) when there is an $x_0 \in S_{\lambda_0}$ such that for each neighbourhood U of (x_0,λ_0) in $X \times \Lambda$ we can find $(x_1,\lambda) \in U$ and $(x_2,\lambda) \in U$ such that $x_1 \neq x_2$ and $M(x_1,\lambda) = M(x_2,\lambda) = 0$. In case one has given a solution branch of (1), i.e. a continuous map $\tilde{x} : \omega \to \Omega$ such that $M(\tilde{x}(\lambda),\lambda) = 0$ for all $\lambda \in \omega$, then $\lambda_0 \in \omega$ will be a bifurcation point at $x_0 = \tilde{x}(\lambda_0)$ if for each neighbourhood U of (x_0,λ_0) in $X \times \Lambda$ one can find $(x,\lambda) \in U$ with $x \neq \tilde{x}(\lambda)$ and $M(x,\lambda) = 0$. By a simple translation this can be reduced to the special case where $\tilde{x}(\lambda) = 0$, $\forall \lambda \in \omega$; then we speak about "bifurcation from the trivial solution".

All restricted bifurcation problems considered in this book will be of the type described above, and will be written down in the form (1). An important remark that can be made at this point is that by taking $\Lambda = C^r(\Omega;Z)$ and by defining $M(x,m) = m(x)$, $\forall x \in \Omega$, $\forall m \in C^r(\Omega;Z)$, the bifurcation problem for (1) coincides with the generic bifurcation problem. Therefore our formalism allows us to handle both generic and restricted problems.

Now suppose that $M(x,\lambda)$ in (1) is continuously differentiable in the variable x. Let $(x_0,\lambda_0) \in X \times \Lambda$ be such that $M(x_0,\lambda_0) = 0$ while $D_xM(x_0,\lambda_0) \in L(X,Z)$ is an isomorphism. Then it is an immediate consequence of the implicit function theorem (see section 2.1) that in a neighbourhood $V \times W$ of (x_0,λ_0) in $X \times \Lambda$ equation (1) will have a unique branch of solutions of the form $\{(x^*(\lambda), \lambda) \mid \lambda \in W\}$. This implies that λ_0 is not a bifurcation point at x_0. For this reason we will restrict our attention to neighbourhoods of such solutions (x_0,λ_0) of (1) at which $D_xM(x_0,\lambda_0)$ is not an isomorphism.

In fact we will ask more about $L = D_xM(x_0,\lambda_0)$ than just that it is not an isomorphism. One of the main hypotheses used throughout the book will be that L is a Fredholm operator; this means that it has a finite-dimensional kernel, while its range is closed and has a finite codimension. This hypothesis will allow us to apply the so-called Liapunov-Schmidt reduction method. This method is based on a splitting of the spaces X and Z, and uses the implicit function theorem to solve part of the equations. The outcome of the method is that there exists a one-to-one correspondance between the solutions of (1) near (x_0,λ_0), and the solutions near $(0,\lambda_0)$ of a similar equation :

$$F(u,\lambda) = 0 \; . \tag{6}$$

Here F is a mapping from ker $L \times \Lambda$ into a complement of the range of L in Z, $F(0,\lambda_0) = 0$ and $D_uF(0,\lambda_0) = 0$. This means that for each λ near λ_0 (6) forms a finite set of scalar equations in a finite number of unknowns (the components of u). So the Liapunov-Schmidt method, when it can be applied, reduces the infinite-dimensional problem (1) to the finite-dimensional problem (6). Several methods have been used to study the bifurcation equation (6) : the Newton polygon, rescaling techniques, the implicit function theorem, or combinations of those. Also topological methods, such as the topological degree or the Liusternik-Schnirelman method can sometimes successfully be used to prove existence of bifurcating solutions. Our main tool in this book will be the implicit function theorem, sometimes in combination with a rescaling of certain variables; we obtain our results by methods which are "elementary and require only the calculus, the implicit function theorem, and a small amount of geometric intuition" (Hale [87]).

To get some feeling for the problems, let us look at a few very simple examples (Hale [80]), for which the bifurcation equation (6) coincides with the original equation (1). We let $X = Z = \Lambda = \mathbb{R}$, and $M(x,\lambda) = x^2-\lambda$. The solution set S and the number of solutions as depending on the parameter λ are shown in the following figures :

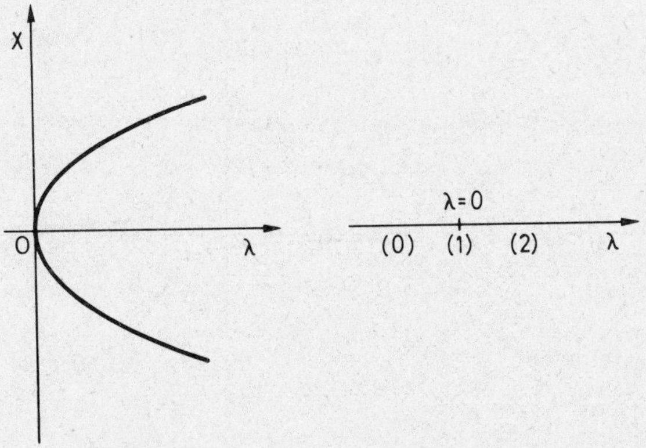

Fig. 1

It is clear that $\lambda = 0$ is a bifurcation point. If we let $M(x,\lambda) = x(x^2-\lambda)$, then $\lambda = 0$ remains the only bifurcation point, and the pictures look as follows :

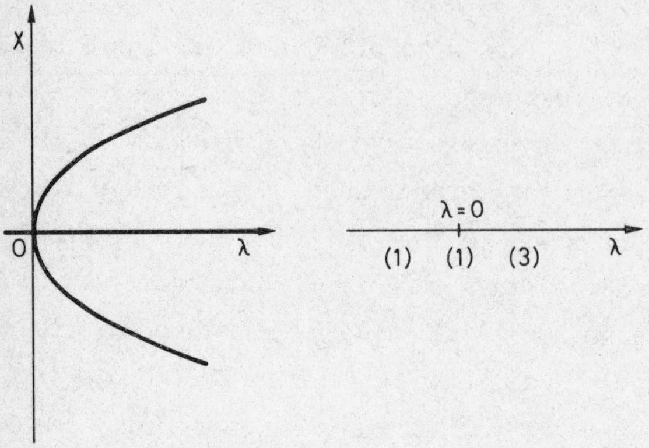

Fig. 2

As a third example, let $X = Z = \mathbb{R}$, $\Lambda = \mathbb{R}^2$ and $M(x,\lambda) = x^3-\lambda_1 x-\lambda_2$. A simple analysis of the graph of the function $x \mapsto M(x,\lambda)$ shows that the equation $M(x,\lambda) = 0$ has exactly one solution when $27\lambda_2^2 > 4\lambda_1^3$, and three distinct solutions when $27\lambda_2^2 < 4\lambda_1^3$. The set of bifurcation points is given by $\{(\lambda_1,\lambda_2)\in\mathbb{R}^2 \mid 27\lambda_2^2 = 4\lambda_1^3\}$, and is represented by the cusp in Fig. 3. These bifurcation points correspond to those values of the parameters for which the equation has a double or a triple solution. Restricting attention to parameters values $\lambda = (\lambda_1,0)$, we obtain again the picture of Fig. 2. If we keep λ_2 fixed at a value different from zero (say $\lambda_2 > 0$); then the solution set looks as in Fig. 4.

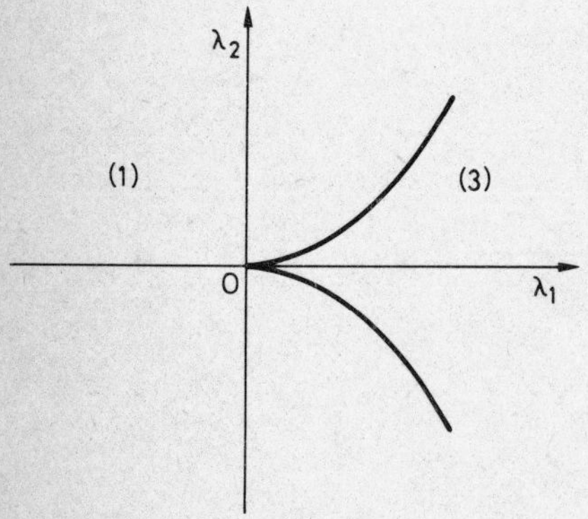

Fig. 3

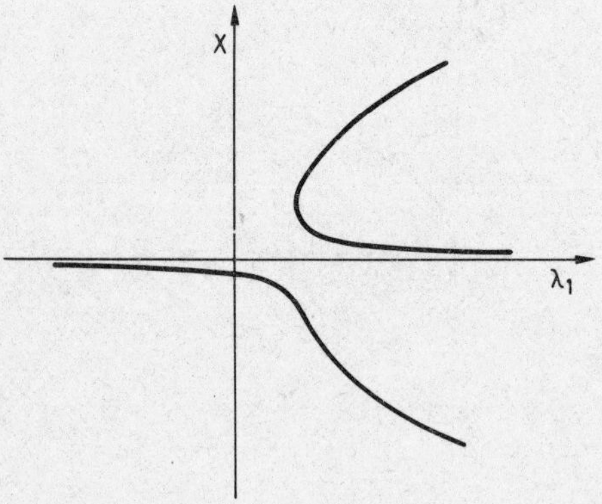

Fig. 4

Let us now return to a generic bifurcation problem, of the form $M(x,m) = m(x) = 0$, where $m \in C^r(\Omega;Z)$ (r sufficiently large) and $x \in \Omega \subset X$. Let $(m_0,x_0) \in C^r(\Omega;Z) \times \Omega$ be such that $L = Dm_0(x_0)$ is a Fredholm operator, with dim ker L = codim R(L) = 1. Then the bifurcation equation (6) takes the form $F(u,m) = 0$, where for each m near m_0, $F(u,m)$ is a scalar function of the scalar variable u. From the general properties of bifurcation functions it follows that

$$F(u,m_0) = cu^k + 0(|u|^k) \quad , \quad \text{as } |u| \to 0 \ , \tag{7}$$

for some $k \geq 2$, and with $c \neq 0$. In a later chapter we will prove that if $k = 2$ in (7), then the bifurcation set is a submanifold of $C^r(\Omega;Z)$, passing through m_0 and having codimension one. For m at one side of the manifold the equation $m(x) = 0$ has no solutions near x_0, for m at the other side there are two solutions; on the manifold itself there is a double solution. When we consider a one-parameter restricted problem, with $\eta : \mathbb{R} \to C^r(\Omega;Z)$ such that $\eta(0) = m_0$ and such that η is transversal to the manifold of bifurcation points, then the restriction of the generic problem to the one-dimensional path described by η gives us precisely a picture as in Fig. 1.

If $k = 3$ in (7), then the bifurcation set is formed by two submanifolds of codimension 1, forming a cusp along their common boundary, which is itself a submanifold of codimension 2. Inside the cusp there are 3 solutions, outside the cusp there is only one solution. When we consider a two-dimensional restricted problem such that the corresponding η is transversal to this generic bifurcation set, then we obtain a bifurcation picture as in Fig. 3. The bifurcation picture of Fig. 2 forms a non-transversal restriction of such generic bifurcation problem.

In general we will call a restricted bifurcation problem which is transversal to the corresponding generic bifurcation set a *generic restricted bifurcation problem*. For such restricted problem the bifurcation set will have the same features as the corresponding generic bifurcation set. For a restricted problem to be generic certain transversality conditions have to be satisfied; we will refer to these conditions as *generic conditions*. They will be generically (i.e. almost always) satisfied if the number of parameters of the restriction is large enough.

Readers who have some acquaintance with singularity theory (see e.g. Golubitsky and Guillemin [72], Brocker [25-26], Martinet [261], Poston and

Stewart [176]) will recognize in the foregoing discussion a number of ideas which appear also in this singularity theory. Indeed, the basic ideas of generic bifurcation theory were very much influenced by the approach of singularity theory, and in recent years there has been an increasing interference between both theories (see e.g. Marsden [155]). Singularity theory was even at the basis of a new area of research in bifurcation theory, which started with the basic papers of Golubitsky and Schaeffer [74-75]; in this work singularity theory is used to classify perturbations of given bifurcation problems. In this book we will not use the methods of singularity theory, although it is undeniable that some ideas from this theory have influenced our presentation.

Now let us come to the main subject of this book, which is the influence of the symmetry properties of the mapping M on its bifurcation properties. In many practical applications one obtains equations which exhibit a certain degree of symmetry, which is due either to some simplifications used in the model, or to some basic symmetry of the problem at hand. Mathematically, this symmetry will be expressed by means of commutation relations between M and a group of symmetry operators; this will give rise to a class of so-called equivariant operators. It is our goal to develop a systematic approach to bifurcation problems for such equivariant operators.

It is clear that the equivariance of M will be reflected into a corresponding symmetry of the solution set S. Indeed, S will be a union of orbits of solutions, each orbit being generated from any of its members by application of the symmetry operators. This forces us to reformulate our basic bifurcation problem, and to consider bifurcation near orbits of solutions. This modified problem will reduce to the classical bifurcation problem if the orbit reduces to one single point, i.e. if we work near a solution which is invariant under the symmetry operators.But then also the linearization at such solution will have the full symmetry of the problem, which in general results in a higher dimensional solution space for this linearized problem. Via the Liapunov-Schmidt method this will result in a higher-dimensional, and therefore more difficult, set of bifurcation equations (6). Fortunately, if the Liapunov-Schmidt reduction is performed in an appropriate way, then also the bifurcation function $F(u,\lambda)$ will be equivariant, and this may compensate to some extent the difficulties arising from the higher dimensions. Group representation theory will play an important role in the analysis of these equivariant bifurcation equations.

Another remark is that equivariant bifurcation problem are in general non-generic, in the sense that small perturbations will destroy the symmetry. Some (e.g. Sattinger [196]) even go a step further and suggest that the non-genericity of equivariant equations is at the origin of bifurcation for general systems : from this point of view, bifurcation is "closely related to symmetry breaking and bifurcation points necessarily have a degree of symmetry. (...) Although systems with symmetries are nongeneric, they nevertheless play pivotal roles as bifurcation points" (Marsden [155]).

In the literature one can find many examples of bifurcation results whose proof relies on some symmetry of the equations; many times this fact is somewhat obscured by an ad hoc formulation of the symmetry properties. It is only by a systematic approach that the underlying structures induced by the symmetry can be revealed. The development of such systematic approach within the framework of modern bifurcation theory has been our guiding motive for writing this book. We hope that our approach will also be useful when dealing with other aspects of bifurcation theory which are not treated here.

SURVEY OF THE CONTENTS

Chapter 2 contains most of the technical material needed further in the book. First we review some results from functional analysis, in particular Fredholm operators and the implicit function theorem. Then we give a number of results on three different particular problems which will be used throughout the book to illustrate our abstract results; these three problems are (i) the determination of periodic solutions of periodic or autonomous differential equations (section 2.2), (ii) some elliptic boundary value problems with Dirichlet data (section 2.3) and, (iii) the buckling problem for plates subject to thrust and normal load, as described by the von Kármán equations (section 2.4). In section 2.5 we define group representations over general Banach spaces, and describe a few results on associated projection operators. Section 2.6 contains some basic results on irreducible representations, and describes the irreducible representations of the groups O(2) and O(3).

In *chapter 3* we introduce the Liapunov-Schmidt method (section 3.1), define equivariant mappings (section 3.2) and study the interference between symmetry and the Liapunov-Schmidt method (section 3.3). We show how the bifurcation equations reduce for symmetry-invariant solutions (section 3.4) and apply the theory to reversible periodic systems (section 3.5). A careful analysis of this example reveals some further properties which can easily be generalized and applied to other problems (section 3.6).

In *chapter 4* we obtain some conditions which ensure that all bifurcating solutions of small perturbations of an equivariant equation will have the symmetry of the perturbation. Such result was proved by Hale and Rodrigues for the Duffing equation [88], and later generalized by Rodrigues and the author [182]. A further generalization is given in section 4.2, while the sections 4.3 and 4.4 contain applications to perturbations of problems with respectively O(2) and O(3) symmetry.

Chapter 5 contains a discussion of generic bifurcation in the important case that the bifurcation equation (6) is a scalar equation in a scalar unknown u. As already mentionned in the introduction, the bifurcation behaviour depends highly on the value of k in (7); in section 5.2 we discuss the case $k = 2$, and in section 5.3 and 5.4 the case $k = 3$. In section 5.5 we show how symmetry leads to nongeneric bifurcation. As an application we study in section 5.6 how the symmetry of the normal load may affect the bifurcation behaviour for the buckling problem of a rectangular plate.

In *chapter 6* we turn to the bifurcation problem at multiple eigenvalues for equivariant mappings. The abstract theory is given in section 6.2, and contains as a special case the classical theorem of Crandall and Rabinowitz [50] for bifurcation from a simple eigenvalue (section 6.3). The remaining sections of chapter 6 contain applications to equations with O(2)-symmetry (section 6.4), to the bifurcation of subharmonic solutions (section 6.5) and to a boundary value problem with O(3)-symmetry (section 6.6).

The main application of the results of section 6.2 is discussed in *chapter 7*, where we study bifurcation for problems with SO(2)-symmetry. An important example of such bifurcation problem is the so-called Hopf bifurcation, which describes the bifurcation of periodic solutions from stationary solutions of autonomous differential equations. Section 7.2 contains the general theory for problems with SO(2)-symmetry, section 7.3 discusses genericity under SO(2)-symmetry, and section 7.4 gives a unified treatment of Hopf bifurcation for both ordinary and functional differential equations. In section 7.5 we give some examples of degenerate Hopf bifurcation; one of the results is the Liapunov Center theorem, which follows here from our Hopf bifurcation analysis.

In the final *chapter 8* we study bifurcation near a compact orbit of solutions of the unperturbed equation. In section 8.2 we reduce this problem to a finite-dimensional one, while in section 8.3 we show how the symmetry of the equations may sometimes imply the existence of particular symmetry-invariant solutions. In the final section 8.4 we discuss the bifurcation equations for a particular example, namely that of periodic perturbations of a conservative oscillation equation.

We conclude this survey with a remark on the organisation of the references within the text. When we want to refer to a particular result, say theorem 5.2.5, then this will be done in one of the following ways :

(i) theorem 5, when the reference appears in section 5.2;

(ii) theorem 2.5, when the reference appears in another section of chapter 5;

(iii) theorem 5.2.5, when the reference appears in another chapter.

A similar system will be used for formulas : formula (7) means formula (7) in the same section, formula (3.7) refers to formula (7) of section 3 of the same chapter, and formula (4.3.7) refers to formula (7) of section 4.3.

Bibliographical note. Since some time the literature on bifurcation theory is growing almost exponentially; therefore the next list of references just intends to give a key to this vaste field of research. There are a number of books, lecture notes and conference proceedings which should provide a good introduction to bifurcation theory and its applications; we mention in particular : Chow and Hale [245], Hale [80],[252], Sattinger [193],[265], Pimbley [173], Iooss [101], Stuart [211], Keller and Antman [114], Berger [21], Krasnosel'skii [134], Ize [103], Rabinowitz [164], Amann, Bazley and Kirchgässner [241] and Iooss and Joseph [255]. Further we want to mention the papers by Crandall and Rabinowitz [50], Dancer [55], Sather [187], Stakgold [209], Westreich [235], Chow, Hale and Mallet-Paret [39], Golubitsky and Schaeffer [74], Rabinowitz [180], Keener and Keller [111], Marsden [155] and Sattinger [266].

2 Mathematical preliminaries

It is the aim of this chapter to provide the mathematical background needed for the treatment in the subsequent chapters. We start with some results from functional analysis, such as the Fredholm alternative for compact operators, the Lax-Milgram theorem, the Krein-Rutman theorem, the implicit function theorem and the rank theorem.

In sections 2, 3 and 4 we describe the three basic examples which will be used in later chapters to illustrate the abstract results. These are the determination of periodic solutions of periodic ordinary differential equations (section 2), elliptic boundary value problems on bounded domains in $\mathbb{R}^n$ (section 3), and the von Kármán equations which describe the buckling problem from plate theory (section 4). Most of the results are stated without proofs; however, we have included a number of intermediate results, which may help to give an idea of how the final conclusions can be obtained. We also give ample bibliographical information on more complete treatments.

Sections 5 and 6 give an elementary introduction to group representation theory. In section 5 we introduce the concept of a representation of a group over a linear space, and prove the existence of an equivariant projection on an invariant subspace when the group is compact. In section 6 we prove some results on irreducible representations, and we discuss the examples of the rotation groups in $\mathbb{R}^2$ and $\mathbb{R}^3$.

As a general remark, which holds throughout this book, let us mention that we consider only *real* spaces, except when the contrary is explicitly stated.

2.1. SOME RESULTS FROM FUNCTIONAL ANALYSIS

2.1.1. Definition. Let X be a topological vector space, and M a closed subspace. Then M has a *topological complement in* X if there exists a closed subspace N of X, which is an algebraic complement of M :

$$X = M + N \quad , \quad M \cap N = \{0\} \ , \tag{1}$$

and such that the map $(x_1, x_2) \mapsto x_1 + x_2$ from the topological product $M \times N$ onto X is a homeomorphism. We say that X is the *direct sum* of M and N, and we write:

$$X = M \oplus N \ . \tag{2}$$

2.1.2. Theorem. A closed subspace M of a topological vector space X has a topological complement if and only if there exists a continuous projection P in X such that

$$M = R(P) \ . \tag{3}$$

Then we have $X = M \oplus N$, with $N = \ker P$, while X/N is topologically isomorphic to M. □

2.1.3. Theorem. Let M be a closed subspace of a topological vector space X.

(i) If X is locally convex, and $\dim M < \infty$, then M has a topological complement in X.

(ii) If $\dim(X/M) < \infty$, then M has a topological complement in X; each algebraic complement of M is also a topological complement. □

2.1.4. Theorem. Let M be a closed subspace of a Hilbert space X. Then M has a topological complement in X, and

$$X = M \oplus M^{\perp} \ . \quad \square \tag{4}$$

In the sequel subspaces with a topological complement will mostly be associated with so-called Fredholm operators.

2.1.5. Definition. Let X and Z be two Banach spaces, and $L : X \to Z$ a bounded linear operator. Then L is called a *Fredholm operator* if the following conditions are satisfied :

(i) $\dim \ker L < \infty$;

(ii) $R(L)$ is closed ;

(iii) $\operatorname{codim} R(L) < \infty$.

The element of $\mathbb{Z}$ given by

$$\text{ind } L = \dim \ker L - \operatorname{codim} R(L) \qquad (5)$$

is called the *index* of the Fredholm operator L.

2.1.6. Theorem. Let X and Z be Banach spaces, and $L \in \mathcal{L}(X,Z)$ a Fredhom operator. Then there exist continuous projections $P \in \mathcal{L}(X)$ and $Q \in \mathcal{L}(Z)$ such that

$$R(P) = \ker L \quad , \quad \ker Q = R(L) \; . \qquad (6)$$

Moreover, the restriction of L to ker P has a bounded inverse $K : R(L) \to \ker P$, satisfying :

$$KLx = (I-P)x \quad , \quad PK(I-Q)z = 0 \quad , \quad LK(I-Q)z = (I-Q)z \; ,$$
$$\forall x \in X \quad , \quad \forall z \in Z \; . \qquad (7)$$

P r o o f. The first part follows from the theorems 2 and 3. The restriction of L to ker P is a continuous linear bijection between the Banach spaces ker P and R(L). So it has a bounded inverse K, by the open mapping theorem (see Rudin [184] , theorem 2.11 and corollary 2.12). □

The result of this theorem will be the main tool used in the Liapunov-Schmidt reduction method given in chapter 3. A particular class of Fredholm operators is given by the compact perturbations of the identity.

2.1.7. Definition. Let X and Z be Banach spaces. An operator $L \in \mathcal{L}(X,Z)$ is called *compact* if L transforms bounded subsets of X into relatively compact subsets of Z.

2.1.8. Lemma. Let $L \in \mathcal{L}(X)$ be compact. Then I-L is a Fredholm operator with zero index. □

2.1.9. Lemma. Let $L \in \mathcal{L}(X,Z)$. Then L is compact if and only if $L^* \in \mathcal{L}(Z^*,X^*)$ is compact. If X = Z, then also

$$\dim \ker(I-L) = \dim \ker(I-L^*) . \quad \square$$

2.1.10. Theorem. Let $L \in \mathcal{L}(X)$ be compact, and $\lambda \neq 0$. Then either the linear operator $\lambda I-L$ has a bounded inverse, or λ is an eigenvalue of L with finite multiplicity.
The set of eigenvalues of L is at most countable, and can only have 0 as an accumulation point.
Further, $\lambda \neq 0$ is an eigenvalue of L if and only if λ is an eigenvalue of L .
We also have :

$$R(\lambda I-L) = (\ker(\lambda I-L^*))^{\perp} \tag{8}$$

and

$$R(\lambda I-L^*) = (\ker(\lambda I-L))^{\perp} . \tag{9}$$

Finally, if $\lambda \neq 0$ is an eigenvalue of L, then there exists an integer $k \geq 1$ such that

$$X = \ker(\lambda I-L)^k \oplus R(\lambda I-L)^k . \quad \square \tag{10}$$

2.1.11. Remark. Theorem 2.1.10 contains the so-called *Fredholm alternative* for compact operators. This can be formulated as follows :
Let $L \in \mathcal{L}(X)$ be compact. Then either the equation

$$x-Lx = y \tag{11}$$

has a unique solution x for each $y \in X$, or the homogeneous equation

$$x-Lx = 0 \tag{12}$$

has a nontrivial solution.
In the first case the operator $(I-L)^{-1}$ is bounded, in the second case the equation (11) has a solution if and only if :

$$\langle x^*, y\rangle = 0 \tag{13}$$

for each solution $x^* \in X^*$ of the homogeneous adjoint equation

$$x^* - L^* x^* = 0 . \tag{14}$$

Now we state two basic theorems on representations of bounded linear functionals over Hilbert spaces.

2.1.12. Riesz representation theorem. For every bounded linear functional F on a Hilbert space H, there is a uniquely determined element $f \in H$ such that $F(x) = (x,f)$ for all $x \in H$; also $\|F\| = \|f\|$. □

2.1.13. The Lax-Milgram theorem. Let $B : H \times H \to \mathbb{R}$ be a bounded *coercive* bilinear form over a Hilbert space H, i.e. :

$$|B(x,y)| \leqslant K\|x\|\|y\| , \qquad \forall x,y \in H \tag{15}$$

and

$$B(x,x) \geqslant \nu\|x\|^2 , \qquad \forall x \in H \tag{16}$$

for some $K > 0$ and $\nu > 0$. Then for every bounded linear functional $F \in H^*$, there exists a unique element $f \in H$ such that

$$F(x) = B(x,f) , \qquad \forall x \in H . \tag{17}$$

Moreover :

$$\nu\|f\| \leqslant \|F\| \leqslant K\|f\| . \quad □ \tag{18}$$

We will also need some results from the theory of ordered Banach spaces, more in particular the Krein-Rutman theorem. More details can be found in Krein & Rutman [136], Krasnoselskii [133], and Amann [5].

2.1.14. Definitions. Let X be a real vector space. A subset $P \subset X$ is called a *cone* if the following is satisfied :

(i) $x+y \in P$, $\forall x,y \in P$;

(ii) $\lambda x \in P$, $\forall \lambda \in \mathbb{R}_+$, $\forall x \in P$;

(iii) $P \cap (-P) = 0$.

Such a cone defines an *ordering* in X, as follows :

$$x \leqslant y \text{ if and only if } y-x \in P .$$

P is called the *positive cone* of the ordering. An *ordered Banach space* is a Banach space X together with a closed positive cone P. We will use the notation (X,P) and the shorthand OBS.

The cone P is *generating* (*reproducing*) if $P-P = X$. A sufficient condition for this is that P has a nonempty interior. The cone P is *total* if $\overline{P-P} = X$.

Let (X,P) and (Z,Q) be two OBS. A linear operator $A : X \to Z$ is called *positive* if $A(P) \subset Q$. If P is total, then the set $L_+(X,Z)$ of all continuous positive linear operators from X into Z forms a closed cone in $L(X,Z)$. We give $L(X,Z)$ the ordering induced by the cone $L_+(X,Z)$; so $A \leqslant B$ if and only if $A(x) \leqslant B(x)$ for each $x \in P$.

Let $e \in P\backslash\{0\}$, and $A : X \to X$ a positive linear operator of X into itself. Then A is called e-*positive* if for every $x \in P\backslash\{0\}$ there exists some $n \in \mathbb{N}\backslash\{0\}$, and constants $0 < \alpha \leqslant \beta$ such that

$$\alpha e \leqslant A^n x \leqslant \beta e .$$

The following is a generalization of the classical Krein-Rutman theorem.

2.1.15. Theorem. Let X be an OBS such that P has nonempty interior. Let $A \in L_+(X)$ be compact and e-positive, for some $e \in P\backslash\{0\}$. Let r(A) be the spectral radius of A :

$$r(A) = \lim_{k \to \infty} \|A^k\|^{1/k} .$$

Then the following is true :

(i) $r(A) > 0$;

(ii) r(A) is a simple eigenvalue of A , having a positive eigenvector;

(iii) r(A) is also a simple eigenvalue of A^*, having a strictly positive eigenvector;

(iv) no other eigenvalues of A correspond to positive eigenvectors; they all satisfy $|\lambda| < r(A)$.

We finish this section by stating a version of the implicit function theorem as we will use it further on. It forms the key step at many places in the further theory. The version quoted below is the one given by Rabinowitz in [180]. For the proof, one can adapt the arguments given in the proof of theorem 10.2.1 of Dieudonné [249].

2.1.16. <u>The implicit function theorem</u>. Suppose :

(a) X, Y and Z are three (real) Banach spaces;

(b) f is a continuous mapping of an open subset A of $X\times Y$ into Z;

(c) if we set

$$A_x = \{y \in Y : (x,y) \in A\}$$

then, for each $x \in X$ such that $A_x \neq \phi$, the map $y \mapsto f(x,y)$ of A_x into Z is Fréchet-differentiable in A_x, and the derivative of this map (denoted by $D_y f$) is continuous in A;

(d) $(x_0,y_0) \in A$ is such that $f(x_0,y_0) = 0$ and $D_y f(x_0,y_0)$ is a linear homeomorphism of Y onto Z.

Then there exist neighbourhoods U of x_0 in X and V of y_0 in Y, and a continuous mapping $y^* : U \to V$ such that

(i) $U\times V \subset A$;

(ii) for each $(x,y) \in U\times V$ we have $f(x,y) = 0$ if and only if $y = y^*(x)$.

If moreover the mapping f is k-times continuously differentiable on A, then also y^* is k-times continuously differentiable on U. □

Two results which are closely related to the implicit function theorem are the inverse function theorem and the rank theorem. We state the inverse function theorem, and prove an infinite-dimensional version of the rank theorem.

2.1.17. <u>The inverse function theorem</u>. Let X and Y be Banach spaces, $\Omega \subset X$ open, $x_0 \in \Omega$, and $f : \Omega \to Y$ a mapping of class C^1. Then there exists an open neighbourhood U of x_0 (with $U \subset \Omega$) such that f(U) is open and f is a C^1-diffeomorphism of U onto f(U), if and only if $Df(x_0)$ is a linear homeomorphism of X onto Y. □

2.1.18. <u>The rank theorem</u>. Let X and Y be Banach spaces, $\Omega \subset X$ open, $x_0 \in \Omega$ and $f : \Omega \to Y$ a mapping of class C^1. Suppose that there exist closed subspaces X_1 and Y_1 of respectively X and Y, such that :

(i) $X = \ker Df(x_0) \oplus X_1$;

(ii) $R(Df(x))$ is closed in Y , $\forall x \in \Omega$;

(iii) $Y = R(Df(x)) \oplus Y_1$, $\forall x \in \Omega$.

Then there exist C^1-diffeomorphisms $\eta : U \subset \Omega \to \eta(U) \subset X$ and $\zeta : V \subset Y \to \zeta(V) \subset Y$, defined respectively in a neighbourhood U of x_0 in X and in a neighbourhood V of $f(x_0)$ in Y, such that $\eta(x_0) = 0$, $D\eta(x_0) = I_X$, $\zeta(f(x_0)) = 0$, $D\zeta(f(x_0)) = I_Y$, $f(U) \subset V$ and

$$(\zeta \circ f \circ \eta^{-1})(x) = Df(x_0)x \quad , \quad \forall x \in \eta(U) . \tag{19}$$

P r o o f. We may suppose that $x_0 = 0$ and $f(0) = 0$. Let $L = Df(0)$, and let $P \in L(X)$ and $Q \in L(Y)$ be projections such that $R(P) = \ker L$, $\ker P = X_1$, $\ker Q = R(L)$ and $R(Q) = Y_1$ (see theorem 2). By the same argument as in the proof of theorem 6 the restriction of L to X_1 has a bounded inverse $K : R(L) \to X_1$, satisfying relations similar to (7). Define $\eta : \Omega \to X$ by

$$\eta(x) = Px + K(I-Q)f(x) \quad , \quad \forall x \in \Omega . \tag{20}$$

Since $\eta(0) = 0$ and $D\eta(0) = P + K(I-Q)L = I_X$, the restriction of η to a suitable neighbourhood U of 0 in Ω is a C^1-diffeomorphism. We may suppose that $\eta(U)$ is convex and such that $x \in \eta(U)$ implies $(I-P)x \in \eta(U)$.
Let $f_1 = f \circ \eta^{-1} : \eta(U) \to Y$. Applying L on the identity :

$$x = P\eta^{-1}(x) + K(I-Q)f_1(x) \quad , \quad \forall x \in \eta(U) \tag{21}$$

we find

$$(I-Q)f_1(x) = Lx \quad , \qquad \forall x \in \eta(U) \; . \tag{22}$$

It follows that $Df_1(x).Ph \in Y_1$ for $x \in \eta(U)$ and $h \in X$. Since by hypothesis $R(Df_1(x)) \cap Y_1 = \{0\}$, we conclude that $Df_1(x)Ph = 0$ for $x \in \eta(U)$ and $h \in X$. Then the mean value theorem gives :

$$f_1(x) = f_1((I-P)x) \quad , \qquad \forall x \in \eta(U) \; .$$

Let $V = \{y \in Y | K(I-Q)y \in \eta(U)\}$ and define $\zeta : V \to Y$ by

$$\zeta(y) = y - Qf_1(K(I-Q)y) \quad , \qquad \forall y \in V \; . \tag{23}$$

Since $K(I-Q)\zeta(y) = K(I-Q)y$ it is straightforward to show that ζ is a C^1-diffeomorphism, with $\zeta(V) = V$ and $\zeta^{-1} : V \to V$ given by :

$$\zeta^{-1}(y) = y + Qf_1(K(I-Q)y) \; , \qquad \forall y \in V \; .$$

From (22) we obtain $K(I-Q)f_1(x) = (I-P)x \in \eta(U)$ for $x \in \eta(U)$, so that $f(U) = f_1(\eta(U)) \subset V$. Then (19) follows from a direct calculation. □

2.1.19. Corollary. Suppose that $f : \Omega \subset X \to Y$ is of class C^1 and such that

$$\text{Rk}\, f(x) = \dim R(Df(x)) = k \quad , \qquad \forall x \in \Omega \; , \tag{24}$$

for some $k \in \mathbb{N}$. Then at each point $x_0 \in \Omega$ the conclusion of the rank theorem holds.

P r o o f. We show that the conditions of the rank theorem are satisfied in a neighbourhood $\Omega' \subset \Omega$ of x_0. The existence of X_1 follows from the fact that $\ker Df(x_0)$ has a finite codimension, by (24). Also condition (ii) follows trivially from (24). In order to show (iii), remark that there exists a closed subspace Y_1 of Y such that (iii) is satisfied for $x = x_0$. From (24) we have $\text{codim}\, Y_1 = k$, and it is sufficient to show that $R(Df(x)) \cap Y_1 = \{0\}$ for all x near x_0.
Let P and Q be as in the proof of the rank theorem. Then $(I-Q)Df(x_0)$ is an

isomorphism between X_1 and $R(Df(x_0))$. Consequently $(I-Q)Df(x)$ remains an isomorphism between X_1 and $R(Df(x_0))$ for all x near x_0. So we have necessarily $\dim Df(x)(X_1) \geq k$, and then the hypothesis implies that $Df(x)(X_1) = Df(x)(X)$. If now $y \in Df(x)(X) \cap Y_1$, then $y = Df(x)h$ for some $h \in X_1$. Since $y \in Y_1$ it follows that $(I-Q)Df(x)h = 0$. Because $(I-Q)Df(x)$ is an isomorphism between X_1 and $R(Df(x_0))$, we conclude that $h = 0$ and $y = 0$. This proves the corollary. □

2.1.20. Corollary. Under the conditions of theorem 18 or corollary 19 each point $x_0 \in \Omega$ has a neighbourhood U such that f(U) is a C^1-submanifold of Y. Also $f^{-1}(f(x_0))$ is a C^1-submanifold of Ω for each $x_0 \in \Omega$.

P r o o f. This follows immediately from the conclusion of the rank theorem and the definition of a submanifold (see e.g. Lang [141]). □

2.1.21. Bibliographical notes. The results about topological complements are taken from Köthe [257]. The theory of Fredholm operators, and more in particular of compact operators, can be found in any textbook on functional analysis; one can see for example Rudin [184], Schechter [198], Lang [140], Taylor [215], Yosida [237]; more detailed results can be found in Kato [108] and Dunford and Schwartz [59].

The Lax-Milgram theorem is the basic result used in the modern Hilbert space approach to boundary value problems for elliptic partial differential equations : see e.g. Schechter [199], Showalter [207], Gilbarg and Trudinger [71], Bers, John and Schechter [22], etc...

Finally, the implicit function theorem (and the associated Banach fixed point theorem) form the start of almost any treatment on nonlinear functional analysis. A nice account can be found in Schwartz [203]. Since it is a local result, it is also frequently used in the theory of differentiable manifolds (see e.g. Lang [141], Chillingworth [37]). In the following sections and chapters it will be our unique tool for solving equations.

A different formulation of the rank theorem can be found in Bourbaki [243], while Dieudonné [249] gives a finite-dimensional version of this theorem.

2.2. PERIODIC SOLUTIONS OF PERIODIC LINEAR ORDINARY DIFFERENTIAL EQUATIONS

In this section we review the Fredholm alternative for the periodic solutions of periodic linear ordinary differential equations.

2.2.1. The set-up of the problem. Let us look for 2π-periodic solutions of the homogeneous linear equation

$$\dot{x} = A(t)x \tag{1}$$

and its non-homogeneous counterpart :

$$\dot{x} = A(t)x + f(t) . \tag{2}$$

Here $A : \mathbb{R} \to L(\mathbb{R}^n)$ and $f : \mathbb{R} \to \mathbb{R}^n$ are 2π-periodic and continuous. We will also have the occasion to consider the (formal) adjoint equation

$$-\dot{y} = A^T(t)y . \tag{3}$$

Since our interest will be in 2π-periodic solutions, let us introduce the following Banach spaces :

$$Z = \{x : \mathbb{R} \to \mathbb{R}^n \mid x \text{ is continuous and } 2\pi\text{-periodic}\}$$

and

$$X = \{x \in Z \mid \dot{x} \in Z\}$$

with the usual C^0, respectively C^1, topology :

$$\|z\|_Z = \sup\{\|z(t)\| \mid t \in \mathbb{R}\} ,$$

and

$$\|x\|_X = \sup\{\|x(t)\| + \|\dot{x}(t)\| \mid t \in \mathbb{R}\} .$$

With the equations (1) and (2) we associate the bounded linear operator

$$L : X \to Z \quad , \quad x(.) \mapsto (Lx)(.) = \frac{dx}{dt}(.) - A(.)x(.) \tag{4}$$

and its formal adjoint

$$L^* : X \to Z \ , \ x(.) \mapsto (L^*x)(.) = -\frac{dx}{dt}(.) - A^T(.)x(.) \ . \tag{5}$$

We want to show that L is a Fredholm operator, and give a description of ker L and R(L). In doing so we will use the following bilinear form over Z :

$$<x,y> = \int_0^{2\pi} (x(t),y(t))dt \qquad , \qquad \forall x,y \in Z \tag{6}$$

where (.,.) is the usual scalar product in $\mathbb{R}^n$.

2.2.2. Some elementary results. The right-hand side of (1) and (2) is continuous in (t,x), and, by the linearity in x, also Lipschitz-continuous in x. As a consequence, the Cauchy problem formed by the equation (1) or (2), together with the initial condition $x(t_0) = x_0$, has a *unique* solution for all $(t_0,x_0) \in \mathbb{R}\times\mathbb{R}^n$; this solution exists and is of class C^1 for all $t \in \mathbb{R}$.

This basic existence and uniqueness result has several consequences. First, if we denote by $\Phi(t,t_0)$ the matrix solution of (1) satisfying the initial condition $X(t_0) = I$ (= the identity matrix), then $\Phi(t,t_0)$ is nonsingular for all $t,t_0 \in \mathbb{R}$, and we have the following relations :

$$\begin{aligned} &\Phi(t_0,t_0) = I \qquad\qquad &, \quad &\forall t_0 \in \mathbb{R} \ , \\ &\Phi(t,t_1)\Phi(t_1,t_0) = \Phi(t,t_0) &, \quad &\forall t,t_0,t_1 \in \mathbb{R} \ . \end{aligned} \tag{7}$$

$\Phi : \mathbb{R}\times\mathbb{R} \to L(\mathbb{R}^n)$ is called the *transition matrix* for equation (1). Indeed, the solution of (1) satisfying the initial condition $x(t_0) = x_0$ is given by

$$x(t;t_0,x_0) = \Phi(t,t_0)x_0 \ ; \tag{8}$$

the analogous solution of (2) is given by

$$x(t;t_0,x_0) = \Phi(t,t_0)x_0 + \int_{t_0}^{t} \Phi(t,s)f(s)ds \ ; \tag{9}$$

this is the so-called *variation-of-constants formula.*

The adjoint equation (3) has similar properties; its transition matrix is given by

$$\Psi(t,t_0) = \Phi^T(t_0,t) \ . \tag{10}$$

2.2.3. Periodic solutions. Another consequence of the uniqueness result is the following criterion for periodicity of solutions : *a solution* $x(t)$ *of (1) or (2) is* 2π*-periodic if and only if* $x(0) = x(2\pi)$. Using (8) this gives the following condition for the initial value of a periodic solution of (1) :

$$x_0 = Cx_0 \ ,$$

where $C \equiv \Phi(2\pi,0)$ is a *monodromy-matrix* for (1). We conclude that

$$\ker L = \{x(.) = \Phi(.,0)x_0 \mid x_0 \in \ker(I-C)\} \ . \tag{11}$$

Since the monodromy-matrix for the adjoint equation (3) is given by

$$C^* = \Phi^T(0,2\pi) = (C^T)^{-1} \tag{12}$$

we also have

$$\begin{aligned} \ker L^* &= \{y(.) = \Phi^T(0,.)y_0 \mid y_0 \in \ker(I-C^*)\} \\ &= \{y(.) = \Phi^T(0,.)y_0 \mid y_0 \in \ker(I-C^T)\} \ . \end{aligned} \tag{13}$$

It follows that

$$\dim \ker L = \dim \ker L^* \leqslant n \ . \tag{14}$$

For the non-homogeneous equation (2) the periodicity condition can be obtained using the variation-of-constants formula. We find :

$$x_0 = Cx_0 + \int_0^{2\pi} \Phi(2\pi,s)f(s)ds \ . \tag{15}$$

The equation (15) has a solution $x_0 \in \mathbb{R}^n$ if and only if :

$$(y_0, \int_0^{2\pi} \Phi(2\pi,s)f(s)ds) = 0 \quad , \quad \forall y_0 \in \ker(I-C^T) \ . \tag{16}$$

Writing $\Phi(2\pi,s) = \Phi(2\pi,0)\Phi(0,s) = C\Phi(0,s)$, and observing that $C^T y_0 = y_0$ for $y_0 \in \ker(I-C^T)$, condition (16) is equivalent to :

$$(y_0, \int_0^{2\pi} \Phi(0,s)f(s)ds) = 0 \quad , \quad \forall y_0 \in \ker(I-C^T)$$

or

$$\int_0^{2\pi} (\Phi^T(0,s)y_0, f(s))ds = 0 \quad , \quad \forall y_0 \in \ker(I-C^T)$$

or still, using the characterization (13) of $\ker L^*$:

$$\langle y,f \rangle = 0 \quad , \quad \forall y \in \ker L^* \ . \tag{17}$$

So (17) is a necessary and sufficient condition for f to belong to R(L). From (14) we can then conclude that L is a Fredholm operator with zero index.

2.2.4. We can also give an explicit form for the projections $P \in L(X)$, $Q \in L(Z)$ and for the operator $K : R(L) \to \ker P$ given by theorem 1.6. Let $\dim \ker L = \dim \ker L^* = p \leq n$, and let

$$\{\phi_1,\ldots,\phi_p\} \quad , \quad \text{respectively } \{\psi_1,\ldots,\psi_p\}$$

be bases for ker L, respectively ker L^*. Since $\langle .,. \rangle$ is a positive definite symmetric bilinear form over Z (and so also over X), we may assume that $\{\phi_i | i=1,\ldots,p\}$ and $\{\psi_i | i=1,\ldots,p\}$ form orthonormal sets with respect to this form :

$$\langle \phi_i,\phi_j \rangle = \langle \psi_i,\psi_j \rangle = \delta_{ij} \quad , \quad \forall i,j = 1,\ldots,p \ . \tag{18}$$

Then we can define P and Q as follows :

$$Px = \sum_{i=1}^{p} \langle x,\phi_i \rangle \phi_i \quad , \quad \forall x \in X \tag{19}$$

and

$$Qz = \sum_{i=1}^{p} <z,\psi_i>\psi_i \qquad , \qquad \forall z \in Z . \tag{20}$$

It is easily checked that P and Q satisfy the necessary conditions. The pseudo-inverse K of L takes the form :

$$Kf = (I-P)[\Phi(.,0)x_0 + \int_0^{\cdot} \Phi(.,s)f(s)ds] \tag{21}$$

where $x_0 = x_0(f)$ is any solution of

$$(I-C)x_0 = \int_0^{2\pi} \Phi(2\pi,s)f(s)ds . \tag{22}$$

2.2.5. <u>Example</u>. Consider the second order scalar equation :

$$\ddot{x} + x = f(t) \tag{23}$$

where f(t) is 2π-periodic and continuous. Taking Z as before, and $X = \{x \in Z \mid \dot{x} \in Z, \ddot{x} \in Z\}$ we can associate with (23) the linear operator

$$L : X \to Z, x(.) \mapsto (Lx)(.) = \ddot{x}(.) + x(.) . \tag{24}$$

Rewriting (23) as two scalar equations of first order, it is immediate to see that :

$$\ker L = \ker L^* = \{a\cos(.) + b\sin(.) \mid (a,b) \in \mathbb{R}^2\} . \tag{25}$$

(Indeed, the homogeneous counterpart of (23), $\ddot{x}+x = 0$, is formally self-adjoint). We also have :

$$R(L) = \{f \in Z \mid \int_0^{2\pi} f(s)\cos s\, ds = \int_0^{2\pi} f(s)\sin s\, ds = 0 \} . \tag{26}$$

In the definition of P and Q we can take :

$$\phi_1(t) = \psi_1(t) = \frac{1}{\sqrt{\pi}} \cos t \quad , \quad \phi_2(t) = \psi_2(t) = \frac{1}{\sqrt{\pi}} \sin t . \tag{27}$$

2.2.6. Example. Our second example is connected with the conservative second order equation

$$\ddot{x} + g(x) = 0 \tag{28}$$

where $g : \mathbb{R} \to \mathbb{R}$ is a C^1-function. Let X and Z be as in the previous example, and define the nonlinear operator

$$M_0 : X \to Z,\ x(.) \mapsto M_0(x)(.) = \ddot{x}(.) + g(x(.)) \ . \tag{29}$$

Then M_0 is continuously (Fréchet)-differentiable. Let now $x_0 \in X$ be any non-constant 2π-periodic solution of (28), i.e. :

$$M_0(x_0) = 0 \tag{30}$$

and let

$$L = DM_0(x_0) \ . \tag{31}$$

Explicitly, L takes the form

$$(Lh)(t) = \ddot{h}(t) + g'(x_0(t))h(t) \quad , \quad \forall t \in \mathbb{R} \ , \forall h \in X \ ; \tag{32}$$

that means, the equation $Lh = 0$ is precisely the variational equation of (28) at the solution x_0. We want to show that L is a Fredholm operator, and determine its kernel and its image.

2.2.7. First, let us remark that when $x(t)$ is any solution of (28), then so is $x(t+\theta)$, for each $\theta \in \mathbb{R}$. Defining :

$$x_{0,M} = \max_{x\in\mathbb{R}} x_0(t) \quad , \quad x_{0,m} = \min_{t\in\mathbb{R}} x_0(t) \tag{33}$$

we may, by an appropriate phase shift, assume that $x_0(0) = x_{0,M}$. Also, the equation (28) has a first integral :

$$H(x,\dot{x}) = \frac{1}{2}\dot{x}^2 + \int_0^x g(s)ds = \frac{1}{2}\dot{x}^2 + G(x) \ . \tag{34}$$

If we set

$$h_0 = \frac{1}{2}\dot{x}_0^2(t) + G(x_0(t)) = H(x_0(t),\dot{x}_0(t)) , \tag{35}$$

then it is easy to see that :

$$h_0 = G(x_{0,M}) = G(x_{0,m}) \tag{36}$$

and

$$G'(x_{0,M}) = g(x_{0,M}) > 0 \tag{37}$$

$$G'(x_{0,m}) = g(x_{0,m}) < 0 \tag{38}$$

while $G(x) < h_0$ for $x \in]x_{0,m},x_{0,M}[$. By (36)-(38) and the implicit function theorem it follows that there are C^2-functions $x_M(h)$ and $x_m(h)$, defined for $h \in]h_0-\delta,h_0+\delta[$, ($\delta > 0$ sufficiently small), such that :

$$\begin{aligned} &(i) \quad G(x_M(h)) = G(x_m(h)) = h \\ &(ii) \quad x_M(h_0) = x_{0,M} \ ; \ x_m(h_0) = x_{0,m} \\ &(iii) \ G(x) < h \text{ for } x \in]x_m(h),x_M(h)[. \end{aligned} \tag{39}$$

If we let $\bar{x}(h)(t)$ be the solution of (28) satisfying the initial conditions :

$$\bar{x}(h)(0) = x_M(h) \quad , \quad \dot{\bar{x}}(h)(0) = 0$$

then $\bar{x}(h)(t)$ is continuously differentiable in h, and it represents a periodic solution of (28). Its least period is given by

$$T(h) = 2\int_{x_m(h)}^{x_M(h)} [\,2(h-G(x))\,]^{-1/2}dx . \tag{40}$$

Let

$$\omega(h) = \frac{2\pi}{T(h)} \tag{41}$$

and define $\tilde{x}(h) \in X$ by :

$$\bar{x}(h)(t) = \tilde{x}(h)(\omega(h)t) \quad , \quad \forall t \in \mathbb{R} \ , \ \forall h \in]h_0-\delta, h_0+\delta[\ . \tag{42}$$

Then the maps $h \mapsto \omega(h)$ and $h \to \tilde{x}(h)$ are continuously differentiable in $]h_0-\delta, h_0+\delta[$, and

$$\begin{aligned} &\omega(h_0) = k \\ &\bar{x}(h_0)(t) = \tilde{x}(h_0)(kt) = x_0(t) \quad , \qquad \forall t \in \mathbb{R} \end{aligned} \tag{43}$$

for some $k \in \mathbb{N}\setminus\{0\}$.

2.2.8. The foregoing gives us a two-parameter family of solutions of (28), of the form :

$$x(t;h,\theta) = \bar{x}(h)(t+\theta) = \tilde{x}(h)(\omega(h)(t+\theta)) \ .$$

Derivation in the parameters gives us two linearly independent solutions of the variational equation, namely :

$$y_1(t) = \frac{\partial}{\partial h} x(t;h,\theta)\big|_{h=h_0,\theta=0}$$

and

$$y_2(t) = \frac{\partial}{\partial \theta} x(t;h,\theta)\big|_{h=h_0,\theta=0} = \dot{x}_0(t) \ .$$

If we assume that

$$\frac{dT}{dh}(h_0) \neq 0 \tag{44}$$

then $y_1(t)$ is non-periodic. We conclude that under the condition (44) :

$$\ker L = \{\alpha\dot{x}_0 \mid \alpha\in\mathbb{R}\} \ . \tag{45}$$

Rewriting the nonhomogeneous equation

$$\ddot{x} + g'(x_0(t))x = f(t) \tag{46}$$

as a system of two first order equations, and applying the preceding theory, one finds that

$$R(L) = \{f \in Z \mid \int_0^{2\pi} f(s)\dot{x}_0(s)ds = 0\} . \tag{47}$$

Finally, we can take for the projections P and Q :

$$P = Q|_X$$

and

$$Qz = \alpha^{-1}\dot{x}_0 \int_0^{2\pi} f(s)\dot{x}_0(s)ds \quad , \quad \forall z \in Z \tag{48}$$

where

$$\alpha = \int_0^{2\pi} \dot{x}_0^2(s)ds . \tag{49}$$

We will use these results in chapter 8.

2.2.9. Bibliographical notes. The main results of this section can be found in any of the numerous textbooks on ordinary differential equations, for example Hale [77], Coddington and Levinson [46], Rouche and Mawhin [183], Knobloch and Kappel [129], etc.

2.3. ELLIPTIC PARTIAL DIFFERENTIAL OPERATORS

In this section we collect some results on boundary value problems for second order elliptic partial differential operators. In particular, we will pay attention to the Dirichlet boundary value problem for the Laplacian. More details and proofs can be found e.g. in Friedman [65],[66], Bers, John and Schechter [22], Gilbarg and Trudinger [71], Folland [64], Miranda [167], Agmon [3], Agmon, Douglis and Nirenberg [2]. We remark that all functions considered are real-valued.

2.3.1. Definition. Let $\Omega \in \mathbb{R}^n$ be an open *bounded* domain. (We assume $n \geq 2$). Consider the linear differential operator

$$L = \sum_{i,j=1}^{n} a_{ij}(x)\frac{\partial^2}{\partial x_i \partial x_j} + \sum_{i=1}^{n} b_i(x)\frac{\partial}{\partial x_i} + c(x) \tag{1}$$

where $a_{ij}, b_i, c \in C^0(\bar{\Omega})$, and $a_{ij} = a_{ji}$. We say that L is *elliptic* when, at each point $x \in \Omega$, we have :

$$\sum_{i,j=1}^{n} a_{ij}(x)\xi_i\xi_j > 0 \quad , \qquad \forall \xi \in \mathbb{R}^n \setminus \{0\} . \tag{2}$$

We say that L is *strictly elliptic* when there exists a constant $\lambda_0 > 0$ such that :

$$\sum_{i,j=1}^{n} a_{ij}(x)\xi_i\xi_j \geqslant \lambda_0 |\xi|^2 \quad , \qquad \forall x \in \Omega \ , \ \forall \xi \in \mathbb{R}^n . \tag{3}$$

In fact, under the conditions imposed on the coefficients of L, this implies that L is *uniformly elliptic*.

2.3.2. Theorem (Weak maximum principle). Let L be elliptic, with $c(x) = 0$, $\forall x \in \Omega$. Suppose that $u \in C^2(\Omega) \cap C^0(\bar{\Omega})$, and that

$$Lu(x) \geqslant 0 \quad , \qquad \forall x \in \Omega . \tag{4}$$

Then the maximum of u in $\bar{\Omega}$ is achieved on $\partial\Omega$, that is :

$$\sup_{x \in \Omega} u(x) = \sup_{x \in \partial\Omega} u(x) . \qquad \square \tag{5}$$

2.3.3. Corollary. Let L be elliptic, and $c(x) \leqslant 0$ for all $x \in \Omega$. Suppose that $u \in C^2(\Omega) \cap C^0(\bar{\Omega})$ and $Lu(x) \geqslant 0$, $\forall x \in \Omega$. Then :

$$\sup_{x \in \Omega} u(x) \leqslant \sup_{x \in \partial\Omega} u^+(x) \tag{6}$$

where $u^+(x) = \max(u(x),0)$. If $Lu(x) = 0$, $\forall x \in \Omega$, then

$$\sup_{x \in \Omega} |u(x)| = \sup_{x \in \partial\Omega} |u(x)| . \qquad \square \tag{7}$$

2.3.4. Theorem. Let L be elliptic, and $c(x) \leqslant 0$ for $x \in \Omega$. Let u and v belong to $C^2(\Omega) \cap C^0(\bar{\Omega})$, and suppose $Lu(x) = Lv(x)$, $\forall x \in \Omega$, and $u(x) = v(x)$,

$\forall x \in \partial\Omega$. Then $u(x) = v(x)$ for all $x \in \Omega$. □

This theorem proves uniqueness for solutions of the *classical Dirichlet problem* for the operator L : given $f \in C^0(\Omega)$ and $\phi \in C^0(\bar{\Omega})$, find $u \in C^2(\Omega) \cap C^0(\bar{\Omega})$ such that

$$\begin{aligned} Lu(x) &= f(x) \quad , \quad \forall x \in \Omega \\ u(x) &= \phi(x) \quad , \quad \forall x \in \partial\Omega \ . \end{aligned} \tag{8}$$

However, the theorem tells nothing about the existence of such solution (see further).

2.3.5. Theorem. Let L be uniformly elliptic and $c(x) = 0$ for all $x \in \Omega$. If $u \in C^2(\Omega) \cap C^0(\bar{\Omega})$ satisfies $Lu(x) \geqslant 0$ for $x \in \Omega$, and if u achieves its maximum in a point of Ω, then $u(x)$ is constant in $\bar{\Omega}$. If $c(x) \leqslant 0$ in Ω, then u cannot achieve a non-negative maximum in Ω unless it is constant. □

This theorem is the *strong maximum principle* of E. Hopf.

2.3.6. Theorem. Let L be strictly elliptic and $c(x) \leqslant 0$ in Ω. Suppose that $u \in C^2(\Omega) \cap C^0(\bar{\Omega})$ and $Lu(x) = f(x)$ for $x \in \Omega$, where f is a given bounded function on Ω. Then :

$$\sup_{x\in\Omega} |u(x)| \leqslant \sup_{x\in\partial\Omega} |u(x)| + C \sup_{x\in\Omega} |f(x)| \tag{9}$$

for some constant C, depending only on diam Ω, λ_0 and $\sup_{x\in\Omega}|b(x)|$. □

Next we formulate a few results which can be obtained from a potential theoretic approach, via the so-called Schauder estimates. In order to formulate these results we need to introduce some function spaces.

2.3.7. Definition. Let $\Omega \subset \mathbb{R}^n$ be an open bounded domain, and consider the space of restrictions to $\bar{\Omega}$ of C^∞-functions u defined in an open neighbourhood of $\bar{\Omega}$. Let $k \in \mathbb{N}$ and $0 \leqslant \alpha \leqslant 1$; consider the following norms and seminorms :

$$\|u\|_k = \sum_{j=0}^{k} \sup_{x\in\Omega} \sup_{|\beta|=j} |D^\beta u(x)| \ , \tag{10}$$

(here $\beta \in \mathbb{N}^n$, $|\beta| = \beta_1+\ldots+\beta_n$, and $D^\beta = D_{x_1}^{\beta_n}\ldots D_{x_n}^{\beta_n}$);

$$H_\alpha(u) = \sup_{\substack{x,y\in\Omega \\ x\neq y}} \frac{|u(x)-u(y)|}{\|x-y\|^\alpha} , \tag{11}$$

$$\|u\|_{k,\alpha} = \|u\|_k + \sup_{|\beta|=k} H_\alpha(D^\beta u) . \tag{12}$$

The completion of this space of C^∞-functions with respect to the norms $\|.\|_k$, respectively $\|.\|_{k,\alpha}$ gives us the Banach spaces $C^k(\bar{\Omega})$ respectively $C^{k,\alpha}(\bar{\Omega})$.

$C^k(\bar{\Omega})$ is the space of functions u in Ω having partial derivatives up to order k, which can be continuously extended up to $\bar{\Omega}$. $C^{k,\alpha}(\bar{\Omega})$ is the space of functions $u \in C^k(\bar{\Omega})$ whose k-th derivatives are uniformly Hölder continuous in $\bar{\Omega}$ with exponent α. For $\alpha = 0$, the space $C^{k,0}(\bar{\Omega})$ coincides with $C^k(\bar{\Omega})$. For $\alpha = 1$, $C^{k,1}(\bar{\Omega})$ is the space of functions $u \in C^k(\bar{\Omega})$ whose k-th derivatives are uniformly Lipschitz continuous in $\bar{\Omega}$.

By $C^{k,\alpha}(\Omega)$ we will denote the space of those functions in $C^k(\Omega)$ whose k-th derivatives are locally Hölder continuous with exponent α, i.e. whose k-th derivatives are Hölder continuous in compact subsets of Ω. $C_0^{k,\alpha}(\Omega)$ is the space of those $u \in C^{k,\alpha}(\Omega)$ having compact support in Ω.

2.3.8. Definition. A bounded domain Ω in $\mathbb{R}^n$ and its boundary $\partial\Omega$ are said to be of class $C^{k,\alpha}$ ($k\in\mathbb{N}$, $0\leqslant\alpha\leqslant 1$) if at each point $x \in \partial\Omega$ there is a neighbourhood B of x in $\mathbb{R}^n$, and a one-to-one mapping Ψ from B onto $D \subset \mathbb{R}^n$ such that

(i) $\Psi(B\cap\Omega) \subset \mathbb{R}_+^n = \{x\in\mathbb{R}^n \mid x_n > 0\}$;

(ii) $\Psi(B\cap\partial\Omega) \subset \partial\mathbb{R}_+^n$;

(iii) $\Psi \in C^{k,\alpha}(B)$, $\Psi^{-1} \in C^{k,\alpha}(D)$.

2.3.9. Definition. Let Ω be a bounded $C^{k,\alpha}$-domain in $\mathbb{R}^n$, and $\phi : \partial\Omega \to \mathbb{R}$. Then ϕ is said to belong to the class $C^{k,\alpha}(\partial\Omega)$ if, at each point $x \in \partial\Omega$, $\phi\circ\Psi^{-1} \in C^{k,\alpha}(D\cap\partial\mathbb{R}_+^n)$, where Ψ and D are as in definition 2.7.8.

2.3.10. Theorem. Let Ψ be a $C^{k,\alpha}$-domain in $\mathbb{R}^n$ ($k\geqslant 1$) and let Ω' be an open set containing $\bar{\Omega}$. Suppose $\phi \in C^{k,\alpha}(\partial\Omega)$. Then there exists a function

$\Phi \in C_0^{k,\alpha}(\Omega')$ such that $\Phi = \phi$ on $\partial\Omega$. Also, if $u \in C^{k,\alpha}(\bar{\Omega})$, then there exists a function $w \in C_0^{k,\alpha}(\Omega')$ such that $w = u$ in Ω and

$$\|w\|_{k,\alpha;\Omega'} \leqslant C\|u\|_{k,\alpha;\Omega} ,$$

where $C = C(k,\Omega,\Omega')$. □

So we can always consider a boundary function $\phi \in C^{k,\alpha}(\partial\Omega)$ as an element of $C^{k,\alpha}(\bar{\Omega})$. This is what we will do in what follows.

2.3.11. Theorem. Let Ω be a $C^{k,\alpha}$ domain in $\mathbb{R}^n$, with $k \geqslant 1$, and let S be a bounded set in $C^{k,\alpha}(\bar{\Omega})$. Then S is precompact in $C^{j,\beta}(\bar{\Omega})$ if $j+\beta < k+\alpha$. That means, the imbedding

$$i : C^{k,\alpha}(\bar{\Omega}) \to C^{j,\beta}(\bar{\Omega})$$

is compact. □

Now we state the two main results of the Schauder theory of elliptic partial differential operators. We assume $\alpha > 0$.

2.3.12. Theorem. Let Ω be a bounded $C^{2,\alpha}$ domain in $\mathbb{R}^n$, and let L be strictly elliptic in Ω, with coefficients belonging to $C^{\alpha}(\bar{\Omega})$. Let $u \in C^{2,\alpha}(\bar{\Omega})$ be a solution of

$$\begin{aligned} Lu(x) &= f(x) \quad , \quad x \in \Omega \\ u(x) &= \phi(x) \quad , \quad x \in \partial\Omega \end{aligned} \qquad (13)$$

where $f \in C^{\alpha}(\bar{\Omega})$ and $\phi \in C^{2,\alpha}(\bar{\Omega})$. Then :

$$\|u\|_{2,\alpha} \leqslant C(\|u\|_0 + \|\phi\|_{2,\alpha} + \|f\|_{0,\alpha}) \qquad (14)$$

for some constant C depending only on n, α, Ω and the coefficients of L. □

Inequality (14) gives the so-called global Schauder estimate for $C^{2,\alpha}(\bar{\Omega})$-solutions of (13).

2.3.13. Theorem (Existence). Let Ω be a bounded $C^{2,\alpha}$ domain in $\mathbb{R}^n$, and let L be strictly elliptic in Ω, with coefficients belonging to $C^\alpha(\bar{\Omega})$ and $c(x) \leq 0$ for all $x \in \Omega$. Then for all $f \in C^\alpha(\bar{\Omega})$ and for all $\phi \in C^{2,\alpha}(\bar{\Omega})$ the Dirichlet problem (13) has a unique solution $u \in C^{2,\alpha}(\bar{\Omega})$. □

Using the preceding results, we can prove the following Fredholm alternative for elliptic operators.

2.3.14. Theorem. Let Ω be a bounded $C^{2,\alpha}$ domain in $\mathbb{R}^n$, and let L be strictly elliptic in Ω, with coefficients belonging to $C^\alpha(\bar{\Omega})$. Then either :

(a) the homogeneous Dirichlet problem :

$$\begin{aligned} Lu(x) &= 0 \quad , \quad x \in \Omega \\ u(x) &= 0 \quad , \quad x \in \partial\Omega \end{aligned} \tag{15}$$

has only the trivial solution $u = 0$, in which case the nonhomogeneous problem (13) has a unique $C^{2,\alpha}(\bar{\Omega})$ solution for each $f \in C^\alpha(\bar{\Omega})$ and for each $\phi \in C^{2,\alpha}(\bar{\Omega})$;

or

(b) the homogeneous problem (15) has nontrivial solutions which form a finite dimensional subspace of $C^{2,\alpha}(\bar{\Omega})$.

P r o o f. Letting $u = \phi+v$, problem (13) becomes equivalent with :

$$\begin{aligned} Lv(x) &= f(x) - L\phi(x) \quad , \quad x \in \Omega \\ v(x) &= 0 \quad , \quad x \in \partial\Omega . \end{aligned}$$

So we can restrict attention to the Dirichlet problem with homogeneous boundary conditions :

$$\begin{aligned} Lu(x) &= f(x) \quad , \quad x \in \Omega \\ u(x) &= 0 \quad , \quad x \in \partial\Omega . \end{aligned} \tag{16}$$

Introducing the Banach space

$$B = \{u \in C^{2,\alpha}(\bar{\Omega}) \mid u(x) = 0, \ \forall x \in \partial\Omega\}$$

we can consider L as a bounded linear operator from $\mathcal{B}$ into $C^{\alpha}(\bar{\Omega})$, and the problem becomes that of determining the solvability of the equation $Lu = f$ with $f \in C^{\alpha}(\bar{\Omega})$.

Let $\sigma \geq \sup_{x\in\Omega} c(x)$, and $L_{\sigma} = L-\sigma$. By theorem 13 the equation $L_{\sigma}u = f$ has a unique solution $u \in \mathcal{B}$. Using theorem 12 and theorem 6 we find :

$$\|u\|_{2,\alpha} \leq C_1(\|u\|_0 + \|f\|_{0,\alpha}) \leq C_1(C_2\|f\|_0 + \|f\|_{0,\alpha}) \leq C\|f\|_{0,\alpha} .$$

That means that the operator $L_{\sigma}^{-1} : C^{\alpha}(\bar{\Omega}) \to \mathcal{B}$ is bounded. Since $C^{2,\alpha}(\bar{\Omega})$ (and so also $\mathcal{B}$) is compactly imbedded in $C^{\alpha}(\bar{\Omega})$, we can consider L_{σ}^{-1} as a compact linear operator from $C^{\alpha}(\bar{\Omega})$ into itself.

Now let $u \in \mathcal{B}$ be a solution of (16). Then $u \in C^{\alpha}(\bar{\Omega})$ and

$$Lu = L_{\sigma}u + \sigma u = f , \tag{17}$$

which, after application of L_{σ}^{-1}, gives :

$$u + \sigma L_{\sigma}^{-1}u = L_{\sigma}^{-1}f . \tag{18}$$

Conversely, let $u \in C^{\alpha}(\bar{\Omega})$ satisfy (18). Since $R(L_{\sigma}^{-1}) = \mathcal{B}$, this implies that in fact $u \in \mathcal{B}$. So we can operate on (18) with L, which gives (17). We conclude that the problem of finding $u \in \mathcal{B}$ satisfying (17) is equivalent to that of finding $u \in C^{\alpha}(\bar{\Omega})$ satisfying (18).

Now L_{σ}^{-1} is a compact operator in $C^{\alpha}(\bar{\Omega})$. Applying the results of section 2.1 to equation (18), we see that (18) has a solution $u \in C^{\alpha}(\bar{\Omega})$ if and only if $\langle u^*, L_{\sigma}^{-1}f\rangle = 0$ for each $u^* \in (C^{\alpha}(\bar{\Omega}))^*$ such that

$$u^* + \sigma(L_{\sigma}^{-1})^*u^* = 0 . \tag{19}$$

Since $\langle u^*, L_{\sigma}^{-1}f\rangle = \langle (L_{\sigma}^{-1})^*u^*, f\rangle$, and since u^* has to satisfy (19), the condition on f becomes : $\langle u^*, f\rangle = 0$ for each $u^* \in (C^{\alpha}(\bar{\Omega}))$ satisfying (19). The space of such u^* is finite-dimensional, with a dimension equal to $\dim\{u \in C^{\alpha}(\bar{\Omega}) \mid u + \sigma L_{\sigma}^{-1}u = 0\}$, which, in turn, is equal to the dimension of the solution space of (15). This proves the theorem. □

2.3.15. Corollary. Under the conditions of theorem 14, the set Σ of real λ for which the homogeneous Dirichlet problem

$$\begin{aligned} Lu(x) - \lambda u(x) &= 0 \quad , \qquad x \in \Omega \\ u(x) &= 0 \quad , \qquad x \in \partial\Omega \end{aligned} \tag{20}$$

has a nontrivial solution is countably infinite, and has no finite accumulation point. Furthermore, if $\lambda \notin \Sigma$, then the problem

$$\begin{aligned} Lu(x) - \lambda u(x) &= f(x) \quad , \qquad x \in \Omega \\ u(x) &= \phi(x) \quad , \qquad x \in \partial\Omega \end{aligned} \tag{21}$$

has a unique solution $u \in C^{2,\alpha}(\bar{\Omega})$ for each $f \in C^{\alpha}(\bar{\Omega})$ and $\phi \in C^{2,\alpha}(\bar{\Omega})$; moreover, this solution satisfies :

$$\|u\|_{2,\alpha} \leq C(\|\phi\|_{2,\alpha} + \|f\|_{0,\alpha}) \tag{22}$$

with C independent of u, f and ϕ. Finally, by theorem 4 or theorem 13, $\Sigma \subset]-\infty,a[$ for $a = \sup\{c(x) \mid x \in \bar{\Omega}\}$. □

Example : for the Laplacian we have $\Sigma \subset]-\infty,0[$.

The foregoing shows that $L : C^{2,\alpha}(\bar{\Omega}) \to C^{\alpha}(\bar{\Omega})$ is a Fredholm operator. A better description of R(L) than the one given in the proof of theorem 14 can be obtained using the so-called L^2- or Hilbert space-approach to the Dirichlet problem. We will briefly review this theory, mostly concentrating on the case of the Laplacian ($L = \Delta$).

2.3.16. Definitions. Let Ω be a bounded open domain in $\mathbb{R}^n$, and $u,v \in L^1_{loc}(\Omega)$. For a multi-index α, we say that $D^{\alpha}u = v$ in the *weak sense*, and that v is the α-th *weak derivative* of u, if

$$\int_{\Omega} uD^{\alpha}\phi dx = (-1)^{|\alpha|}\int_{\Omega} v\phi dx \quad , \qquad \forall \phi \in C_0^{\infty}(\Omega) . \tag{23}$$

We denote by $W^k(\Omega)$ the space of functions which are k-times weakly differentiable.

Let $W^{k,p}(\Omega)$ be the space of functions $u \in W^k(\Omega)$ such that $D^\alpha u \in L^p(\Omega)$ for all $|\alpha| \leq k$. This is a Banach space, using the norm :

$$\|u\|_{W^{k,p}} = \left(\int_\Omega \sum_{|\alpha|\leq k} |D^\alpha u|^p dx\right)^{1/p} . \qquad (24)$$

We denote by $W_0^{k,p}(\Omega)$ the subspace of $W^{k,p}(\Omega)$ obtained by the completion of $C_0^\infty(\Omega)$ with respect to the norm (24).

In the case $p = 2$ one also uses the notation $H^k(\Omega) = W^{k,2}(\Omega)$ and $H_0^k(\Omega) = W_0^{k,2}(\Omega)$. These are Hilbert spaces, with inner product :

$$(u,v)_k = \sum_{|\alpha|\leq k} \int_\Omega D^\alpha u D^\alpha v dx . \qquad (25)$$

Finally, we will write $(.,.)$ for $(.,.)_0$, and $\|.\|_k$ for $\|.\|_{H^k(\Omega)}$.

There exist a whole scale of imbedding theorems between these so-called *Sobolev-spaces* (see e.g. Adams [1]). We just mention the following results; an elegant proof, using periodic extensions, can be found in Bers, John and Schechter [22].

2.3.17. Theorem. Let $u \in H_0^k(\Omega)$ and $0 \leq m < k - \frac{n}{2}$. Then $u \in C^m(\bar{\Omega})$, and :

$$\|u\|_{C^m(\bar{\Omega})} \leq C\|u\|_k \quad , \qquad \forall u \in H_0^k(\Omega) . \quad \square \qquad (26)$$

2.3.18. Theorem. Let $0 \leq j < k$. Then $H_0^k(\Omega)$ is compactly imbedded in $H_0^j(\Omega)$. □

2.3.19. Theorem. The norm $\|.\|_k$ in $H_0^k(\Omega)$ is equivalent with the norm :

$$|u|_k = \left(\sum_{|\alpha|=k} \int_\Omega |D^\alpha u|^2 dx\right)^{1/2} . \qquad (27)$$

P r o o f. It is sufficient to prove the inequality

$$\|\phi\|_0 \leq C|\phi|_1 \qquad (28)$$

for each $\phi \in C_0^\infty(\Omega)$. Considering ϕ as the product of ϕ with a test function which is equal to 1 in Ω, we have, for $x \in \Omega$:

$$|\phi(x)|^2 = |\int_{-\infty}^{x_i} \frac{\partial\phi}{\partial x_i} dx_i|^2 \leq C_1 \int_{-\infty}^{x_i} |\frac{\partial\phi}{\partial x_i}|^2 dx_i$$
$$\leq C_1 \int_{-\infty}^{+\infty} |\frac{\partial\phi}{\partial x_i}|^2 dx_i$$

by the Schwarz inequality. Integration over Ω gives (28). □

2.3.20. Lemma. Assume that $\partial\Omega$ is of class C^k. If $u \in H_0^k(\Omega) \cap C^{k-1}(\bar{\Omega})$, then

$$\frac{\partial^j u}{\partial \nu^j}(x) = 0 \quad , \quad \forall x \in \partial\Omega \quad , \quad 0 \leq j \leq k-1 \ , \tag{29}$$

where $\partial/\partial\nu$ is the derivative in the direction of the outward normal to $\partial\Omega$. □

2.3.21. Lemma. Let $\partial\Omega$ be of class C^k, and $u \in C^k(\bar{\Omega})$. If also (29) is satisfied, then $u \in H_0^k(\Omega)$. □

2.3.22. The generalized Dirichlet problem. Let $u \in C^2(\Omega) \cap C^1(\bar{\Omega})$ be a classical solution of the Dirichlet problem :

$$\begin{aligned} -\Delta u(x) &= f(x) \quad , \quad x \in \Omega \\ u(x) &= 0 \quad , \quad x \in \partial\Omega \ . \end{aligned} \tag{30}$$

Multiplication by some $\phi \in C_0^\infty(\Omega)$, and integration over Ω gives us :

$$B(\phi,u) = (\phi,f) \quad , \quad \forall\phi \in C_0^\infty(\Omega) \tag{31}$$

where

$$B(\phi,u) = \sum_{i=1}^{n} (D_i\phi, D_i u) \ . \tag{32}$$

However, (31) makes sense for all $u \in H^1(\Omega)$. We will consider any $u \in H^1(\Omega)$ satisfying (31) as a generalized solution of $-\Delta u = f$ in Ω.

As for the boundary conditions, if u is a classical solution as described above, then it follows from lemma 21 that $u \in H_0^1(\Omega)$. So we can give the following definition :

A function $u \in H_0^1(\Omega)$ satisfying (31) is called a *generalized solution* of the Dirichlet problem (30). The problem of finding such a function is called the *generalized Dirichlet problem* for (30).

2.3.23. Theorem. Let $\partial\Omega$ be of class C^1. If $u \in C^1(\bar{\Omega}) \cap C^2(\Omega)$ is a classical solution of the Dirichlet problem (30), then u is a generalized solution of the Dirichlet problem. Conversely, if u is a generalized solution of the Dirichlet problem (30), and if $u \in C^2(\Omega) \cap C^0(\bar{\Omega})$, then u is a classical solution of (30). □

It is clear that condition (31) implies that the same relation remains valid for each $\phi \in H_0^1(\Omega)$. The following estimate forms the basis for the proof of the existence of generalized solutions.

2.3.24. Theorem (Gårding's inequality). B(u,v), as defined by (32), is a bounded coercive bilinear form over $H_0^1(\Omega)$ (cfr. theorem 1.13), i.e., there exists some $\nu > 0$ such that

$$B(u,u) \geqslant \nu\|u\|_1^2 \quad , \qquad \forall u \in H_0^1(\Omega) . \tag{33}$$

P r o o f. This follows immediately from theorem 19 and $B(u,u) = |u|_1^2$. □

2.3.25. Existence theorem. For each $f \in L^2(\Omega)$ there exists a unique solution of the generalized Dirichlet problem for (30). That means, there exists a unique $u \in H_0^1(\Omega)$ such that

$$B(v,u) = (v,f) \quad , \qquad \forall v \in H_0^1(\Omega) . \tag{34}$$

P r o o f. The linear functional $F(v) = (v,f)$, $v \in H_0^1(\Omega)$, is bounded. The theorem then follows from the Lax-Milgram theorem (theorem 1.13). □

The problem, and the corresponding existence result can also be formulated as follows. Fix some $u \in H_0^1(\Omega)$. The map $v \mapsto B(v,u)$ from $H_0^1(\Omega)$ into $\mathbb{R}$ is then bounded and linear. By the Riesz representation theorem we can associate with this functional a unique element of $H_0^1(\Omega)$, which we will denote by Lu. Thus $L : H_0^1(\Omega) \to H_0^1(\Omega)$ is such that

$$B(v,u) = (v,Lu)_1 \qquad , \qquad \forall u,v \in H_0^1(\Omega) \ . \tag{35}$$

Since B is bounded, L is a bounded linear operator over $H_0^1(\Omega)$.

Let now $f \in H_0^1(\Omega)$. By the Lax-Milgram theorem and the coercivity of B, there exists a unique $u \in H_0^1(\Omega)$ such that $B(u,v) = (v,f)_1$ for all $v \in H_0^1(\Omega)$; moreover, $\|u\|_1 \leq \frac{1}{\nu}\|f\|_1$. Since then $Lu = f$, this means that the operator L defined by (35) has a bounded inverse $L^{-1} : H_0^1(\Omega) \to H_0^1(\Omega)$.

With each $f \in L^2(\Omega)$ we can associate a unique element $\pi(f) \in H_0^1(\Omega)$ such that

$$(v,f) = (v,\pi(f))_1 \qquad , \qquad \forall v \in H_0^1(\Omega) \ . \tag{36}$$

The map $\pi : L^2(\Omega) \to H_0^1(\Omega)$ is linear and bounded. The generalized Dirichlet problem (34) then takes the form

$$Lu = \pi(f)$$

and its unique solution is given by $u = L^{-1}\pi(f)$.

Remark. Similar results hold for generalized solutions of the Dirichlet problem

$$\begin{aligned} -\Delta u(x) - \lambda u(x) &= f(x) \qquad , \qquad x \in \Omega \\ u(x) &= 0 \qquad , \qquad x \in \partial\Omega \end{aligned} \tag{37}$$

when $\lambda \leq 0$. Indeed, the corresponding bilinear form

$$B_\lambda(v,u) = B(v,u) - \lambda(v,u) \tag{38}$$

satisfies also the inequality (33).

Let us now consider problem (37) for general $\lambda \in \mathbb{R}$. We have the following result.

2.3.26. Theorem. There is a Fredholm alternative for the generalized solutions of the Dirichlet problem (37). More precisely : either there exists for each $f \in L^2(\Omega)$ a unique $u \in H_0^1(\Omega)$ such that

$$B_\lambda(v,u) = (v,f) \qquad , \qquad \forall v \in H_0^1(\Omega) \tag{39}$$

or the problem

$$B_\lambda(v,u) = 0 \qquad , \qquad \forall v \in H_0^1(\Omega) \tag{40}$$

has non-trivial solutions, which form a finite-dimensional subspace of $H_0^1(\Omega)$. If $\{u_j \mid j=1,2,\ldots,p\}$ is a base for this subspace, then a necessary and sufficient condition for (39) to have a solution is that $(f,u_j) = 0$, for all $j = 1,\ldots,p$.

P r o o f. Define a linear mapping $J : H_0^1(\Omega) \to H_0^1(\Omega)$ by $J = \pi \circ i$, where $i : H_0^1(\Omega) \to L^2(\Omega)$ is the inclusion map. Since i is compact (theorem 18), the same holds for J. Problem (39) is then equivalent with :

$$Lu - \lambda Ju = \pi(f) \ ,$$

or, since L has a bounded inverse, with :

$$u - \lambda L^{-1}Ju = L^{-1}\pi(f) \ . \tag{41}$$

Now $L^{-1}J$ is compact, and we have a Fredholm alternative for (41). Equation (41) has a solution $u \in H_0^1(\Omega)$ if and only if

$$(u^*, L^{-1}\pi(f))_1 = 0 \tag{42}$$

for each solution $u^* \in H_0^1(\Omega)$ of the homogeneous adjoint equation :

$$u^* - \lambda J^*(L^{-1})^*u^* = 0 \ . \tag{43}$$

Let $u^* \in H_0^1(\Omega)$ be a solution of (43), and let $v^* = (L^*)^{-1}u^* = (L^{-1})^*u^*$. Then, since $B(u,v) = B(v,u)$ by (32), we have for each $v \in H_0^1(\Omega)$:

$$\begin{aligned} B(v,v^*) &= B(v^*,v) = (v^*,Lv)_1 = (L^*v^*,v)_1 = (u^*,v)_1 = \lambda(J^*v^*,v)_1 \\ &= \lambda(v^*,Jv)_1 = \lambda(v^*,v) = \lambda(v,v^*) \ , \end{aligned} \tag{44}$$

i.e. $v^* \in H_0^1(\Omega)$ is a generalized solution of (40).

Conversely, if $v^* \in H_0^1(\Omega)$ is a generalized solution of (40), then the same relations show that $u^* = L^* v^*$ is a solution of (43).

Then we also have :

$$(u^*, L^{-1}\pi(f))_1 = (v^*, \pi(f))_1 = (v^*, f) \ . \qquad (45)$$

We conclude that (41) has a solution $u \in H_0^1(\Omega)$ if and only if

$$(v^*, f) = 0 \qquad (46)$$

for each generalized solution v^* of (40). This proves the theorem.

Remark that if $v^* \in H_0^1(\Omega)$ and $(v^*, f) = 0$ for all $f \in L^2(\Omega)$, then necessarily $v^* = 0$. This shows that the two possibilities given by the Fredholm alternative exclude each other. □

2.3.27. <u>Theorem</u>. If $\partial\Omega$ is of class C^2, then all generalized solutions of (37) belong to $H^2(\Omega)$. If $\partial\Omega$ is of class C^{2+k} $(k \geq 0)$, and $f \in H^k(\Omega)$, then all generalized solutions of (37) belong to $H^{2+k}(\Omega)$. If $\partial\Omega$ is of class C^∞ and $f \in C^\infty(\bar{\Omega})$, then also $u \in C^\infty(\bar{\Omega})$. □

2.3.28. Suppose now that $\partial\Omega$ is of class $C^{2,\alpha}$ and $f \in C^\alpha(\bar{\Omega})$. If the Dirichlet problem (37) has a classical solution, then this classical solution is also a generalized solution. The same holds for the classical solution of the homogeneous equation

$$\begin{aligned} -\Delta u(x) - \lambda u(x) &= 0 \ , & x &\in \Omega \\ u(x) &= 0 \ , & x &\in \partial\Omega \ . \end{aligned} \qquad (47)$$

It follows from theorem 26 that necessarily $(u,f) = 0$ for each classical solution of (47). Defining the operator $L_\lambda : \{u \in C^{2,\alpha}(\bar{\Omega}) \mid u(x) = 0, \ x \in \partial\Omega\} \to C^{0,\alpha}(\bar{\Omega})$ by $L_\lambda u = -\Delta u - \lambda u$, this proves that

$$R(L_\lambda) \subset \{f \in C^{0,\alpha}(\bar{\Omega}) \mid (f,u) = 0, \ \forall u \in \ker L_\lambda\} \ . \qquad (48)$$

However, we know from the proof of theorem 14 that $\dim \ker L_\lambda = \operatorname{codim} R(L_\lambda)$. So we can conclude that in fact we have equality in (49) :

$$R(L_\lambda) = \{f \in C^{0,\alpha}(\bar{\Omega}) \mid (f,u) = 0,\ \forall u \in \ker L_\lambda\} . \tag{49}$$

Also, each generalized solution of (47) will be a classical solution. Indeed, if not, there would be some $u \in H_0^1(\Omega)$ with $u \notin \ker L_\lambda$, and such that $(f,u) = 0$ for $f \in R(L_\lambda)$. However this contradicts (49).

2.3.29. Under the conditions of the preceding section, let us reconsider the eigenvalue problem (47). Assume again that $\partial\Omega$ is of class $C^{2,\alpha}$. Let $\Sigma = \{\lambda \in \mathbb{C} \mid \ker L_\lambda \neq \{0\}\}$. Since each classical solution of (47) is also a generalized solution, it follows that $\Sigma \subset \sigma(L)$, where L is the operator defined by (35). Since L is self-adjoint, we have $\Sigma \subset \mathbb{R}$. It follows from corollary 15 that $\Sigma \subset]0,\infty[$, and that for each $\lambda \in \Sigma$, we have $\dim \ker L_\lambda < \infty$. Also, Σ has no accumulation points. So we can order the elements of Σ as follows :

$$0 < \lambda_1 < \lambda_2 < \ldots < \lambda_n < \lambda_{n+1} < \ldots \to \infty .$$

We want to prove that $\dim \ker L_{\lambda_1} = 1$, and that there exists some $u_1 \in \ker L_{\lambda_1}$ such that $u_1(x) > 0$, $\forall x \in \Omega$. Our tool will be the generalized Krein-Rutman theorem 1.15.

2.3.30. For each $f \in C^\alpha(\bar{\Omega})$, denote by Kf the unique solution of

$$\begin{aligned} -\Delta u(x) &= f(x) \quad , \qquad x \in \Omega \\ u(x) &= 0 \quad , \qquad x \in \partial\Omega . \end{aligned} \tag{50}$$

Considering K as a map from $C^\alpha(\bar{\Omega})$ into itself, we know already (proof of theorem 14) that K is compact.

We give $C^\alpha(\bar{\Omega})$ an order structure by introducing the following cone of positive elements :

$$C_+^\alpha(\bar{\Omega}) = \{v \in C^\alpha(\bar{\Omega}) \mid v(x) \geq 0,\ \forall x \in \bar{\Omega}\} . \tag{51}$$

This cone has a nonempty interior; for example, the function $v(x) \equiv 1$,

$\forall x \in \bar{\Omega}$ belongs to the interior of $C_+^\alpha(\bar{\Omega})$. Also, the operator K is positive, as a consequence of the maximum principle (theorem 5). Indeed, if $f(x) \geqslant 0$, $\forall x \in \bar{\Omega}$, and $u = Kf$, then $\Delta u(x) \leqslant 0$, $\forall x \in \Omega$. It follows from the boundary condition in (50) and the maximum principle that $u(x) \geqslant 0$, $\forall x \in \Omega$.

However, there is more. Let e(x) be the solution of

$$\begin{aligned} -u(x) &= 1 \quad , \quad \forall x \in \Omega \\ u(x) &= 0 \quad , \quad \forall x \in \partial\Omega . \end{aligned} \tag{52}$$

We will prove that K is e-positive (see definition 1.14).

Let $v \in C_+^\alpha(\bar{\Omega}) \setminus \{0\}$. Since $v(x) \leqslant \|v\|_0 = \sup\{\|v(x)\| \mid x \in \bar{\Omega}\}$, it follows from the maximum principle that

$$Kv \leqslant \|v\|_0 e .$$

Since $v(x) \not\equiv 0$, there is some closed ball $\bar{B} \subset \Omega$ such that $v(x) \geqslant \beta > 0$, $\forall x \in \bar{B}$. Let $v_1 \in C_0^\infty(B) \setminus \{0\}$ be such that $0 \leqslant v_1(x) \leqslant \beta$, $\forall x \in B$ and let $u_1 = Kv_1$. It follows from the proof of the maximum principle that, under the conditions on the boundary which we have assumed ($\partial\Omega \in C^{2,\alpha}$), the following holds :

if $u \in C^2(\Omega) \cap C^0(\bar{\Omega})$, $\Delta u(x) \geqslant 0$, and for some $x_0 \in \partial\Omega$:

$$u(x_0) > u(x) \quad , \quad \forall x \in \Omega$$

then the outer normal derivative of u at x_0 satisfies :

$$\frac{\partial u}{\partial \nu}(x_0) > 0 .$$

Applying this result on the functions u_1 and e, we find : $\frac{\partial u_1}{\partial \nu}(x) < 0$ and $\frac{\partial e}{\partial \nu}(x) < 0$ for all $x \in \partial\Omega$. By continuity there exists a number $\alpha_1 > 0$ such that :

$$\frac{\partial}{\partial \nu}(u_1 - \alpha e)(x) < 0 \quad , \quad \forall x \in \partial\Omega,\ \forall \alpha \in [0, \alpha_1] .$$

Since $(u_1-\alpha e)(x) = 0$, $\forall x \in \partial\Omega$, it follows again by continuity that there is a neighbourhood U of $\partial\Omega$ in $\bar{\Omega}$ such that $(u_1-\alpha e)(x) > 0$ for all $x \in \bar{U}\backslash\partial\Omega$, and for all $\alpha \in [0,\alpha_1]$. Moreover, since $u_1(x) > 0$ for $x \in \Omega$, we can take $\alpha > 0$ sufficiently small such that $(u_1-\alpha e)(x) > 0$ for all $x \in \Omega\backslash U$. This shows that there is some $\alpha > 0$ such that $\alpha e \leqslant Kv$. So K is e-positive.

Since $\lambda \in \Sigma$ if and only if $1/\lambda$ is an eigenvalue of K, theorem 1.15 gives us the following result.

2.3.31. Theorem. Let $\partial\Omega$ be of class $C^{2,\alpha}$, and

$$\lambda_1 = \inf \{\lambda \mid \lambda \in \Sigma\} . \tag{53}$$

Then λ_1 is a simple characteristic value :

$$\dim \ker L_{\lambda_1} = 1 , \tag{54}$$

and

$$\ker(K - \frac{1}{\lambda_1} I)^n = \ker(K - \frac{1}{\lambda_1} I) \quad , \quad \forall n \geqslant 1 . \tag{55}$$

Also there exists a $u_1 \in \ker L_{\lambda_1}$ such that

$$u_1(x) > 0 \quad , \quad \forall x \in \Omega . \tag{56}$$

P r o o f. Except for the last statement, the theorem follows from the generalized Krein-Rutman theorem. By the same theorem there exists some $u_1 \in \ker L_{\lambda_1}$ such that $u_1(x) \geqslant 0$, $\forall x \in \bar{\Omega}$, $u_1(x) \not\equiv 0$. But $-\Delta u_1(x) = \lambda_1 u_1(x) \geqslant 0$, and (56) follows from the maximum principle. □

2.4. THE VON KARMAN EQUATIONS

In this section we have brought together a number of results on the von Kármán equations, which describe the equilibrium state for thin elastic plates subject to forces and stresses along its boundary, and to normal loading. Depending on the physical situation the equations take a slightly different form, and also different boundary conditions are possible. We consider the case of a flat plate with arbitrary shape, clamped along its edges or simply supported; we are particularly interested in the cases of rectangular and circular plates. We also consider the buckling problem for cylindrical plates.

2.4.1. The von Kármán equations. Consider a thin flat elastic plate of arbitrary shape; let $\bar{\Omega} \subset \mathbb{R}^2$ be the bounded region occupied by the plate in its undeformed state. Let the plate be subjected to compressive (or stretching) forces along its edges, which are clamped, and to some normal loading. Later we will allow the forces along the edges and the loading to depend on parameters. The equilibrium state of the plate is then described by two functions $f : \bar{\Omega} \to \mathbb{R}$ and $w : \bar{\Omega} \to \mathbb{R}$ which satisfy in Ω the following system of partial differential equations :

$$\Delta^2 f = -\frac{1}{2}[w,w] \tag{1}$$

$$\Delta^2 w = [w,f] + [w,F_0] + p \tag{2}$$

subject to the boundary conditions :

$$f = f_x = f_y = w = w_x = w_y = 0 \quad \text{along } \partial\Omega . \tag{3}$$

In these equations Δ^2 is the biharmonic operator, and

$$[u,v] = u_{xx}v_{yy} + u_{yy}v_{xx} - 2u_{xy}v_{xy} . \tag{4}$$

These equations are obtained after a number of transformations and rescalings. Physically, $F = F_0+f$ represents the so-called Airy-stress function; F_0 is that part of the stress-function produced in the plate when the normal loading is zero and when the plate is artificially prevented from buckling; f is the excess stress-function produced by the buckling of the plate. The function w represents the deflections of the plate out of its equilibrium plane, and p is a measure for the normal load (and so is a given function). Also F_0 is supposed to be given : it represents in fact all the external forces along the edges. We will suppose that F_0 is bounded in $\bar{\Omega}$, together with its derivatives up to second order.

2.4.2. Definition. A *classical solution* of the equations (1)-(3) is a pair of functions (f,w), both belonging to $C^4(\Omega) \cap C^1(\bar{\Omega})$, and satisfying (1)-(3) pointwise.

We want to use the same approach to the von Kármán problem (1)-(3) as we used in section 3 for the generalized Dirichlet problem. Let us first state the following results on L^p- and Sobolev spaces.

2.4.3. Generalized Hölder inequality. Let $u_i \in L^{p_i}(\Omega)$, $1 \leq p_i \leq +\infty$, $i = 1,\ldots,m$, and

$$\sum_{i=1}^{m} \frac{1}{p_i} = 1 \ .$$

Then $\prod_{i=1}^{m} u_i \in L^1(\Omega)$, and :

$$\int_\Omega |u_1 u_2 \ldots u_m| dx \leq \|u_1\|_{p_1} \|u_2\|_{p_2} \ldots \|u_m\|_{p_m} \ . \quad \square \tag{5}$$

2.4.4. Kondrachov compactness theorem. Let Ω be a bounded domain in $\mathbb{R}^n$. Then the following imbeddings are compact :

(i) if $kp < n$:

$$W_0^{j+k,p}(\Omega) \to W_0^{j,q} \ , \text{ for all } j \in \mathbb{N} \text{ and all } q < \frac{np}{n-kp} \ ;$$

(ii) if $kp = n$:

$$W_0^{j+k,p}(\Omega) \to W_0^{j,q} \ , \text{ for all } j \in \mathbb{N} \text{ and all } q < \infty \ ;$$

(iii) if $kp > n$:

$$W_0^{k,p}(\Omega) \to C^{m,\alpha}(\bar{\Omega}) \ , \text{ for } 0 \leq m+\alpha < k - \frac{n}{p} \ . \quad \square$$

2.4.5. Generalized solutions. Suppose (f,w) form a classical solution of (1)-(3), belonging to $C^4(\Omega) \cap C^2(\bar{\Omega})$, and assume that the boundary $\partial\Omega$ is sufficiently smooth (say of class C^2). Multiplying (1) by some $\phi \in C_0^\infty(\Omega)$ and integrating over Ω gives, after some integrations by part

$$\int_\Omega (\Delta f)(\Delta\phi) = \int_\Omega (w_{yy}w_x\phi_x - w_{xy}w_x\phi_y) \quad , \quad \forall\phi \in C_0^\infty(\Omega) \; . \tag{6}$$

Analogously we find from equation (2) :

$$\int_\Omega (\Delta w)(\Delta\eta) = \int_\Omega (F_{xy}w_y - F_{yy}w_x)\eta_x + (F_{xy}w_x - F_{xx}w_y)\eta_y + \int_\Omega p\eta \; , \quad \forall\eta \in C_0^\infty(\Omega) \tag{7}$$

where, as before, $F = F_0 + f$.

Furthermore, we know that f and w belong to $W_0^{2,2}(\Omega)$ (see lemma 3.21); consequently, the relations (6) and (7) remain meaningfull for all ϕ and η in $W_0^{2,2}(\Omega)$. This is immediate for the left-hand sides of the equations. As for the right-hand sides, we have from the theorems 3 and 4 that $w \in W_0^{1,4}$ and, for example :

$$\begin{aligned} \left|\int_\Omega F_{xy}w_y\eta_x\right| &\leq \|F\|_{2,2}\|w\|_{1,4}\|\eta\|_{1,4} \\ &\leq C\|F\|_{2,2}\|w\|_{1,4}\|\eta\|_{2,2} \end{aligned} \tag{8}$$

with similar inequalities for the other terms. This suggests the following definition.

2.4.6. Definition. A *generalized solution* of the equations (1)-(3) is a pair of functions f and w belonging to $W_0^{2,2}(\Omega)$, and such that the relations (6) and (7) are satisfied for each $\phi \in W_0^{2,2}(\Omega)$ and each $\eta \in W_0^{2,2}(\Omega)$.

The following result can now be proven (see Berger [17]) :

2.4.7. Theorem. Any classical solution of (1)-(3) is a generalized solution. Conversely, any generalized solution of (1)-(3) is a classical solution in Ω and at all sufficiently smooth portions of $\partial\Omega$. □

This allows us to concentrate on the generalized problem (6)-(7), which we will reformulate now as an operator equation in $W_0^{2,2}(\Omega)$.

2.4.8. Let $H = W_0^{2,2}(\Omega)$. We have, for each $\phi \in C_0^\infty(\Omega)$:

$$|\phi|_{2,2}^2 = \sum_{|\alpha|=2} \int_\Omega (D^\alpha\phi)^2 = \int_\Omega (\Delta\phi)^2 \; . \tag{9}$$

By theorem 3.19 the norm $|\cdot|_{2,2}$ is equivalent to the usual norm in H. Since $C_0^\infty(\Omega)$ is dense in H, (9) shows that we can use the following inner product and associated norm in H :

$$(u,v) = \int_\Omega (\Delta u)(\Delta v) \quad , \qquad \forall u,v \in H \; , \tag{10}$$

and

$$\|u\| = \left(\int_\Omega (\Delta u)^2\right)^{1/2} \quad , \qquad \forall u \in H \; . \tag{11}$$

Let now $F_0 \in W^{2,2}(\Omega)$ and $p \in L^2(\Omega)$ be fixed. For each f and w in H, the right-hand sides of (6) and (7) define bounded linear functionals of ϕ and η. Using the Riesz representation theorem this allows us to define elements $B(f,w)$, Aw and $\tilde{p}$ in H such that :

$$(B(f,w),\phi) = \int_\Omega (f_{xy}w_y - f_{yy}w_x)\phi_x + \int_\Omega (f_{xy}w_x - f_{xx}w_y)\phi_y \; , \qquad \forall \phi \in H \tag{12}$$

$$Aw = B(F_0,w) \tag{13}$$

$$(\tilde{p},\phi) = \int_\Omega p\phi \quad , \qquad \forall \phi \in H \; . \tag{14}$$

2.4.9. <u>Theorem</u>. The generalized solutions of (1)-(3) coincide with the solutions $(f,w) \in H\times H$ of the equations

$$\begin{aligned} w &= Aw + B(f,w) + \tilde{p} \\ f &= -\tfrac{1}{2} B(w,w) \end{aligned} \tag{15}$$

which are equivalent to

$$f = -\tfrac{1}{2} B(w,w) \; , \tag{16}$$

$$w - Aw + C(w) = \tilde{p} \; , \tag{17}$$

where $C : H \to H$ is given by

$$C(w) = \frac{1}{2}B(B(w,w),w). \qquad \square \tag{18}$$

2.4.10. Lemma. The operator $A : H \to H$ is a bounded, selfadjoint and compact linear operator.

P r o o f. Estimates similar to (8) show that

$$|(Aw,\phi)| \leq \|F_0\|_{2,2}\|w\|_{1,4}\|\phi\|_{1,4} \leq C\|w\|_H\|\phi\|_H ,$$

which implies the boundedness and compactness of A, since H is compactly imbedded in $W_0^{1,4}(\Omega)$. The selfadjointness follows from $(Aw,v) = (Av,w)$, which is easily verified for $v,w \in C_0^\infty(\Omega)$. $\square$

2.4.11. Lemma. The operator $B : H \times H \to H$ is bounded, bilinear and compact. Moreover, the mapping $(u,v,w) \mapsto (B(u,v),w)$ defines a symmetric multilinear form over H :

$$(B(u,v),w) = (B(v,u),w) = B(u,w),v) \quad , \quad \forall u,v,w \in H . \qquad \square \tag{19}$$

2.4.12. Lemma. The map $C : H \to H$ is compact and cubic :

$$C(\alpha w) = \alpha^3 C(w) \quad , \quad \forall \alpha \in \mathbb{R} \quad , \quad \forall w \in H \tag{20}$$

and is the gradient of the C^∞-functional $\sigma : H \to \mathbb{R}$ defined by

$$\sigma(w) = \frac{1}{4}(C(w),w) \quad , \quad \forall w \in H . \qquad \square \tag{21}$$

The proofs are similar to the proof of lemma 10. Remarkt that the symmetry relation (19) implies that

$$(C(w),w) = \|B(w,w)\|^2 \quad , \quad \forall w \in H . \tag{22}$$

We also have :

$$D_w C(w).u = \frac{1}{2}B(B(w,w),u) + B(B(u,w),w) \quad , \quad \forall u,w \in H . \tag{23}$$

2.4.13. Example. Consider a circular plate subjected to a uniform compressive pressure at its boundary. We can assume that the radius is equal to 1, so we have :

$$\Omega = \{(x,y) \mid x^2+y^2 < 1\} .$$

We will assume that the compressive pressure is proportional to a real parameter $\lambda \geqslant 0$; then we can take :

$$F_0(x,y) = -\tfrac{1}{2}\lambda(x^2+y^2) . \tag{24}$$

(If $\lambda < 0$, this represents a uniform tension along the boundary). The equations take the form :

$$\begin{aligned} \Delta^2 f &= -\tfrac{1}{2}[w,w] \\ \Delta^2 w &= [w,f] - \lambda\Delta w + p \end{aligned} \tag{25}$$

together with the clamped plate boundary conditions :

$$f = f_x = f_y = 0 = w = w_x = w_y \text{ along } \partial\Omega . \tag{26}$$

The operator A is defined by

$$\int_\Omega (Aw)\Delta^2\phi = -\int_\Omega w\Delta\phi \quad , \quad \forall w \in H \quad , \quad \forall\phi \in C_0^\infty(\Omega) . \tag{27}$$

It follows that

$$(A\phi,\phi) = -\int_\Omega \phi\Delta\phi = \int_\Omega \nabla\phi.\nabla\phi \geqslant 0 \quad , \quad \forall\phi \in C_0^\infty(\Omega) .$$

So A is a positive operator, i.e. :

$$(Aw,w) \geqslant 0 \qquad , \quad \forall w \in H . \tag{28}$$

Since A is compact and self-adjoint, its spectrum consists of real eigenvalues converging to zero. So the characteristic values of A, i.e. those $\lambda \in \mathbb{R}$ for which the equation

$$w - \lambda A w = 0 \tag{29}$$

has a nontrivial solution $w \in H$, form a sequence $\lambda_1 < \lambda_2 < \dots < \lambda_n < \dots$ converging to $+\infty$. Each of these characteristic values has a finite multiplicity. We now determine some of these characteristic values.

2.4.14. Characteristic values corresponding to radially symmetric eigenfunctions. Since Ω is a smooth domain, solutions of (29) correspond to classical solutions of the following boundary value problem :

$$\begin{aligned} &\Delta^2 w + \lambda \Delta w = 0 && \text{in } \Omega \\ &w = w_x = w_y = 0 && \text{in } \partial\Omega . \end{aligned} \tag{30}$$

Since both the domain Ω and the equation (30) are rotationally invariant, it follows from the representation theory for the rotation group as given in section 6 that eigenfunctions of (30) either are radially symmetric, or are linear combinations of functions of the form $a_k(r)\cos k\theta$ and $a_k(r)\sin k\theta$, for some $k = 1,2,\dots$ (Here (r,θ) are the polar coordinates of the point $x \in \Omega$). In the first case the corresponding characteristic value is generically simple, in the second case its multiplicity is at least two, generically it is equal to two.

In order to determine the characteristic values of (30) correspondong to radially symmetric eigenfunctions, let $w = w(r)$ be a radially symmetric solution of (30). Then it is easily seen that $\psi(r) = \frac{dw}{dr}(r)$ satisfies

$$\begin{cases} \dfrac{d^2\psi}{dr^2} + \dfrac{1}{r}\dfrac{d\psi}{dr} + (\lambda - \dfrac{1}{r^2})\psi = 0 \\ \psi(0) = 0 \quad , \quad \psi(1) = 0 . \end{cases} \tag{31}$$

By the positiveness of A we only have to consider the case $\lambda > 0$. Putting $\psi(r) = \phi(\sqrt{\lambda} r)$, the equation (31) and the first boundary condition $\psi(0) = 0$ show that $\phi(x) = J_1(x)$, the first order Bessel function. Then the second boundary condition $\psi(1) = 0$ takes the form

$$J_1(\sqrt{\lambda}) = 0 , \tag{32}$$

and gives us the characteristic values of A corresponding to radially symmetric solutions. These characteristic values are simple and form a sequence $\tilde{\lambda}_1 < \tilde{\lambda}_2 < \ldots < \tilde{\lambda}_n < \ldots$ tending to $+\infty$; $\tilde{\lambda}_1$ corresponds to an eigenfunction which satisfies $w(x) > 0$, $\forall x \in \Omega$. All other characteristic values of (30) are at least double; one can prove that they are all strictly greater than $\tilde{\lambda}_1$; so $\lambda_1 = \tilde{\lambda}_1$.

2.4.15. The simply supported plate. Consider again the von Kármán equations for the buckling of a plate, but let us assume now that the plate is simply supported instead of being clamped. The equations (1) and (2) remain, but there are different boundary conditions; the general theory for this case has been described in Berger & Fife [18].

Here we will only consider the particular case of a rectangular plate, for which the theory can be considerably simplified. Let the domain occupied by the plate in its undeformed state be

$$\Omega = \{(x,y) \mid -\ell<x<\ell,\ -1<y<1\} .$$

We assume that a compressive thrust T is applied to the edges $x = \pm\ell$, while also a normal load acts on the plate. After some rescalings, the von Kármán equations for the equilibrium state of the plate take the form (see Bauer & Reiss [14]) :

$$\begin{aligned} \Delta^2 f &= -\tfrac{1}{2}[w,w] \\ \Delta^2 w &= [w,f] - \lambda w_{xx} + p \end{aligned} \tag{33}$$

together with the boundary conditions :

$$f = \Delta f = 0 \quad , \quad w = \Delta w = 0 \quad \text{along } \partial\Omega . \tag{34}$$

Here λ is proportional to the thrust T, and p is proportional to the normal load; the other entities have the same meaning as in the case of the clamped plate.

2.4.16. Definition. A *classical solution* of the problem (33)-(34) is a pair of functions (f,w) belonging to $C^4(\Omega) \cap C^2(\bar{\Omega})$ and satisfying (33) and (34) pointwise.

Before defining generalized solutions and putting the problem in an operator form, let us remark that the domain Ω has a smooth boundary, except at the four corners; even at these corners the singularity is not too bad, since $\partial\Omega$ (corners included) satisfies a Lipschitz condition, that is, $\partial\Omega$ is of class $C^{0,1}$. Under this condition the imbedding and compactness theorem 4 remains valid if we replace $W_0^{m,p}(\Omega)$ by $W^{m,p}(\Omega)$ (see Adams [1]). Also $W^{m,p}(\Omega)$ is the completion of $C^\infty(\bar{\Omega})$ under the $W^{m,p}(\Omega)$-norm.

2.4.17. <u>A function space</u>. Let us define now a Hilbert space which will be the underlying function space for the generalized form of the problem (33)-(34). Consider the space of all functions $u \in C^\infty(\bar{\Omega})$ satisfying $u(x) = 0$ for $x \in \partial\Omega$. Let H be the completion of this space under the $W^{2,2}(\Omega)$-norm. It is clear that

$$W_0^{2,2}(\Omega) \subset H \subset W^{2,2}(\Omega) .$$

Moreover, since $W^{2,2}(\Omega)$ is continuously imbedded in $C^{0,\alpha}(\bar{\Omega})$ for any $0 \leqslant \alpha < 1$, H will consist of continuous functions which are zero along the boundary $\partial\Omega$. In fact :

$$H = \{u \in W^{2,2}(\Omega) \mid u(x) = 0, \ \forall x \in \partial\Omega\} . \qquad (35)$$

2.4.18. <u>Lemma</u>. The inner product

$$(u,v)_H = \int_\Omega (\Delta u)(\Delta v) \quad , \quad \forall u,v \in H \qquad (36)$$

and the corresponding norm

$$\|u\|_H = \left(\int_\Omega (\Delta u)^2\right)^{1/2} \quad , \quad \forall u \in H \qquad (37)$$

determine a topology in H equivalent to the $W^{2,2}(\Omega)$-topology.

P r o o f. It is immediate that $\|u\|_H \leqslant C\|u\|_{2,2}$ for some constant C. In order to prove the coercivity of $\|.\|_H$, remark first that

$$\int_\Omega (\Delta u)^2 = \sum_{|\alpha|=2} \int_\Omega |D^\alpha u|^2 \quad , \quad \forall u \in H . \qquad (38)$$

Indeed, (38) can be explicitly verified for $u \in C^\infty(\Omega) \cap H$; the boundary terms appearing after integration by parts cancel out because of the special geometry of the domain Ω and the fact that $u = 0$ along $\partial\Omega$.

It remains to show that there is a constant $C > 0$ such that

$$\sum_{|\alpha|=2} \int_\Omega |D^\alpha u|^2 \geq C\left(\int_\Omega u^2 + \sum_{|\alpha|=1} \int_\Omega |D^\alpha u|^2\right) \quad , \quad \forall u \in H , \tag{39}$$

(i.e. $|u|^2_{2,2} \geq C\|u\|^2_{1,2}$). If no such constant exists, then there is a sequence $\{u_n \mid n \in \mathbb{N}\}$ in H such that $\|u_n\|_{1,2} = 1$, $\forall n \in \mathbb{N}$, and $|u_n|_{2,2} \to 0$ as $n \to \infty$. This implies that $\{u_n\}$ is bounded in $W^{2,2}$; since $W^{2,2}$ is compactly imbedded in $W^{1,2}$, we may assume that $\{u_n\}$ converges to some u in $W^{1,2}$. Since $|u_n|_{2,2} \to 0$, $\{u_n\}$ also converges in $W^{2,2}$, and its limit u has vanishing second derivatives. So u must be linear, but then $u \in H$ implies that $u(x) = 0$, $\forall x \in \Omega$. This, however, contradicts the fact that $\|u_n\|_{1,2} = 1$ for all $n \in \mathbb{N}$. □

From now on we equip H with the topology given by the lemma.

2.4.19. Generalized solutions. Let (f,w) be a classical solution of the problem (33)-(34). Then f and w both belong to H. Let ϕ and η be $C^\infty(\bar{\Omega})$ functions vanishing along $\partial\Omega$. If we multiply (33) by ϕ, and (34) by η, and integrate over Ω, then we find :

$$\int_\Omega (\Delta f)(\Delta\phi) = -\frac{1}{2}\int_\Omega (w_{xy}w_y - w_{yy}w_x)\phi_x - \frac{1}{2}\int_\Omega (w_{xy}w_x - w_{xx}w_y)\phi_y \tag{40}$$

$$\int_\Omega (\Delta w)(\Delta\eta) = \int_\Omega (f_{xy}w_y - f_{yy}w_x)\eta_x + \int_\Omega (f_{xy}w_x - f_{xx}w_y)\eta_y$$
$$+ \lambda\int_\Omega w_x\eta_x + \int_\Omega p\eta \; . \tag{41}$$

Using the compact imbedding of $W^{2,2}(\Omega)$ into $W^{1,4}(\Omega)$, we can, as in subsection 7, prove that the relations (40) and (41) remain valid for each ϕ and η in H. This suggests the following definition :

2.4.20. Definition. A *generalized solution* of the problem (33)-(34) is a pair of functions (f,w), both belonging to H, such that (40) and (41) are satisfied for each ϕ and η in H.

2.4.21. Theorem. Every classical solution of (33)-(34) is also a generalized solution.

Conversely, every generalized solution is a classical solution in $\Omega \cup \partial'\Omega$, where $\partial'\Omega$ is equal to $\partial\Omega$ with the corners deleted. □

The main (and most difficult) part of the proof consists in showing that each generalized solution is of class C^4 in $\Omega \cup \partial'\Omega$. The proof uses a so-called bootstrapping argument, and can be found in Knightly & Sather [124] for the particular case under consideration here, in Berger & Fife [18] for the general case.

2.4.22. Theorem. The generalized solutions of (33)-(34) coincide with the solutions $f \in H$ and $w \in H$ of the pair of equations :

$$w - \lambda Aw + C(w) = \tilde{p} \tag{42}$$

$$f = -\frac{1}{2}B(w,w) \; . \tag{43}$$

Here A is a compact selfadjoint positive linear operator of H in itself, B is a compact, continuous bilinear operator mapping $H \times H$ into H, and finally C is a compact, cubic C^∞ map of H into itself. □

The proof, and the definition of the operators are analogous to the case of the clamped plate. So we have :

$$(Aw,\phi) = \int_\Omega w_x \phi_x \qquad , \qquad \forall\phi \in H \; ; \tag{44}$$

$$(B(u,v),\phi) = \int_\Omega (u_{xy}v_y - u_{yy}v_x)\phi_x + \int_\Omega (u_{xy}v_x - u_{xx}v_y)\phi_y \; , \quad \forall\phi \in H \; ; \tag{45}$$

$$C(u) = \frac{1}{2}B(B(u,u),u) \; ; \tag{46}$$

$$(\tilde{p},\phi) = \int_\Omega p\phi \qquad , \qquad \forall\phi \in H \; . \tag{47}$$

2.4.23. The linearized problem. The linear problem

$$w - \lambda Aw = 0 \tag{48}$$

has, by the compactness, selfadjointness and positivity of A, only nontrivial solutions $w \in H$ for a sequence of positive values for λ, tending to $+\infty$. Let us determine these characteristic values. By theorem 21, equation (48) is equivalent with the classical boundary value problem :

$$\Delta^2 w + \lambda w_{xx} = 0 \qquad \text{in } \Omega$$
$$w = \Delta w = 0 \qquad \text{in } \partial\Omega \ . \tag{49}$$

This problem can easily be solved by using a Fourier expansion for w :

$$w(x,y) = \sum_{m,n=1}^{\infty} a_{mn} \sin\frac{m\pi}{2\ell}(x+\ell) \sin\frac{n\pi}{2}(y+1) \ . \tag{50}$$

Bringing this in (49) one observes that (49) has only the trivial solution $w = 0$ except when

$$\lambda = \lambda_{mn} = \frac{\pi^2(m^2+n^2\ell^2)^2}{4\ell^2 m^2} \ . \tag{51}$$

The characteristic value λ_{mn} corresponds to the eigenfunction

$$\phi_{mn}(x,y) = C_{mn} \sin\frac{m\pi}{2\ell}(x+\ell) \sin\frac{n\pi}{2}(y+1) \tag{52}$$

where the constant C_{mn} can be chosen in such a way that the set $\{\phi_{mn} \mid m,n = 1,2,\ldots\}$ forms a complete orthonormal system in H :

$$C_{mn} = \frac{4\ell^{3/2}}{\pi^2(m^2+n^2\ell^2)} \ . \tag{53}$$

It is clear that each characteristic value λ_{mn} has finite multiplicity : if we fix some (m,n), then the set

$$\sigma(m,n) = \{(m',n') \mid \lambda_{m'n'} = \lambda_{mn}\} \tag{54}$$

is finite. If we define :

$$L_{mn} = I - \lambda_{mn} A \tag{55}$$

then

$$\ker L_{mn} = \text{span}\{\phi_{m'n'} \mid (m',n') \in \sigma(m,n)\} \tag{56}$$

and

$$R(L_{mn}) = (\ker L_{mn})^{\perp}$$
$$= \{w \in H \mid (w,\phi_{m'n'}) = 0,\ \forall (m',n') \in \sigma(m,n)\} \ . \qquad (57)$$

2.4.24. First characteristic value. Now we prove the following result :

The lowest characteristic value for (49) is simple, except when $\ell^2 = m(m+1)$ for some $m = 1,2,\ldots$ In that case the lowest characteristic value has multiplicity two.

P r o o f. It is immediate from (51) that $\lambda_{mn} > \lambda_{m1}$ if $n > 1$, and that for each m. Now

$$\lambda_{m1} = \frac{\pi^2}{4\ell^2}\left(m + \frac{\ell^2}{m}\right)^2 \ .$$

The function $x \mapsto x + \frac{\ell^2}{x}$ attains its minimum in $x > 0$ at the point $x = \ell$. So we can expect the lowest characteristic value λ_0 to be equal to $\lambda_{m,1}$ for $m = [\ell]$ or $m = [\ell+1]$. If $\ell = m$, then $\lambda_0 = \lambda_{m,1}$, and λ_0 is simple. If $m < \ell < m+1$, then $\lambda_{m,1} = \lambda_{m+1,1}$ if and only if $\ell^2 = m(m+1)$. In that case λ_0 has multiplicity two. In all other cases λ_0 is simple.

The special importance of the first characteristic value follows from the next results.

2.4.25. Lemma. Let $H = W_0^{2,2}(\Omega)$, respectively as in 2.4.17. Let $C : H \to H$ be as in (18), respectively (46). Let $w \in H$ be a solution of (29), respectively (48), for some $\lambda \in \mathbb{R}$. Suppose also that $(C(w),w) = 0$. Then $w = 0$.

P r o o f. We have in both cases that $(C(w),w) = \|B(w,w)\|^2$ (cfr. (22)), so $(C(w),w) = 0$ implies that $B(w,w) = 0$, or, by the definition of B and since w is sufficiently smooth :

$$\int_\Omega [w,w]\,\phi = 0 \qquad , \qquad \forall \phi \in C_0^\infty(\Omega) \ .$$

Consequently :

$$w_{xx}w_{yy} - w_{xy}^2 = 0 \qquad \text{in } \Omega$$
$$w(x) = 0 \qquad , \qquad \forall x \in \partial\Omega \ .$$

Considering the surface $w = w(x,y)$, this implies that the Gaussian curvature of this surface is identically zero. This means that the surface is developable and contains a family of straight lines. Because of the boundary condition on w this is only possible for $w = 0$. □

2.4.26. Theorem. Let λ_0 be the lowest characteristic value for the problem (29), respectively (48). Then the equation

$$w - \lambda Aw + C(w) = 0 \tag{58}$$

has, for $\lambda \leq \lambda_0$, only the solution $w = 0$.

P r o o f. λ_0 has the following variational characterization (see Courant & Hilbert [48], Weinberger [234]) :

$$\frac{1}{\lambda_0} = \sup_{\substack{w \in H \\ w \neq 0}} \frac{(Aw,w)}{(w,w)} \ . \tag{59}$$

Let $w \neq 0$ be a solution of (58), for some $\lambda \leq \lambda_0$. Then (22) implies that $(w - Aw,w) \leq 0$. By (59) this is only possible if $\lambda = \lambda_0$ and $w \in \ker L_{\lambda_0}$. But then (58) gives $(C(w),w) = 0$, which, by lemma 25 gives $w = 0$, a contradiction. □

2.4.27. The cylindrical plate. As a last example we consider a cylindrical plate which is simply supported, axially compressed, and subject to a radial loading. The von Kármán-Donnel equations which govern the equilibrium state of such system reduce after some rescaling to :

$$\begin{aligned} \Delta^2 f &= -\tfrac{1}{2}[w,w] + \alpha w_{xx} \ , \\ \Delta^2 w &= [w,f] - \lambda w_{xx} - \alpha f_{xx} + p \ . \end{aligned} \tag{60}$$

These equations hold in the domain $\Omega = \{(x,y) \mid -\pi\ell < x < \pi\ell\}$, together with the boundary condition

$$w = \Delta w = 0 \quad , \quad f = \Delta w = 0 \ \text{ along } \partial\Omega \tag{61}$$

and the periodicity condition

$$w(x,y) = w(x,y+2\pi) \quad , \; f(x,y) = f(x,y+2\pi) \quad , \quad \forall (x,y) \in \Omega \; . \tag{62}$$

The constant α is proportional to the curvature of the cylinder. In our setting x measures distances parallel to the axis, while y measures distances along the circumference of the cylinder; this explains the periodicity condition (62).

One can use the same approach to these equations as we did before for flat plates. The basic Hilbert space will be the closure in the $W^{2,2}(\Omega_0)$-norm of the $C^\infty(\bar{\Omega})$-functions which are zero along $\partial\Omega$ and 2π-periodic in y. (Here $\Omega_0 = \{(x,y) \mid -\pi\ell < x < \pi\ell,\ 0 \leq y \leq 2\pi\}$.) The abstract equations in H for the generalized solutions (which are also classical solutions) take the form

$$f = -\tfrac{1}{2}B(w,w) - \alpha Aw \;, \; w = B(w,f) + \lambda Aw + \alpha Af + \tilde{p} \; . \tag{63}$$

Elimination of f gives

$$(w - \lambda Aw + \alpha^2 A^2 w) + \alpha Q(w) + C(w) = \tilde{p} \; , \tag{64}$$

where the quadratic and compact operator $Q : H \to H$ is given by

$$Q(w) = B(w,Aw) + \tfrac{1}{2}AB(w,w) \; . \tag{65}$$

2.4.28. Characteristic values. The characteristic values for the cylindrical plate problem are those $\lambda \in \mathbb{R}$ for which the linearized equation

$$L_\lambda w \equiv (I - \lambda A + \alpha^2 A^2)w = 0 \tag{66}$$

has a nontrivial solution. We have

$$L_\lambda = (I-\mu_-(\lambda)A)(I-\mu_+(\lambda)A) \quad , \quad \mu_\pm(\lambda) = \tfrac{1}{2}(\lambda \pm \sqrt{\lambda^2 - 4\alpha^2}) \; , \tag{67}$$

which implies that

$$\ker L_\lambda = \ker(I-\mu_-(\lambda)A) \cup \ker(I-\mu_+(\lambda)A) \; . \tag{68}$$

Consequently, $\lambda \in \mathbb{R}$ is a characteristic value if and only if $\mu_+(\lambda)$ or $\mu_-(\lambda)$

are a characteristic value of A, that is, if and only if $\lambda = \mu + \alpha^2/\mu$ for some characteristic value μ of A.

The characteristic values μ of A are determined by the boundary value problem :

$$\begin{aligned} &\Delta^2 w + \mu w_{xx} = 0 && \text{in } \Omega \\ &w = \Delta w = 0 && \text{in } \partial\Omega , \end{aligned} \tag{69}$$

which we have to solve for solutions $w = w(x,y)$ which are 2π-periodic in y. Using Fourier expansions one finds :

$$\mu_{mn} = \frac{(m^2+4\ell^2n^2)^2}{4\ell^2m^2} \quad , \quad m = 1,2,\ldots \; ; \; n = 0,1,2,\ldots \tag{70}$$

The characteristic values $\mu_{m,0}$ for $m = 1,2,\ldots$ are simple, and correspond to the eigenfunction :

$$\phi_{m,0}(x,y) = C_{m,0} \, \sin\frac{m}{2\ell}(x+\pi\ell) \; . \tag{71}$$

For $n > 0$, μ_{mn} is a characteristic value of A of multiplicity two, corresponding to the eigenfunctions :

$$\phi_{m,n}(x,y) = C_{m,n} \, \sin\frac{m}{2\ell}(x+\pi\ell) \cos ny \tag{72}$$

and

$$\psi_{m,n}(x,y) = C_{m,n} \, \sin\frac{m}{2\ell}(x+\pi\ell) \sin ny \; . \tag{73}$$

The constants $C_{m,n}$ are normalization constants.

2.4.29. Bibliographical notes. There exists an extensive literature on the buckling problem for plates. An account of the physical motivation for the von Kårmån equations can be found in Landau and Lifschitz [139] and Leipholz [142]; the basic ideas leading to these equations were introduced by von Kårmån [107]. A general theory of the von Kårmån equations can be found in Berger and Fife [17],[18]. Further references are Ambrosetti [6], Antman [8], Bauer, Keller and Reiss [12],[14], Berger [20],[21], Dickey [58], Fife [62],

Friedrichs and Stoker [68], Keener [109], Keener and Keller [110], Keller, Keller and Reiss [113], Knightly [121],[122], Knightly and Sather [123],[124], [125],[126],[127], Koiter [131], Magnus and Poston [152], Matkowsky and Putnick [158], Reiss [181], Sather [190],[191], Shearer [206], Wolkowisky [236]. Recently a very nice discussion of the von Kàrmàn equations has been given in the lecture notes of Ciarlet and Rabier [246]. A somewhat different approach can be found in Duvaut and Lions [60].

2.5. GROUP REPRESENTATIONS AND EQUIVARIANT PROJECTIONS

This section contains some general results on representations of groups over Banach spaces. Most of these results are taken from Rudin [184] (in particular sections 5.11 to 5.18), where full proofs can be found. Finite-dimensional representations will be considered in the next section.

2.5.1. Definition. A *topological group* is a group G on which a topology is defined such that :

(i) every point of G forms a closed set;

(ii) the map

$$\phi : G\times G \to G, \ (s,t) \mapsto \phi(s,t) = st^{-1} \qquad (1)$$

is continuous.

The topology of G is a Hausdorff topology, and is completely determined by any neighbourhood base of the identity element $e \in G$.

2.5.2. Definition. Let X be a topological vector space, and G a topological group. A *representation of* G *over* X is a map $\Gamma : G \to L(X)$ such that :

(i) Γ is a group homomorphism :

$$\Gamma(e) = I \quad , \quad \Gamma(st) = \Gamma(s)\Gamma(t) \quad , \quad \forall s,t \in G \; ; \qquad (2)$$

(ii) Γ is strongly continuous :

$$\lim_{t\to e} \Gamma(t)x = x \quad , \quad \forall x \in X \; . \qquad (3)$$

2.5.3. Lemma. Suppose that X is a Fréchet-space, G a locally compact topological group, and $\Gamma : G \to L(X)$ a representation of G over X.

Then the map $(s,x) \mapsto \Gamma(s)x$ from $G \times X$ into X is continuous.

P r o o f. For each $x \in X$ the map $s \mapsto \Gamma(s)x$ from G into X is continuous, by the definition of a representation. Fixing a compact neighbourhood U of $s_0 \in G$, this implies that $\{\Gamma(s)x \mid s \in U\}$ is bounded in X, for each $x \in X$. It follows from the Banach-Steinhaus theorem that $\{\Gamma(s) \mid s \in U\}$ forms an equicontinuous set of bounded linear operators. This proves the lemma. □

2.5.4. The groups used in this book will in general be *compact* topological groups. On such groups it is possible to define an invariant measure; in order to introduce this measure, we need some concepts from integration theory.

If X is any set, then a *σ-algebra* over X is a collection Σ of subsets of X, such that

(i) $X \in \Sigma$;

(ii) $A \in \Sigma \Rightarrow A^c \in \Sigma$;

(iii) $A_n \in \Sigma , \forall n \in \mathbb{N} \Rightarrow \bigcup_{n \in \mathbb{N}} A_n \in \Sigma$.

A set X together with a σ-algebra Σ over X is called a *measurable space* (X,Σ). In particular, if X is a compact or locally compact Hausdorff space, then the smallest σ-algebra Σ_B containing all open subsets of X, is called the σ-algebra of *Borel sets* in X.

A *positive measure* over (X,Σ) is a map $\mu : \Sigma \to [0,\infty]$ which is countably additive :

$$A_j \in \Sigma , A_i \cap A_j = \phi , i \neq j \Rightarrow \mu(\bigcup_{j \in \mathbb{N}} A_j) = \sum_{j \in \mathbb{N}} \mu(A_j) .$$

Such positive measure is *finite* if $\mu(X) < \infty$. In case $\mu(X) = 1$, μ is called a *probability measure*. In case X is a locally compact Hausdorff space, a *positive Borel measure* is a positive measure over (X,Σ_B). Such Borel measure is *regular* if

$$\begin{aligned} \mu(E) &= \sup\{\mu(K) \mid K \subset E \text{ compact}\} \\ &= \inf\{\mu(G) \mid E \subset G \text{ open}\} \quad , \quad \forall E \in \Sigma_B. \end{aligned}$$

2.5.5. Definition. Let μ be a positive measure over a measure space (Q,Σ), X a locally convex topological vector space, and $f : Q \to X$ a mapping such that $x^* \circ f \in L^1(\mu)$, for each $x^* \in X^*$. If there exists a vector $x \in X$ such that

$$x^*(x) = \int_Q (x^* \circ f) d\mu \quad , \qquad \forall x^* \in X^* , \tag{4}$$

then x is called the *Pettis-integral of f over Q with respect to the measure* μ ; we use the notation :

$$x = \int_Q f d\mu . \tag{5}$$

2.5.6. Theorem. Suppose X is a Fréchet space, and μ is a Borel probability measure on a compact Hausdorff space Q. Then the integral $\int_Q f d\mu$ exists for every continuous $f : Q \to X$. Moreover,

$$\int_Q f d\mu \in \overline{co}\, f(Q) ,$$

the closed convex hull of $f(Q)$ in X. □

2.5.7. Let G be a compact topological group. Then we denote by $C(G)$ the Banach space of all complex-valued continuous functions on G, with the supremum norm. For given $s \in G$, we then define $L_s : C(G) \to C(G)$ and $R_s : C(G) \to C(G)$ by :

$$(L_s f)(t) = f(st) \quad , \quad (R_s f)(t) = f(ts) \quad , \quad \forall t \in G , \ \forall f \in C(G) . \tag{6}$$

2.5.8. Theorem. On every compact topological group G there exists a unique Borel probability measure m which is left-invariant, in the sense that :

$$\int_G f dm = \int_G (L_s f) dm \quad , \qquad \forall s \in G , \ \forall f \in C(G) . \tag{7}$$

This measure is also right-invariant

$$\int_G f dm = \int_G (R_s f) dm \quad , \qquad \forall s \in G , \ \forall f \in C(G) , \tag{8}$$

and it satisfies the relation

$$\int_G f(t)dm(t) = \int_G f(t^{-1})dm(t) \qquad , \qquad \forall f \in C(G) \ . \tag{9}$$

This measure is called the *Haar measure* on G. □

2.5.9. Theorem. Let X be a Fréchet space, and Y a closed subspace of X, having a topological complement. Let G be a compact topological group, and $\Gamma : G \to L(X)$ a representation of G over X, such that

$$\Gamma(s)(Y) \subset Y \qquad , \qquad \forall s \in G \ .$$

Then there exists a continuous projection Q of X onto Y which is *equivariant* with respect to Γ :

$$\Gamma(s)Q = Q\Gamma(s) \qquad , \qquad \forall s \in G \ . \tag{10}$$

When $P \in L(X)$ is a continuous projection with $R(P) = Y$, then one can construct an equivariant projection Q from P by

$$Qx = \int_G \Gamma(s^{-1})P\Gamma(s)x\,dm(s) \ , \qquad \forall x \in X \ , \tag{11}$$

where m is the Haar measure of G, and the integral is defined in the sense of definition 5. □

2.5.10. Example. Let G be a finite group

$$G = \{s_i \mid i = 1,\ldots,N\} \ . \tag{12}$$

Using the discrete topology, G becomes a compact topological group. The corresponding Haar measure is given by

$$m(E) = \frac{1}{N} \times \text{(number of elements in E)} \quad , \quad \forall E \in G \ . \tag{13}$$

Under the hypotheses of theorem 9, the formula (11) for an equivariant projection takes the form :

$$Q = \frac{1}{N} \sum_{i=1}^{N} \Gamma(s_i^{-1})P\Gamma(s_i) \ . \tag{14}$$

For this particular case the proof is much easier than for the general situation of theorem 9 (see Vanderbauwhede [225]).

2.5.11. <u>Example</u>. Let X be a Hilbert space, G a compact topological group, and $\Gamma : G \to L(X)$ a *unitary* representation :

$$\Gamma^{-1}(s) = \Gamma(s^{-1}) = \Gamma^*(s) \qquad , \qquad \forall s \in G . \tag{15}$$

If Y is a closed subspace of X which is invariant under the representation (i.e. such that $\Gamma(s)(Y) \subset Y$, $\forall s \in G$), then the orthogonal projection Q of X onto Y is equivariant with respect to Γ.

P r o o f. Let $y \in Y$, $z \in Y^{\perp}$ and $s \in G$. Then

$$\langle y,\Gamma(s)z\rangle = \langle\Gamma^*(s)y,z\rangle = \langle\Gamma(s^{-1})y,z\rangle = 0 ,$$

since $\Gamma(s^{-1})y \in Y$. This shows that also $Y^{\perp}$ is invariant under Γ. For each $x \in X$ we have :

$$\Gamma(s)x = \Gamma(s)Qx + \Gamma(s)(I-Q)x .$$

Since, by the foregoing, we have $\Gamma(s)Qx \in Y$ and $\Gamma(s)(I-Q)x \in Y$, it follows that $Q\Gamma(s)x = \Gamma(s)Qx$. □

2.5.12. <u>Example</u>. In the subsequent chapters we will frequently meet the following situation, which is somewhat similar to the preceding example.

Let X be a (real) Banach space, and $\langle .,. \rangle : X \times X \to \mathbb{R}$ a continuous, symmetric and positive definite bilinear form over X. Let G be a compact topological group, and Γ a representation of G over X, such that

$$\langle\Gamma(s)x,\Gamma(s)y\rangle = \langle x,y\rangle \qquad , \qquad \forall s \in G , \forall x,y \in X . \tag{16}$$

Let Y be a finite-dimensional subspace of X, which is invariant under Γ. Let $\{e_i \mid i = 1,\ldots,n\}$ be a basis for Y, which we can assume to be orthonormal with respect to the bilinear form $\langle .,. \rangle$. Define a projection Q of X onto Y by

$$Qx = \sum_{i=1}^{n} \langle x,e_i \rangle e_i \quad , \qquad \forall x \in X . \tag{17}$$

Defining $Y^{\perp} = \{x \in X \mid \langle x,y \rangle = 0, \forall x \in Y\}$ one easily sees that $Qx = 0$ is equivalent to $x \in Y^{\perp}$, i.e. $\ker Q = Y^{\perp}$. Since, for each $s \in G$, $\Gamma(s)$ is an automorphism of X, the restriction of $\Gamma(s)$ to Y is an isomorphism of Y onto $\Gamma(s)(Y) \subset Y$; we conclude that $\Gamma(s)Y = Y$, since Y is finite-dimensional. Then the same argument as in example 11 shows that $\Gamma(s)(Y^{\perp}) \subset Y^{\perp}$, and that the projection Q given by (17) is equivariant.

We conclude this section with the following theorem, whose proof is very similar to the proof of theorem 9, as given in Rudin [184].

2.5.13. Theorem. Let X be a Fréchet space, G a compact topological group, and $\Gamma : G \to L(X)$ a representation of G over X. Let

$$X_0 = \{x \in X \mid \Gamma(s)x = x, \forall s \in G\} . \tag{18}$$

Then X_0 has a topological complement in X, and $P_0 \in L(X)$, defined by

$$P_0 x = \int_G \Gamma(s)x\,dm(s) \quad , \qquad \forall x \in X , \tag{19}$$

is a continuous projection of X onto X_0. Moreover

$$P_0\Gamma(s) = \Gamma(s)P_0 = P_0 \quad , \qquad \forall s \in G . \tag{20}$$

P r o o f. It is clear that X_0 is closed. By theorem 2.1.2 it is sufficient to prove that P_0, as defined by (29), is a continuous projection, and that $R(P_0) = X_0$.

First, the integral in (19) exists since the map $s \mapsto \Gamma(s)x$ from G into X is continuous, for each $x \in X$. It is clear that P_0 is linear. Since G is compact, the set $\{\Gamma(s)x \mid s \in G\}$ is compact, for each $x \in X$. By the Banach-Steinhaus theorem (see Rudin [184], theorem 2.6) it follows that $\{\Gamma(s) \mid s \in G\}$ is an equicontinuous family of linear operators in X.

Let V be any neighbourhood of 0 in X. Choose a convex neighbourhood W of 0 such that $\bar{W} \subset V$. There exists a neighbourhood U of 0 such that

$$\Gamma(s)(U) \subset W \quad , \qquad \forall s \in G .$$

Then it follows from theorem 6, the definition of P_0 and the convexity of W, that

$$P_0(U) \subset \bar{W} \subset V .$$

This shows that P_0 is continuous.

It is clear that $P_0 x = x$ for $x \in X_0$. Now remark that it follows from the definition of the Pettis-integral that the integral commutes with bounded linear operators. Using the left invariance of the Haar measure, we have :

$$P_0 x = \int_G \Gamma(ts)x\,dm(s) = \Gamma(t)P_0 x \quad , \quad \forall t \in G , \forall x \in X .$$

This shows that $R(P_0) \subset X_0$; since $P_0(X_0) = X_0$, we have in fact $R(P_0) = X_0$.

Finally, it follows from the right invariance of the Haar measure that

$$P_0\Gamma(t)x = \int_G \Gamma(s)\Gamma(t)x\,dm(s) = P_0 x = \Gamma(t)P_0 x \quad , \forall t \in G , \forall x \in X .$$

This proves (20). □

2.6. IRREDUCIBLE GROUP REPRESENTATIONS

In this section we give an introduction to the theory of irreducible group representations, and then determine the irreducible representations of the groups SO(2), O(2), SO(3) and O(3). These groups will be frequently encountered in the subsequent chapters.

Let us remark that we will concentrate on real representations, in contrast to the complex representations considered almost exclusively in classical handbooks on group theory (see e.g. Hamermesh [92], Miller [166], Husain [99]). Our approach to the representations of the group SO(3) is an adaption to real representations of the approach in Miller [166]. A nice account of the theory of spherical harmonics can be found in Müller [168].

We consider representations of the form

$$\Gamma : G \to L(\mathbb{R}^n) ,$$

where G is a topological group. The natural number n is called the *dimension* of the representation.

2.6.1. Lemma. Let Γ be an n-dimensional representation of G, and let $T \in L(\mathbb{R}^n)$ be non-singular. Then

$$\Gamma_1 : G \to L(\mathbb{R}^n) \ , \ s \mapsto \Gamma_1(s) = T^{-1}\Gamma(s)T \tag{1}$$

also defines a representation of G. □

2.6.2. Definition. Two n-dimensional representations Γ and Γ_1 of a topological group G are *similar* or *equivalent* if there exists some non-singular $T \in L(\mathbb{R}^n)$ such that

$$\Gamma_1(s) = T^{-1}\Gamma(s)T \quad , \quad \forall s \in G \ . \tag{2}$$

Similarity defines an equivalence relation among the representations of a group G. This reduces the study of all representations of G to the study of equivalence classes of representations. For compact topological groups, each such equivalence class contains an *orthogonal representation*, that is, a representation such that $\Gamma(s)$ is orthogonal for each $s \in G$.

2.6.3. Theorem. Each representation of a compact group is equivalent with an orthogonal representation.

P r o o f. Denote by $\langle .,. \rangle$ the inner product in $\mathbb{R}^n$, and by m the Haar measure of the group G. Define a positive definite, symmetric bilinear form $B : \mathbb{R}^n \times \mathbb{R}^n \to \mathbb{R}$ by :

$$B(x,y) = \int_G \langle \Gamma(s)x, \Gamma(s)y \rangle dm(s) \quad , \quad \forall x,y \in \mathbb{R}^n \ . \tag{3}$$

It follows from the invariance of the Haar measure that

$$B(\Gamma(t)x, \Gamma(t)y) = B(x,y) \quad , \quad \forall t \in G \ , \ \forall x,y \in \mathbb{R}^n \ . \tag{4}$$

We have $B(x,y) = \langle Ax,y \rangle$, for some symmetric positive definite $A \in L(\mathbb{R}^n)$. Also, there is a symmetric positive definite $T \in L(\mathbb{R}^n)$ such that $T^2 = A$. Then $B(x,y) = \langle Tx,Ty \rangle$, and (4) can be written in the form :

$$\langle T\Gamma(t)T^{-1}x, T\Gamma(t)T^{-1}y \rangle = \langle x,y \rangle \quad , \quad \forall t \in G \ , \ \forall x,y \in \mathbb{R}^n \ . \tag{5}$$

This shows that $T\Gamma(t)T^{-1}$ is orthogonal, for each $t \in G$. □

2.6.4. Definition. A set $M = \{M_k \mid k \in K\} \subset L(\mathbb{R}^n)$ of linear operators over $\mathbb{R}^n$ is said to be *reducible* if there exists a subspace V of $\mathbb{R}^n$ such that

(i) $0 < \dim V < n$;

(ii) $M_k(V) \subset V \quad , \quad \forall k \in K$.

If M is not reducible, we say that M is *irreducible*.

If $\mathbb{R}^n = \sum_{j=1}^{N} \oplus V_j$, such that for each $j = 1,\ldots,N$:

(i) $0 < \dim V_j \leq n$;

(ii) $M_k(V_j) \subset V_j \quad , \quad \forall k \in K$;

(iii) $M_j = \{M_k|_{V_j} \mid k \in K\}$ is irreducible ;

then we say that M is *completely reducible*.

2.6.5. Definition. A representation Γ of a group G is called *reducible* (respectively *irreducible, completely reducible*) when the set of linear operators $\{\Gamma(s) \mid s \in G\}$ is reducible (respectively irreducible, completely reducible).

2.6.6. Lemma. Each orthogonal representation of a group G is completely reducible.

P r o o f. If V is a proper subspace of $\mathbb{R}^n$ which is invariant under Γ, then the argument used in example 2.5.11 shows that also $V^\perp$ is invariant under Γ. Applying this argument a finite number of times one obtains a complete reduction of Γ. □

The foregoing reduces the study of all representations of a compact topological group to the study of all irreducible orthogonal representations. The following result is fundamental for the study of such representations.

2.6.7. Theorem (Schur's lemma). Suppose that $M \subset L(\mathbb{R}^m)$ and $N \subset L(\mathbb{R}^n)$ are irreducible, and that $A \in L(\mathbb{R}^n,\mathbb{R}^m)$ is such that for each $M \in M$ there is some $N \in N$, and for each $N \in N$ there is some $M \in M$ such that

$$MA = AN .$$

Then either $A = 0$, or $n = m$ and A is non-singular.

P r o o f. Let $V = \ker A \subset \mathbb{R}^n$. By the hypothesis, we can find for each $N \in \mathcal{N}$ some $M \in \mathcal{M}$ such that

$$ANx = MAx = 0 \qquad , \qquad \forall x \in V .$$

This shows that $N(V) \subset V$, $\forall N \in \mathcal{N}$. Since $\mathcal{N}$ is irreducible, we conclude that $V = \{0\}$ or $V = \mathbb{R}^n$. In the first case $n \leqslant m$, and A is injective; in the second case $A = 0$.

Similarly, if $V' = R(A) \subset \mathbb{R}^m$, $M \in \mathcal{M}$ and $x \in \mathbb{R}^n$, then we have :

$$MAx = ANx \in V' .$$

So $M(V') \subset V'$ for each $M \in \mathcal{M}$. Then the irreducibility of $\mathcal{M}$ implies that $V' = \{0\}$ or $V' = \mathbb{R}^m$. In the first case $A = 0$, in the second case $n \geqslant m$ and A is surjective. Combining both results proves the theorem. □

2.6.8. <u>Corollary</u>. Let Γ_1 and Γ_2 be two irreducible representations of a group G, with $\dim \Gamma_1 = m$ and $\dim \Gamma_2 = n$. Let $A \in L(\mathbb{R}^n, \mathbb{R}^m)$ be such that

$$\Gamma_1(s)A = A\Gamma_2(s) \qquad , \qquad \forall s \in G . \tag{6}$$

Then either $A = 0$, or $n = m$ and A is non-singular. In this last case Γ_1 and Γ_2 are equivalent. □

2.6.9. <u>Corollary</u>. Let Γ_1 and Γ_2 be two n-dimensional representations of a group G, and assume Γ_2 is irreducible. If there exists a non-zero $A \in L(\mathbb{R}^n)$ such that (6) holds, then A is non-singular, and Γ_1 is an irreducible representation equivalent to Γ_2. □

2.6.10. <u>Corollary</u>. Let $\Gamma : G \to L(\mathbb{R}^n)$ be an irreducible representation, and let $A \in L(\mathbb{R}^n)$ be such that

$$A\Gamma(s) = \Gamma(s)A \qquad , \qquad \forall s \in G . \tag{7}$$

Suppose also that A has at least one real eigenvalue $c \in \mathbb{R}$.

Then $A = cI$.

P r o o f. It is sufficient to apply corollary 8 to $A - cI$. □

2.6.11. Lemma. Let $\Gamma : G \to L(\mathbb{R}^n)$ be an irreducible representation. Then either $n = 1$ and $\Gamma(s)x = x$, $\forall x \in \mathbb{R}$, $\forall s \in G$ (i.e. Γ is the trivial representation), or

$$\{x \in \mathbb{R}^n \mid \Gamma(s)x = x,\ \forall s \in G\} = \{0\} . \tag{8}$$

P r o o f. The set at the left-hand side of (8) is an invariant subspace for Γ, and so equals $\mathbb{R}^n$ or $\{0\}$. This proves the lemma. □

2.6.12. Definition. For each $n = 1,2,\ldots$ we denote by $O(n)$ the group of orthogonal linear operators on $\mathbb{R}^n$:

$$O(n) = \{R \in L(\mathbb{R}^n) \mid RR^T = R^T R = I\} . \tag{9}$$

By $SO(n)$ we denote the subgroup of $O(n)$ containing those $R \in O(n)$ which are orientation preserving, i.e. which satisfy $\det R = 1$. Both $O(n)$ and $SO(n)$ are compact topological groups if we give them the topology induced by the usual topology on $L(\mathbb{R}^n)$.

It follows from lemma 2.5.3 that if $\Gamma : O(n) \to L(\mathbb{R}^m)$ is a representation of $O(n)$ over $\mathbb{R}^m$, then Γ is continuous. The same holds for representations of $SO(n)$. We will study in particular the representations of $O(2)$, $SO(2)$, $O(3)$ and $SO(3)$.

2.6.13. Representations of O(2) and SO(2). Starting with $SO(2)$, consider the following mapping :

$$\phi : \mathbb{R} \to L(\mathbb{R}^2) , \quad \alpha \mapsto \phi(\alpha) = \begin{pmatrix} \cos\alpha & \sin\alpha \\ -\sin\alpha & \cos\alpha \end{pmatrix} . \tag{10}$$

This is a C^∞-mapping, satisfying

$$\phi(0) = I , \quad \phi(\alpha+\beta) = \phi(\alpha)\phi(\beta) , \quad \forall \alpha,\beta \in \mathbb{R} . \tag{11}$$

Also, $\phi(\mathbb{R}) = SO(2)$, and $\phi^{-1}(\phi(\alpha)) = \{\alpha+2\pi k \mid k\in\mathbb{Z}\}$. So we can consider ϕ as a homeomorphism between $\mathbb{R}$ (mod 2π), with the quotient topology, and SO(2). In fact, ϕ can be used to give SO(2) a differential structure. If now $\Gamma : SO(2) \to L(\mathbb{R}^n)$ is a representation, then $\tilde{\Gamma} = \Gamma\circ\phi : \mathbb{R} \to L(\mathbb{R}^n)$ is a continuous mapping such that :

$$\text{(i)} \quad \tilde{\Gamma}(\alpha+2\pi) = \tilde{\Gamma}(\alpha) \quad , \quad \forall\alpha \in \mathbb{R} ;$$

$$\text{(ii)} \quad \tilde{\Gamma}(0) = I ;$$

$$\text{(iii)} \quad \tilde{\Gamma}(\alpha+\beta) = \tilde{\Gamma}(\alpha)\tilde{\Gamma}(\beta) \quad , \quad \forall\alpha,\beta \in \mathbb{R} . \tag{12}$$

Conversely, given a continuous mapping $\tilde{\Gamma} : \mathbb{R} \to L(\mathbb{R}^n)$ satisfying (i)-(iii), then there exists a unique representation $\Gamma : SO(2) \to L(\mathbb{R}^n)$ such that $\tilde{\Gamma} = \Gamma\circ\phi$. Therefore, Γ is uniquely determined by $\Gamma\circ\phi$, which we will itself denote by Γ, and also call a representation of SO(2). So, a representation Γ of SO(2) is a continuous map $\Gamma : \mathbb{R} \to L(\mathbb{R}^n)$ satisfying (i)-(iii) above.

As for O(2), define $\sigma \in O(2)$ by :

$$\sigma e_1 = e_1 \quad , \quad \sigma e_2 = -e_2 . \tag{13}$$

Then $O(2) = \phi(\mathbb{R}) \cup (\sigma\circ\phi)(\mathbb{R})$. Also :

$$\sigma^2 = I \quad \text{and} \quad \sigma\circ\phi(\alpha) = \phi(-\alpha)\circ\sigma \quad , \quad \forall\alpha \in \mathbb{R} . \tag{14}$$

A representation of O(2) is uniquely determined by a representation $\Gamma : \mathbb{R} \to L(\mathbb{R}^n)$ of SO(2), together with an operator $\Gamma_\sigma = \Gamma(\sigma) \in L(\mathbb{R}^n)$ such that

$$\Gamma_\sigma^2 = I \quad \text{and} \quad \Gamma_\sigma.\Gamma(\alpha) = \Gamma(-\alpha).\Gamma_\sigma \quad , \quad \forall\alpha \in \mathbb{R} . \tag{15}$$

2.6.14. Lemma. Let $\Gamma : \mathbb{R} \to L(\mathbb{R}^n)$ be a representation of SO(2). Then Γ is continuously differentiable. If we define

$$A = \frac{d\Gamma}{d\alpha}(0) , \tag{16}$$

then we have :

(i) $\frac{d\Gamma}{d\alpha}(\alpha) = A\Gamma(\alpha)$, $\forall\alpha \in \mathbb{R}$;

(ii) Γ is irreducible if and only if $\mathbb{R}^n$ is irreducible for A.

Moreover, if Γ is an orthogonal representation, then we also have :

$$\text{(iii)}\ A^T = -A\ . \tag{17}$$

P r o o f. If we can show that Γ is differentiable at $\alpha = 0$, then the differentiability everywhere and (i) follow immediately from (12). In order to show the differentiablity at $\alpha = 0$, define $B : \mathbb{R} \to L(\mathbb{R}^n)$ by :

$$B(\beta) = \int_0^\beta \Gamma(\alpha)d\alpha = \beta\int_0^1 \Gamma(s\beta)ds = \beta\tilde{B}(\beta) \quad , \quad \forall\beta \in \mathbb{R}\ .$$

Then $\tilde{B} : \mathbb{R} \to L(\mathbb{R}^n)$ is continuous, and $\tilde{B}(0) = I$. Consequently $\tilde{B}(\beta)$ is non-singular for $|\beta|$ sufficiently small. Using (12), it is easily verified that

$$(\Gamma(\alpha)-I)B(\beta) = (\Gamma(\beta)-I)B(\alpha) \quad , \quad \forall\alpha,\beta \in \mathbb{R}\ .$$

Dividing by $\alpha\beta$, we find, for $\alpha \neq 0$, $\beta \neq 0$ and $|\beta|$ sufficiently small :

$$\frac{1}{\alpha}[\,\Gamma(\alpha)-I\,] = \frac{1}{\beta}[\,\Gamma(\beta)-I\,]\tilde{B}(\alpha)(\tilde{B}(\beta))^{-1}\ . \tag{18}$$

Since the right-hand side of (18) has a limit for $\alpha \to 0$, the same holds for the left-hand side, proving the existence of $\frac{d\Gamma}{d\alpha}(0)$.

It follows from (i) and $\Gamma(0) = I$ that $\Gamma(\alpha) = \exp(A\alpha)$, $\forall\alpha \in \mathbb{R}$. Since a subspace V of $\mathbb{R}^n$ is invariant under $\exp(A\alpha)$ if and only if V is invariant under A, we have (ii). Finally, if Γ is orthogonal then $\Gamma^T(\alpha) = \Gamma^{-1}(\alpha) = \Gamma(-\alpha)$. Differentiation at $\alpha = 0$ gives (iii). □

2.6.15. <u>Theorem</u>. Let $\Gamma : \mathbb{R} \to L(\mathbb{R}^n)$ be an irreducible orthogonal representation of SO(2). Then either

(i) $n = 1$, and

$$\Gamma(\alpha) = \Gamma^{(0)}(\alpha) \overset{\text{def}}{=} 1 \quad , \quad \forall\alpha \in \mathbb{R}\ , \tag{19}$$

or

(ii) $n = 2$, and there exists some $k \in \mathbb{Z}\setminus\{0\}$ such that

$$\Gamma(\alpha) = \Gamma^{(k)}(\alpha) \stackrel{\text{def}}{=} \begin{pmatrix} \cos k\alpha & \sin k\alpha \\ -\sin k\alpha & \cos k\alpha \end{pmatrix} \quad , \quad \forall\alpha \in \mathbb{R} . \tag{20}$$

P r o o f. Define $A \in L(\mathbb{R}^n)$ by (16). Since Γ is irreducible, $\mathbb{R}^n$ is irreducible for A. It follows from (17) that A^2 is a symmetric operator, and consequently has real eigenvalues. Since A^2 commutes with A, Schur's lemma implies that $A^2 = \lambda I$, for some $\lambda \in \mathbb{R}$.

Let $\mu \in \mathbb{C}$ be an eigenvalue of A; then $\exp(\mu\alpha)$ is an eigenvalue of $\Gamma(\alpha) = \exp(A\alpha)$. Since $\Gamma(2\pi) = I$, it follows that $\exp(2\pi\mu) = 1$, i.e. $\mu = ik$ for some $k \in \mathbb{Z}$. We conclude that $A^2 = -k^2 I$ for some $k \in \mathbb{Z}$.

First suppose that 0 is an eigenvalue of A, i.e. there is some $u_0 \in \mathbb{R}^n$, $u_0 \neq 0$ such that $Au_0 = 0$. Then $V = \text{span}\{u_0\}$ is invariant under A. By the irreducibility hypothesis it follows that $V = \mathbb{R}^n$, $n = 1$ and $A = 0$. This proves (19).

Next suppose that $A^2 = -k^2 I$ for some $k \in \mathbb{Z}\setminus\{0\}$. Take any $u_0 \in \mathbb{R}^n$, $u_0 \neq 0$, and let $v_0 = Au_0$. Then v_0 is linearly independent from u_0, since otherwise A would have a real eigenvalue, which can only be zero, and we would have $A^2 = 0$. Since $Av_0 = A^2 u_0 = -k^2 u_0$ it follows that $V = \text{span}\{u_0, v_0\}$ is invariant under A. Since $\mathbb{R}^n$ is irreducible for A, we have $V = \mathbb{R}^n$ and $n = 2$. Let $\{e_1, e_2\}$ be the canonical basis of $\mathbb{R}^2$. Then $\langle Ae_i, e_j\rangle = \langle e_i, A^T e_j\rangle = -\langle e_i, Ae_j\rangle$, for $i,j = 1,2$. It follows that

$$Ae_1 = -\lambda e_2 \quad , \quad Ae_2 = \lambda e_1$$

for some $\lambda \in \mathbb{R}$. The condition $A^2 = -k^2 I$ then implies that $\lambda = k$ (possibly after replacing k by -k). Since $\Gamma(\alpha) = \exp(A\alpha)$ this gives :

$$\left.\begin{aligned} \Gamma(\alpha)e_1 &= \cos k\alpha . e_1 - \sin k\alpha . e_2 , \\ \Gamma(\alpha)e_2 &= \sin k\alpha . e_1 + \cos k\alpha . e_2 , \end{aligned}\right. \qquad \forall\alpha \in R .$$

This proves (20). □

2.6.16. Lemma. For each $k \neq 0$ the representations $\Gamma^{(k)}$ and $\Gamma^{(-k)}$ given by theorem 15 are equivalent.

P r o o f. The foregoing proof shows that we obtain $\Gamma^{(-k)}$ from $\Gamma^{(k)}$ by interchanging the role of e_1 and e_2. More formally, if we define $T \in L(\mathbb{R}^2)$ by $Te_1 = e_2$, $Te_2 = e_1$, then $T^{-1}\Gamma^{(k)}(\alpha)T = \Gamma^{(-k)}(\alpha)$, $\forall\alpha \in \mathbb{R}$. $\square$

2.6.17. Theorem. Let $\Gamma : \mathbb{R} \to L(E)$ be an irreducible representation of SO(2) over a finite-dimensional vector space E. Then one has either dim E = 1 and $\Gamma(\alpha)u = u$, $\forall\alpha \in \mathbb{R}$, $\forall u \in E$, or dim E = 2, and E has a basis $\{u_1,u_2\}$ such that

$$\begin{aligned} \Gamma(\alpha)u_1 &= \cos k\alpha . u_1 - \sin k\alpha . u_2 \\ \Gamma(\alpha)u_2 &= \sin k\alpha . u_1 + \cos k\alpha . u_1 , \end{aligned} \tag{21}$$

for some $k \in \mathbb{N}\setminus\{0\}$. $\square$

2.6.18. Lemma. A finite-dimensional representation Γ of O(2) is irreducible if and only if the restriction of Γ to SO(2) is irreducible.

P r o o f. Γ is determined by a representation $\Gamma : \mathbb{R} \to L(\mathbb{R}^n)$ of SO(2), and by $\Gamma_\sigma \in L(\mathbb{R}^n)$ such that (15) is satisfied. By lemma 14, Γ is differentiable. If we define A by (16), then Γ is irreducible if and only if $\mathbb{R}^n$ is irreducible for $\{A,\Gamma_\sigma\}$. We have to show that $\mathbb{R}^n$ is irreducible for $\{A,\Gamma_\sigma\}$ if and only if $\mathbb{R}^n$ is irreducible for A.

It is clear that if $\mathbb{R}^n$ is irreducible for A, then $\mathbb{R}^n$ is also irreducible for $\{A,\Gamma_\sigma\}$. Conversely, suppose $\mathbb{R}^n$ is irreducible for $\{A,\Gamma_\sigma\}$. Differentiation of (15) at $\alpha = 0$ gives

$$\Gamma_\sigma A = -A\Gamma_\sigma . \tag{22}$$

This implies that A^2 commutes with A and Γ_σ. Then, as before, it follows from Schur's lemma and $\Gamma(2\pi) = I$ that $A^2 = -k^2 I$ for some $k \in \mathbb{Z}$. Moreover, since $\Gamma_\sigma^2 = I$, Γ_σ must have an eigenvalue equal to +1 or -1. Let $u_0 \in \mathbb{R}^n$, $u_0 \neq 0$ be such that $\Gamma_\sigma u_0 = \pm u_0$. Define $v_0 = Au_0$ and $V = \text{span}\{u_0,v_0\}$.

We have $Av_0 = A^2u_0 = -k^2u_0$ and $\Gamma_\sigma v_0 = \Gamma_\sigma Au_0 = -A\Gamma_\sigma u_0 = \mp Au_0 = \mp v_0$. This shows that V is invariant under A and Γ_σ; consequently $V = \mathbb{R}^n$. Now there are two possibilities. First, assume that v_0 is linearly dependent of u_0; then A has a real eigenvalue, which can only be zero, with eigenvector u_0. Then

$n = 1$, since $\dim V = 1$, and V is irreducible for A. If u_0 and v_0 are linearly independent, then $n = 2$ and the relations $v_0 = Au_0$, $Av_0 = -k^2u_0$ $(k \neq 0)$ show that $V = \mathbb{R}^2$ is irreducible for A. This proves the lemma. □

2.6.19. Theorem. Let $\Gamma : O(2) \to L(E)$ be an irreducible representation of $O(2)$ over a finite-dimensional vectorspace E. Then we have the following possibilities :

(i) $n = 1$, $\Gamma(\alpha)u = u$, $\Gamma_\sigma u = u$, $\forall\alpha \in \mathbb{R}$, $\forall u \in E$;

(ii) $n = 1$, $\Gamma(\alpha)u = u$, $\Gamma_\sigma u = -u$, $\forall\alpha \in \mathbb{R}$, $\forall u \in E$;

(iii) $n = 2$, and E has a basis $\{u_1,u_2\}$ such that

$$\Gamma(\alpha)u_1 = \cos k\alpha . u_1 - \sin k\alpha . u_2 ,$$
$$\Gamma(\alpha)u_2 = \sin k\alpha . u_1 + \cos k\alpha . u_2 , \qquad \forall\alpha \in \mathbb{R}$$
$$\Gamma_\sigma u_1 = u_1 , \quad \Gamma_\sigma u_2 = -u_2 ,$$

for some $k \in \mathbb{N}\backslash\{0\}$.

P r o o f. The proof of the foregoing lemma shows that either $n = 1$ or $n = 2$. In case $n = 1$, then $A = 0$ and $\Gamma_\sigma = \pm 1$, which gives the possibilities (i) and (ii). In case $n = 2$, both $+1$ and -1 are eigenvalues of Γ_σ; also $A^2 = -k^2I$ for some $k \in \mathbb{N}\backslash\{0\}$. Let $u_1 \in E$, $u_1 \neq 0$ be such that $\Gamma_\sigma u_1 = u_1$. Define $u_2 \in E$ such that $Au_1 = -ku_2$; then $Au_2 = ku_1$, giving us the case (iii). □

Next we turn to the representations of the groups $O(3)$ and $SO(3)$.

2.6.20. Lemma. Let $\Gamma : O(3) \to L(\mathbb{R}^n)$ be a representation of the group $O(3)$. Then Γ is irreducible if and only if its restriction to the subgroup $SO(3)$ is irreducible. If this is the case, then

$$\Gamma_- = \Gamma(-I) = \pm I .$$

P r o o f. If the restriction of Γ to $SO(3)$ is irreducible, then also Γ is irreducible. As for the converse, suppose that Γ is irreducible. Then $\Gamma_- = \Gamma(-I)$ commutes with $\Gamma(R)$, for each $R \in O(3)$. Moreover, $\Gamma_-^2 = I$, and consequently Γ_- has real eigenvalues ± 1. By corollary 10, it follows that $\Gamma_- = \pm I$. Since $O(3) = SO(3) \cup \{-R \mid R \in SO(3)\}$ it is then clear that the irreducibility of Γ implies the irreducibility of the restriction of Γ to SO(3). □

This lemma allows us to restrict attention to the subgroup SO(3).

2.6.21. The Lie group SO(3). We will show now that SO(3) forms a 3-dimensional smooth submanifold of the 9-dimensional space $L(\mathbb{R}^3)$. (For the general theory of differentiable manifolds, see e.g. Lang [141]). Using the group structure of SO(3), it is sufficient to show that there is a neighbourhood U of I in $L(\mathbb{R}^3)$ such that $SO(3) \cap U$ is a submanifold. Since the determinant function is continuous on $L(\mathbb{R}^3)$, we can take U sufficiently small such that $SO(3) \cap U = O(3) \cap U$.

Let $L_s(\mathbb{R}^3)$ and $L_a(\mathbb{R}^3)$ be the subspaces of symmetric, respectively anti-symmetric operators :

$$L_s(\mathbb{R}^3) = \{R \in L(\mathbb{R}^3) \mid R^T = R\} \ , \ L_a(\mathbb{R}^3) = \{R \in L(\mathbb{R}^3) \mid R^T = -R\} \ .$$

Then $\dim L_s(\mathbb{R}^3) = 6$, $\dim L_a(\mathbb{R}^3) = 3$, and $L(\mathbb{R}^3) = L_s(\mathbb{R}^3) \oplus L_a(\mathbb{R}^3)$.

Define $\eta : L(\mathbb{R}^3) \to L(\mathbb{R}^3)$ by :

$$\eta(R) = \frac{1}{2}(R^T R - I) + \frac{1}{2}(R^T - R) \quad , \quad \forall R \in L(\mathbb{R}^3) \ .$$

Then $\eta(I) = 0$, and $D\eta(I)$ is the identity in $L(\mathbb{R}^3)$. Consequently, the restriction of η to an appropriate open neighbourhood U of I in $L(\mathbb{R}^3)$ is a C^∞-diffeomorphism from U onto an open neighbourhood $\eta(U)$ of 0 in $L(\mathbb{R}^3)$. Moreover, $O(3) \cap U = \eta^{-1}(L_a(\mathbb{R}^3) \cap \eta(U))$. This shows that $O(3) \cap U$ is a three-dimensional smooth submanifold of $L(\mathbb{R}^3)$. The tangent space to SO(3) in the point I is given by $D\eta^{-1}(0)(L_a(\mathbb{R}^3)) = L_a(\mathbb{R}^3)$.

Let V be an open neighbourhood of I in $L(\mathbb{R}^3)$ such that $\{A.B \mid A \in V, B \in V\} \subset U$. Then it is clear that the mapping

$$(A,B) \mapsto (\eta^{-1}(A).\eta^{-1}(B))$$

is a C^∞-mapping from $\eta(V) \times \eta(V)$ into $\eta(U)$. So also its restriction to $(\eta(V) \cap L_a(\mathbb{R}^3)) \times (\eta(V) \cap L_a(\mathbb{R}^3))$ is a C^∞-mapping into $\eta(U) \cap L_a(\mathbb{R}^3)$. This shows that the mapping $(R_1,R_2) \mapsto R_1.R_2$ from $SO(3) \times SO(3)$ into SO(3) is smooth. Then an easy application of the implicit function theorem shows that the mapping $R \mapsto R^{-1}$ from SO(3) into itself is also smooth. This proves that

SO(3) is a 3-dimensional Lie group.

2.6.22. Lemma. Let $\Gamma : SO(3) \to L(\mathbb{R}^n)$ be a finite-dimensional representation of the Lie group SO(3). Then Γ is a C^∞-mapping.

P r o o f. Using the fact that Γ is a group representation, it is sufficient to proof that the restriction of Γ to an appropriate neighbourhood U of I in SO(3) is smooth. For that purpose, we will use a special system of local coordinates, which we introduce now.

Let $\{e_1,e_2,e_3\}$ be the canonical basis of $\mathbb{R}^3$, and define a basis $\{J_1,J_2,J_3\}$ of $L_a(\mathbb{R}^3)$ by

$$\begin{aligned} &J_1.e_1 = 0 \quad , \quad J_1.e_2 = -e_3 \;, \quad J_1.e_3 = e_2 \;; \\ &J_2.e_1 = -e_3 \;, \quad J_2.e_2 = 0 \quad , \quad J_2.e_3 = e_1 \;; \\ &J_3.e_1 = -e_2 \;, \quad J_3.e_2 = e_1 \quad , \quad J_3.e_3 = 0 \quad . \end{aligned} \tag{23}$$

For i = 1,2,3, define $\phi_i : \mathbb{R} \to SO(3)$ by

$$\phi_i(\alpha) = \exp(\alpha J_i) \qquad , \qquad \forall \alpha \in \mathbb{R} \; . \tag{24}$$

Finally, define $\psi : \mathbb{R}^3 \to SO(3)$ by

$$\psi(\alpha_1,\alpha_2,\alpha_3) = \phi_1(\alpha_1)\phi_2(\alpha_2)\phi_3(\alpha_3) \quad , \quad \forall \alpha_i \in \mathbb{R},\; i = 1,2,3 \; . \tag{25}$$

ψ is a C^∞-mapping, with $\psi(0) = I$, and $D\psi(0).(\tilde{\alpha}_1,\tilde{\alpha}_2,\tilde{\alpha}_3) = \tilde{\alpha}_1 J_1+\tilde{\alpha}_2 J_2+\tilde{\alpha}_3 J_3$. So $D\psi(0)$ is an isomorphism between $\mathbb{R}^3$ and the tangent space $L_a(\mathbb{R}^3)$ to SO(3) at I; consequently ψ defines local coordinates near I on SO(3). It remains to show that $\Gamma\circ\psi : \mathbb{R}^3 \to L(\mathbb{R}^n)$ is C^∞.

We have :

$$\Gamma(\psi(\alpha_1,\alpha_2,\alpha_3)) = \Gamma(\phi_1(\alpha_1)).\Gamma(\phi_2(\alpha_2)).\Gamma(\phi_3(\alpha_3)) \; .$$

Now the argument used in the proof of lemma 14 shows that for each i = 1,2,3, the mapping $\alpha_i \mapsto \Gamma(\phi_i(\alpha_i))$ from $\mathbb{R}$ into $L(\mathbb{R}^n)$ is differentiable. This implies that also $\Gamma\circ\psi$ is differentiable. □

2.6.23. The Lie algebra $so(3) = L_a(\mathbb{R}^3)$. Fix $A,B \in L_a(\mathbb{R}^3)$ and $\alpha \in \mathbb{R}$. Then the mapping

$$\beta \mapsto \exp(A\alpha)\exp(B\beta)\exp(-A\alpha)$$

defines a 1-parameter subgroup of SO(3); differentiating at $\beta = 0$ we obtain a smooth mapping from $\mathbb{R}$ into $L_a(\mathbb{R}^3)$ given by

$$\alpha \mapsto \exp(A\alpha)B.\exp(-A\alpha) \ .$$

Differentiating again at $\alpha = 0$ we find an element of $L_a(\mathbb{R}^3)$, which we denote by $[A,B]$, and call the Lie bracket of A and B. So $[.,.] : L_a(\mathbb{R}^3)\times L_a(\mathbb{R}^3) \to L_a(\mathbb{R}^3)$ is defined by

$$[A,B] = \frac{d}{d\alpha}(\frac{d}{d\beta}\exp(A\alpha)\exp(B\beta)\exp(-A\alpha)|_{\beta=0})|_{\alpha=0} \ ,$$

and is explicitly given by

$$[A,B] = AB - BA \ . \tag{26}$$

An explicit calculation shows that for the J_i (i=1,2,3) defined by (23) we have :

$$[J_1,J_2] = J_3 \quad , \quad [J_2,J_3] = J_1 \quad , [J_3,J_1] = J_2 \ . \tag{27}$$

$L_a(\mathbb{R}^3)$, equipped with the Lie bracket, forms the Lie algebra of the group SO(3); as such, it is sometimes denoted by so(3).

2.6.24. Lemma. Let $\Gamma : SO(3) \to L(\mathbb{R}^n)$ be a representation. For i = 1,2,3 define $L_i \in L(\mathbb{R}^n)$ by

$$L_i = \frac{d}{d\alpha}\Gamma(\phi_i(\alpha))|_{\alpha=0} \ . \tag{28}$$

Then

$$[L_1,L_2] = L_3 \quad , \quad [L_2,L_3] = L_1 \quad , \quad [L_3,L_1] = L_2 \ . \tag{29}$$

P r o o f. By the differentiability of Γ we have for each smooth curve $\zeta : \mathbb{R} \to SO(3)$ such that $\zeta(0) = I$:

$$\frac{d}{d\alpha}\Gamma(\zeta(\alpha))\Big|_{\alpha=0} = D\Gamma(I).\frac{d\zeta}{d\alpha}(0) ,$$

where $D\Gamma(I) \in L(L_a(\mathbb{R}^3), L(\mathbb{R}^n))$ and $\frac{d\zeta}{d\alpha}(0) \in L_a(\mathbb{R}^3)$. In particular :

$$L_i = D\Gamma(I).J_i \qquad , \qquad i = 1,2,3 . \tag{30}$$

It follows that

$$[L_1,L_2] = \frac{d}{d\alpha}\Big(\frac{d}{d\beta}\Gamma(\phi_1(\alpha))\Gamma(\phi_2(\beta))\Gamma(\phi_1(-\alpha))\Big|_{\beta=0}\Big)\Big|_{\alpha=0}$$

$$= \frac{d}{d\alpha}\Big(\frac{d}{d\beta}\Gamma(\phi_1(\alpha)\phi_2(\beta)\phi_1(-\alpha))\Big|_{\beta=0}\Big)\Big|_{\alpha=0}$$

$$= D\Gamma(I).[J_1,J_2] = D\Gamma(I).J_3 = L_3 . \qquad \square$$

2.6.25. Lemma. Let $\Gamma : SO(3) \to L(\mathbb{R}^n)$ be a representation. Then Γ is irreducible if and only if $\mathbb{R}^n$ is irreducible under $\{L_1,L_2,L_3\}$.

P r o o f. Each $R \in SO(3)$ can be written as an appropriate compositon of operators of the form $\phi_i(\alpha_i)$ (i=1,2,3). Since $\phi_i(\alpha+\beta) = \phi_i(\alpha)\phi_i(\beta)$ it follows that

$$\frac{d}{d\alpha}\Gamma(\phi_i(\alpha)) = L_i\Gamma(\phi_i(\alpha)) \qquad , \qquad \forall\alpha \in \mathbb{R} , \ i = 1,2,3 ,$$

and consequently $\Gamma(\phi_i(\alpha)) = \exp(L_i\alpha)$. It follows that Γ is irreducible if and only if $\mathbb{R}^n$ is irreducible under $\{\Gamma(\phi_i(\alpha)) \mid i=1,2,3, \ \alpha\in\mathbb{R}\}$, which in turn holds if and only if $\mathbb{R}^n$ is irreducible under $\{L_i \mid i=1,2,3\}$. $\square$

2.6.26. The Casimir operator. Let now $\Gamma : SO(3) \to L(\mathbb{R}^n)$ be an orthogonal irreducible representation. Then the L_i (i=1,2,3) are anti-symmetric.

Consider the restriction of Γ to $\{\phi_3(\alpha) \mid \alpha\in\mathbb{R}\}$, which is a subgroup of SO(3) isomorphic to SO(2). So this restriction of Γ forms a representation of SO(2). It follows from the theory of such representations that we can write $\mathbb{R}^n$ as a direct sum of one- and two-dimensional subspaces, on which the representation of SO(2) is irreducible. The one-dimensional subspaces are spanned by vectors u such that $L_3u = 0$, the two-dimensional ones by vectors u, v such that $L_3u = -qv$, $L_3v = qu$, for some $q = 1,2,3,\ldots$

Consider such nonzero vectors u, v for which $L_3u = -qv$, $L_3v = qu$. Define u^+, v^+, u^- and v^- by

$$u^+ = L_2u - L_1v \quad , \quad v^+ = L_1u + L_2v \ , \tag{31}$$

and

$$u^- = -L_2u - L_1v \quad , \quad v^- = L_1u - L_2v \ . \tag{32}$$

Then it follows from (29) that

$$L_3u^+ = -(q+1)v^+ \quad , \quad L_3v^+ = (q+1)u^+ \ , \tag{33}$$

and

$$L_3u^- = -(q-1)v^- \quad , \quad L_3v^- = (q-1)u^- \ . \tag{34}$$

Now we define the so-called Casimir operator

$$C = -L_1^2 - L_2^2 - L_3^2 \ . \tag{35}$$

It is a symmetric operator, and consequently its eigenvalues are real. Moreover, it is easily seen from (29) that C commutes with each of the L_i. Since $\mathbb{R}^n$ is irreducible under $\{L_i \mid i=1,2,3\}$, it follows from corollary 10 that $C = \lambda I$ for some $\lambda \in \mathbb{R}$.

Let ℓ be the maximal value of q appearing in the decomposition of $\mathbb{R}^n$ discussed above, and let u, v be a corresponding pair of nonzero vectors such that $L_3u = -\ell v$, $L_3v = \ell u$. Then (33) implies that $L_2u - L_1v = 0$ and $L_1u + L_2v = 0$. But then

$$\begin{aligned} Cu &= -L_2(L_2u-L_1v) - L_1(L_1u+L_2v) + L_3v - L_3^2u \\ &= \ell(\ell+1)u \ . \end{aligned}$$

We conclude that $C = \ell(\ell+1)I$.

2.6.27. Lemma. Let $\Gamma : SO(3) \to L(\mathbb{R}^n)$ be an orthogonal irreducible representation. Then there exists some $u_0 \neq 0$ such that $L_3u_0 = 0$.

P r o o f. Let s be the minimal value of q appearing in the decomposition of $\mathbb{R}^n$ discussed above, where $s = 0$ if there is some $u \neq 0$ such that $L_3u = 0$. We have $0 \leq s \leq \ell$, and we must prove that $s = 0$. Suppose that $0 < s \leq \ell$. Then there are nonzero vectors u, v such that $L_3u = -sv$, $L_3v = su$, while the definition of s and (34) imply that $-L_2u - L_1v = 0$ and $L_1u - L_2v = 0$. But then

$$\begin{aligned} Cu &= L_2(-L_2u-L_1v) - L_1(L_1u-L_2v) - L_3v - L_3^2u \\ &= s(s-1)u \ . \end{aligned}$$

Since $u \neq 0$ we must have $s(s-1) = \ell(\ell+1)$, which is impossible for $0 < s \leq \ell$. □

2.6.28. Theorem. Let $\Gamma : SO(3) \to L(\mathbb{R}^n)$ be an irreducible representation. Then $n = 2\ell+1$ for some integer ℓ. Up to equivalence, Γ is uniquely determined by ℓ.

P r o o f. We may assume that Γ is orthogonal. By lemma 27, we can find some $u_0 \in \mathbb{R}^n$, $u_0 \neq 0$ such that $L_3u_0 = 0$. Define $u_1 = L_2u_0$, $v_1 = L_1u_0$, and more generally :

$$u_{q+1} = L_2u_q - L_1v_q \quad , \quad v_{q+1} = L_1u_q + L_2v_q \quad , \quad \forall q \in \mathbb{N} \ . \tag{36}$$

Then

$$L_3u_q = -qv_q \quad , \quad L_3v_q = qu_q \quad , \quad \forall q \in \mathbb{N} \ . \tag{37}$$

If $(u_q,v_q) \neq (0,0)$, then the vectors $\{u_0,u_1,v_1,\ldots,u_q,v_q\}$ are linearly independent. Consequently, there exists some $\ell \in \mathbb{N}$ such that $(u_\ell,v_\ell) \neq (0,0)$ and $(u_{\ell+1},v_{\ell+1}) = (0,0)$. Let $U = \mathrm{span}\{u_0,u_1,v_1,\ldots,u_\ell,v_\ell\}$. Then U is a $(2\ell+1)$-dimensional subspace of $\mathbb{R}^n$, which is clearly invariant under L_3. Also, the same argument as used before shows that $C = \ell(\ell+1)I$. Moreover, $L_1u_0 \in U$, $L_2u_0 \in U$, and we have for $0 < q \leq \ell$:

$$-L_2u_q - L_1v_q = Cu_{q-1} + L_3^2u_{q-1} - L_3v_{q-1}$$

$$= [\ell(\ell+1)-q(q-1)]\,u_{q-1}\ , \tag{38}$$

$$L_1u_q - L_2v_q = [\ell(\ell+1)-q(q-1)]\,v_{q-1}\ .$$

Together with (36) this shows that U is invariant under L_1 and L_2. By lemma 25, this implies that $U = \mathbb{R}^n$, and $n = 2\ell+1$. Since the action of L_1, L_2 and L_3 on the basis $\{u_0,u_1,v_1,\ldots,u_\ell,v_\ell\}$ is uniquely determined by the foregoing relations, also the second part of the theorem is proved. □

2.6.29. Corollary. A representation $\Gamma : SO(3) \to L(\mathbb{R}^n)$ is irreducible if and only if

$$\dim\{u\in\mathbb{R}^n \mid L_3u = 0\} = 1\ . \tag{39}$$

P r o o f. This follows from lemma 27 and the proof of theorem 28. Remark also that

$$\{u\in\mathbb{R}^n \mid L_3u = 0\} = \{u\in\mathbb{R}^n \mid u = \Gamma(\phi_3(\alpha))u,\ \forall\alpha\in\mathbb{R}\}\ . \qquad \square$$

2.6.30. Spherical harmonics. For each $\ell \in \mathbb{N}$ one can realize a $(2\ell+1)$-dimensional representation of SO(3) by using spherical harmonics, which we introduce now.

Let H_ℓ be the space of homogeneous polynomials of degree ℓ in three scalar variables $H_\ell : \mathbb{R}^3 \to \mathbb{R}$, which are also harmonic :

$$\Delta H_\ell(x) = 0 \qquad , \qquad \forall x\in \mathbb{R}^3\ . \tag{40}$$

Let S^2 be the unit sphere in $\mathbb{R}^3$:

$$S^2 = \{\theta = (x_1,x_2,x_3)\in\mathbb{R}^3 \mid x_1^{\,2}+x_2^{\,2}+x_3^{\,2} = 1\}\ . \tag{41}$$

Finally, let

$$U_\ell = \{Y_\ell = H_\ell|_{S^2} \mid H_\ell\in H_\ell\}\ . \tag{42}$$

An element Y_ℓ of U_ℓ is called a spherical harmonic of order ℓ in 3 variables; it is the restriction to the unit sphere of a harmonic homogeneous polynomial of degree ℓ.

We can define a representation $\Gamma^{(\ell)} : SO(3) \to L(U_\ell)$ of SO(3) over the finite-dimensional space U_ℓ as follows :

$$(\Gamma^{(\ell)}(R)Y_\ell)(\theta) = Y_\ell(R^{-1}\theta) \quad , \quad \forall\theta \in S^2, \forall R \in SO(3), \forall Y_\ell \in U_\ell \ . \tag{43}$$

We have the following result.

2.6.31. Theorem. We have for each $\ell \in \mathbb{N}$:

(i) $\dim U_\ell = 2\ell+1$;

(ii) the representation $\Gamma^{(\ell)} : SO(3) \to L(U_\ell)$ defined by (43) is irreducible.

P r o o f. It is clear that $\dim U_\ell = \dim H_\ell$. Writing $H_\ell \in H_\ell$ in the form :

$$H_\ell(x) = \sum_{j=0}^{\ell} x_3{}^j A_{\ell-j}(x_1,x_2) \ , \tag{44}$$

where $A_i(x_1,x_2)$ is a homogeneous polynomial of degree i in x_1 and x_2, we find

$$\Delta H_\ell(x) = \sum_{j=2}^{\ell} j(j-1)x_3^{j-2}A_{\ell-j}(x_1,x_2) + \sum_{j=0}^{\ell-2} x_3{}^j \Delta_2 A_{\ell-j}(x_1,x_2) \ ,$$

where $\Delta_2 = \frac{\partial^2}{\partial x_1{}^2} + \frac{\partial^2}{\partial x_1{}^2}$ is the Laplacian in 2 variables. Bringing this in (40) and equating coefficients of equal powers of x_3, we find :

$$\Delta_2 A_{\ell-j} = -(j+2)(j+1)A_{\ell-j-2} \quad , \quad j = 0,\dots,\ell-2 \ . \tag{45}$$

This shows that there are no conditions on the coefficients appearing in A_ℓ and $A_{\ell-1}$, while $A_{\ell-2},\dots,A_0$ are uniquely determined by A_ℓ and $A_{\ell-1}$. Since A_ℓ and $A_{\ell-1}$ contain respectively $\ell+1$ and ℓ coefficients, we conclude that $\dim H_\ell = 2\ell+1$.

Now consider the subspace :

$$V = \{Y_\ell \in U_\ell \mid \Gamma(\phi_3(\alpha))Y_\ell = Y_\ell,\ \forall\alpha\in\mathbb{R}\} . \tag{46}$$

If $Y_\ell = H_\ell|S^2$, with $H_\ell \in \mathcal{H}_\ell$ of the form (44), then $Y_\ell \in V$ if and only if :

$$A_i(x_1\cos\alpha + x_2\sin\alpha,\ -x_1\sin\alpha + x_2\cos\alpha) = A_i(x_1,x_2),$$
$$\forall i = 0,\dots,\ell\ ,\ \forall\alpha \in \mathbb{R} . \tag{47}$$

This implies that

$$\begin{aligned} A_i(x_1,x_2) &= a_i(x_1{}^2+x_2{}^2)^{i/2} && \text{if} \quad i = \text{even} \\ &= 0 && \text{if} \quad i = \text{odd} . \end{aligned}$$

Since the A_i also have to satisfy (45), this shows that dim V = 1. Then it follows from Corollary 29 that $\Gamma^{(\ell)}$ is irreducible. □

2.6.32. Remark. The representation $\Gamma^{(\ell)}$ of SO(3) defined by (43) can be extended to a representation of O(3), by using (43) for all $R \in O(3)$. Then $\Gamma^{(\ell)}(-I) = (-1)^\ell I$.

3 Symmetry and the Liapunov–Schmidt method

The main objective of this chapter is to introduce the Liapunov-Schmidt reduction method, and to show how this reduction can be performed in a way which is compatible with the symmetries which may be present in the problem under consideration. The Liapunov-Schmidt method results in the so-called bifurcation equations, which form a finite set of equations, equivalent to the original problem. Our main result shows that if the reduction is done properly, then the bifurcation equations inherit the symmetry properties of the original problem.

In section 1 we describe the Liapunov-Schmidt method under very general hypotheses. In section 2 we introduce the concept of an equivariant mapping; the definition of such mappings contains all the ingredients of what we will consider as a mapping with symmetry properties. Section 3 contains the main result referred to above. In section 4 we restrict attention to symmetry-invariant solutions, and we show that for such solutions there is a further reduction of the bifurcation equations. In section 5 we use our formalism to describe some properties of reversible systems of ordinary differential equations (also called "systems with property E"); these results were found earlier by Hale [77]. Some aspects of the approach for such systems can be put in a more abstract form, as is done in section 6. This section also contains some further examples : subharmonic solutions of periodic differential equations, oscillations in conservative systems, Dirichlet boundary value problems, and the von Kármán equations for the buckling problem of rectangular and cylindrical plates.

3.1. THE LIAPUNOV-SCHMIDT METHOD

In this section we describe the Liapunov-Schmidt method. Application of this method allows in most cases to reduce an infinite-dimensional problem to a finite-dimensional one. Sometimes this approach is referred to as the alternative method, although this last method is in fact somewhat more general than the method described here (see e.g. Cesari [31], Hale [78]).

3.1.1. <u>The hypotheses</u>. Let X, Z and Γ be real Banach spaces, Ω an open neighbourhood of the origin in X, and ω an open neighbourhood of the origin in Λ. Let $M : \Omega\times\omega \subset X\times\Lambda \to Z$ be a C^1-map, such that $M(0,0) = 0$. We want to study the solution set of the equation

$$M(x,\lambda) = 0 \tag{1}$$

in a neighbourhood of (0,0) in X×Λ.

Define

$$L = D_x M(0,0) \ . \tag{2}$$

Our basic assumption will be the following :

(C) ker L has a topological complement in X, while R(L) is closed and has a topological complement in Z.

By definition 2.1.5 and theorem 2.1.3 (C) will be verified under the following hypothesis :

(F) L is a Fredholm operator.

Not only does (F) imply (C), but (F) also has an important advantage on the general hypothesis (C) : under the hypothesis (F) the result of the Liapunov-Schmidt reduction will be a finite number of scalar equations in a finite number of scalar unknowns, and depending on the parameter λ. In all applications considered further on (F) will be satisfied.

3.1.2. Under the hypothesis (C) it follows from theorem 2.1.2 that there exist continuous projections $P \in L(X)$ and $Q \in L(Z)$ such that

$$\ker L = R(P) \quad , \quad R(L) = \ker Q \ . \tag{3}$$

We can write $x \in \Omega$ in the form $x = u+v$, where $u = Px \in \ker L = X_P$ and $v = (I-P)x \in \ker P = X_{I-P}$. (In general, if $P \in L(X)$ is a projection, we write X_P for the image R(P) of P). Then we can rewrite the equation (1) in the form :

$$(I-Q)M(u+v,\lambda) = 0 \tag{4.a}$$
$$QM(u+v,\lambda) = 0 . \tag{4.b}$$

3.1.3. Lemma. Assume (C). Then there exist a neighbourhood U of the origin in X_P, a neighbourhood V of the origin in X_{I-P}, a neighbourhood ω_0 of the origin in Λ, and a continuously differentiable mapping $v^* : U\times\omega_0 \to V$ such that :

(i) $U\times V \subset \Omega \quad , \quad \omega_0 \subset \omega$;

(ii) for each $(u,v,\lambda) \in U\times V\times\omega_0$ (4a) is satisfied if and only if $v = v^*(u,\lambda)$.

Moreover :

$$v^*(0,0) = 0 \tag{5}$$

and

$$D_u v^*(0,0) = 0 . \tag{6}$$

P r o o f. The map $\psi : X_P\times X_{I-P}\times\Lambda \to R(L)$, defined in a neighbourhood of the origin by

$$\psi(u,v,\lambda) = (I-Q)M(u+v,\lambda)$$

is continuously differentiable, $\psi(0,0,0) = 0$ and $D_v\psi(0,0,0) = L|_{\ker P}$, which is an isomorphism between ker P and R(L). Then the result follows from the implicit function theorem. □

3.1.4. Remark. Define $N : \Omega\times\omega \to Z$ by

$$N(x,\lambda) = Lx - M(x,\lambda) . \tag{7}$$

Then N is a C^1-map, with

$$N(0,0) = 0 \quad , \quad D_x N(0,0) = 0 , \tag{8}$$

and equation (1) can be rewritten in the form :

$$Lx = N(x,\lambda) \ . \tag{9}$$

By the same argument as used in the proof of theorem 2.1.6, the hypothesis (C) implies that the restriction of L to ker P has a bounded inverse $K : R(L) \to \ker P$, satisfying relations such as (2.1.7). It is then easily seen that (9) is equivalent to the equations :

$$v = K(I-Q)N(u+v,\lambda) \tag{10.a}$$
$$0 = Q\,N(u+v,\lambda) \ . \tag{10.b}$$

The result of the preceding lemma then also holds for the equation (10.a); in fact, the two solutions $v^*(u,\lambda)$ coincide.

Sometimes it is interesting to write the equations in the form (10). For example, if the original problem can be written in the form (9), without N being necessarily of class C^1 it may still be possible to prove that the right-hand side of (10.a) defines a contraction in a sufficiently small neighbourhood of the origin. Then (10.a) can still be solved in a unique way for v as a function of u and λ. Also, in case one would like to generalize the Liapunov-Schmidt method to more general topological vector spaces X and Z, it may in certain cases be possible to prove the existence of a pseudo-inverse K for L such that (9) is equivalent to (10). Then (10.a) may be solved by a fixed point theorem. For more details on generalized inverses, one can consult Z. Nashed [262].

3.1.5. The result of lemma 3 allows us to define C^1-mappings $x^* : U\times\omega_0 \to \Omega$ and $F : U\times\omega_0 \to R(Q)$ by :

$$x^*(u,\lambda) = u + v^*(u,\lambda) \tag{11}$$

and

$$F(u,\lambda) = QM(u+v^*(u,\lambda),\lambda) = M(u+v^*(u,\lambda),\lambda) \ . \tag{12}$$

Then

$$x^*(0,0) = 0 \quad , \quad D_u x^*(0,0) = I_{\ker L} \tag{13}$$

and

$$F(0,0) = 0 \quad , \quad D_u F(0,0) = 0 . \tag{14}$$

The following theorem summarizes the essential result of the Liapunov-Schmidt reduction.

3.1.6. Theorem. Assume (C). Let U, V and ω_0 be the neighbourhoods given by lemma 3. Let x^* and F be the mappings defined by (11) and (12).

Then for all $u \in U$, $x \in U\times V \subset X$ and $\lambda \in \omega_0$ the following statements are equivalent :

(i) $Px = u$ and $M(x,\lambda) = 0$;

(ii) $x = x^*(u,\lambda)$ and

$$F(u,\lambda) = 0 . \quad \square \tag{15}$$

3.1.7. Conclusion. Theorem 6 gives us a one-to-one relation between the solution set of (1) and the solution set of (15), both restricted to appropriate neighbourhoods of the origin. This reduces the bifurcation problem for (1) to the bifurcation problem for (15). Since $u \in \ker L$ and $F(u,\lambda) \in R(Q)$ this is a considerable reduction, both of the domain and the range of the nonlinear operators. In case (F) is satisfied (15) is a finite dimensional problem : for each $\lambda \in \omega_0$ (15) is equivalent to m scalar equations in n scalar unknowns, where $n = \dim \ker L$ and $m = \operatorname{codim} R(L)$.

Equation (15) is called the *bifurcation equation* corresponding to (1). Remark that by (14) the *bifurcation function* F is completely degenerated as far as a direct application of the implicit function theorem is concerned.

In the sequel we will consider several examples for which we can analyse the structure of the bifurcation equation and give methods to solve the equation. For a survey of some simple cases we refer to Hale [80], Lichnewsky (Exposé nr. 2 in [204]) and Vainberg and Trenogin [224].

3.1.8. Bibliographical notes. Although the Liapunov-Schmidt method is explained in almost any text on bifurcation theory (see the references given in chapter 1) we can refer more in particular to Hale [78], Cesari [31], Vainberg and Trenogin [223],[224]. The method itself originated from the work of Liapunov [143] and Schmidt [202] on nonlinear integral equations.

3.2. EQUIVARIANT MAPPINGS

3.2.1. An example. Consider the following simple example of a problem showing some symmetry invariance : find the 2π-periodic solutions of the autonomous ordinary differential equation

$$\dot{x} = f(x,\lambda) \tag{1}$$

where $f : \mathbb{R}^n \times \Lambda \to \mathbb{R}^n$ is, for example, a C^1-function. Since (1) is autonomous, each 2π-periodic solution $x : \mathbb{R} \to \mathbb{R}^n$ of (1) will generate a whole family of such solutions, obtained from x by a phase shift over an arbitrary $\theta \in \mathbb{R}$:

$$x_\theta : \mathbb{R} \to \mathbb{R}^n \quad , \quad t \mapsto x_\theta(t) = x(t+\theta) \ . \tag{2}$$

However, since x is 2π-periodic, θ and $\theta + 2k\pi$ will give the same result.

In order to put this in an abstract form, we introduce the spaces

$$Z = \{z : \mathbb{R} \to \mathbb{R}^n \mid z \text{ is continuous and } 2\pi\text{-periodic}\}$$

and $X = \{x \in Z \mid x \text{ is } C^1\}$, and the operator

$$M : X \times \Lambda \to Z \quad , \quad x(.) \mapsto M(x,\lambda)(.) = \dot{x}(.) - f(x(.),\lambda) \ . \tag{3}$$

Then our problem can be written in the form

$$M(x,\lambda) = 0 \ . \tag{4}$$

Moreover, for each $\theta \in \mathbb{R}$ we can define a bounded linear operator $\Gamma(\theta) : Z \to Z$ by

$$\Gamma(\theta)z(t) = z(t+\theta) \quad , \quad \forall z \in Z \ , \ \forall t \in \mathbb{R} \ . \tag{5}$$

It is clear that $\Gamma(\theta+2k\pi) = \Gamma(\theta)$; so Γ is in fact a representation over Z of the rotation group $SO(2)$ introduced in section 2.6. If (x,λ) is a solution of (4), then the same holds for $(\Gamma(\theta)x,\lambda)$, for each $\theta \in \mathbb{R}$. Also, the operator M defined by (3) satisfies

$$M(\Gamma(\theta)x,\lambda) = \Gamma(\theta)M(x,\lambda) \quad , \quad \forall\theta \in \mathbb{R} \ , \ \forall(x,\lambda) \in X\times\Lambda \ . \tag{6}$$

This example motivates the following definition.

3.2.2. Definition. Let X, Z and Λ be real Banach spaces, $\Omega\times\omega \subset X\times\Lambda$ open and $M : \Omega\times\omega \to Z$. Let G be a compact topological group. Let $\Gamma : G \to L(X)$ and $\tilde{\Gamma} : G \to L(Z)$ be representations of G over X, respectively Z. Then we say that M is *equivariant with respect to* $(G,\Gamma,\tilde{\Gamma})$ if :

$$\text{(i)} \quad \Gamma(s)(\Omega) = \Omega \qquad , \quad \forall s \in G \ ;$$

$$\text{(ii)} \quad M(\Gamma(s)x,\lambda) = \tilde{\Gamma}(s)M(x,\lambda) \quad , \quad \forall s \in G \ , \ \forall(x,\lambda) \in \Omega\times\omega \ . \tag{7}$$

3.2.3. Remark. Although in many applications X will be a subspace of Z, we did not make such an assumption here. This forces us to consider two different representations (Γ and $\tilde{\Gamma}$) of G. Again, in many applications Γ will be obtained from $\tilde{\Gamma}$ by an appropriate restriction. However, our formalism allows different representations, even if $X \subset Z$. As we will see in section 5, this is especially important for the treatment of reversible systems.

3.2.4. Some immediate properties. Let M be equivariant with respect to $(G,\Gamma,\tilde{\Gamma})$. Then we have the following :

(i) if $(x,\lambda) \in \Omega\times\omega$ is a solution of $M(x,\lambda) = 0$, then so is $(\Gamma(s)x,\lambda)$, for all $s \in G$;

(ii) if G_1 is a closed subgroup of G, and

$$\Gamma_1 = \Gamma|_{G_1} \quad , \quad \tilde{\Gamma}_1 = \tilde{\Gamma}|_{G_1} \ ,$$

then M is also equivariant with respect to $(G_1,\Gamma_1,\tilde{\Gamma}_1)$;

(iii) if M is of class C^1, and if we define L as in 3.1.2, then :

$$L\Gamma(s) = \tilde{\Gamma}(s)L \quad , \quad \forall s \in G \ , \tag{8}$$

i.e. : L is equivariant with respect to $(G,\Gamma,\tilde{\Gamma})$;

(iv) under the conditions of (iii), we also have :

$$\Gamma(s)(\ker L) = \ker L \quad , \quad \forall s \in G , \tag{9}$$

and

$$\tilde{\Gamma}(s)(R(L)) = R(L) \quad , \quad \forall s \in G ; \tag{10}$$

(v) define

$$X_0 = \{x \in X \mid \Gamma(s)x = x, \ \forall s \in G\} \tag{11}$$

and

$$Z_0 = \{z \in Z \mid \tilde{\Gamma}(s)z = z, \ \forall s \in G\} ; \tag{12}$$

then, for each $(x,\lambda) \in (\Omega \cap X_0)\times\omega$:

$$M(x,\lambda) \in Z_0 . \tag{13}$$

3.2.5. Remarks. (i) The relations (9) and (10), in combination with theorem 2.5.9, will be at the basis of the theory in the next section.

(ii) If we consider only solutions $x \in X_0$ of $M(x,\lambda) = 0$, this results, by property (v) above, in a reduction of the equation; we only have to consider the symmetry-invariant part of the equation. In section 4 we will show that a similar situation arises for the bifurcation equation resulting from a Liapunov-Schmidt reduction.

(iii) As an example of such a reduction, consider the problem introduced at the beginning of this section. In this case X_0 consists of all constant functions, and the reduced problem if that of finding the singular points of (1). The reduced equation is :

$$f(x,\lambda) = 0 . \tag{14}$$

3.2.6. Bibliographical note. Hypotheses similar to (7) were used by Loginov and Trenogin [148],[149], Sattinger [195],[247], and Dancer [57]; in these references one has $X \subset Z$, and Γ is taken equal to $\tilde{\Gamma}$. The definition as given here (using different representations) was first introduced in [225], for the case of finite groups.

3.3. THE LIAPUNOV-SCHMIDT METHOD FOR EQUIVARIANT EQUATIONS

3.3.1. The hypotheses. In this section we reconsider the Liapunov-Schmidt reduction for the equation

$$M(x,\lambda) = 0 , \tag{1}$$

as given in section 3.1, under the supplementary condition that M is equivariant with respect to some triple $(G,\Gamma,\tilde{\Gamma})$. We resume the hypotheses on M :

(H) (i) $M : \Omega\times\omega \subset X\times\Lambda \to Z$ is of class C^1, and

$$M(0,0) = 0 . \tag{2}$$

(X, Z and Λ are real Banach spaces, Ω is a neighbourhood of the origin in X, and ω is a neighbourhood of the origin in Λ).

(ii) The linear operator

$$L = D_x M(0,0) \tag{3}$$

satisfies the hypothesis (C) of section 3.1; this implies that we can apply the Liapunov-Schmidt reduction.

(iii) M is equivariant with respect to some $(G,\Gamma,\tilde{\Gamma})$, with G compact.

3.3.2. Theorem. Let (H) be satisfied. Then the projections P and Q satisfying (3.1.3) can be chosen in such a way that also :

$$\Gamma(s)P = P\Gamma(s) \qquad , \qquad \forall s \in G \tag{4}$$

and

$$\tilde{\Gamma}(s)Q = Q\tilde{\Gamma}(s) \qquad , \qquad \forall s \in G . \tag{5}$$

P r o o f. This follows from (2.9), (2.10) and theorem 2.5.9. □

3.3.3. Important remark. From now on we will always assume that, under the hypotheses (H), the projections P and Q are chosen such that also (4) and (5) are satisfied. *This will not be mentionned explicitly.*

Remark also that the pseudo-inverse $K : R(L) \to \ker P$ of L depends on the choice of P. In case P satisfies (4) we can prove the following.

3.3.4. Lemma. Assume (H), and let $P \in L(X)$ be a projection on ker L, such that (4) is satisfied. Then we have on R(L) :

$$K\tilde{\Gamma}(s) = \Gamma(s)K \qquad , \qquad \forall s \in G . \tag{6}$$

P r o o f. Let $z = Lx \in R(L)$, then, since $KL = I-P$:

$$K\tilde{\Gamma}(s)z = K\tilde{\Gamma}(s)Lx = KL\Gamma(s)x = (I-P)\Gamma(s)x$$
$$= \Gamma(s)(I-P)x = \Gamma(s)KLx = \Gamma(s)Kz .$$

This proves (6). □

3.3.5. Lemma. Assume (H). Then the neighbourhoods U and V given by lemma 1.3 can be chosen in such a way that

$$\Gamma(s)(U) = U \quad , \quad \Gamma(s)(V) = V \quad , \quad \forall s \in G . \tag{7}$$

Then the mapping $v^* : U\times\omega_0 \to V$ given by the same lemma satisfies :

$$v^*(\Gamma(s)u,\lambda) = \Gamma(s)v^*(u,\lambda) \quad , \quad \forall s \in G , \forall (u,\lambda) \in U\times\omega_0 . \tag{8}$$

P r o o f. Let U, V, ω_0 and $v^*(u,\lambda)$ be as in lemma 1.3. First, we can shrink U and ω_0 in such a way that $\Gamma(s)v^*(u,\lambda) \in V$ for each $(u,\lambda) \in U\times\omega_0$ and for each $s \in G$. This follows from the fact that $\Gamma(s)v^*(0,0) = 0$ for all $s \in G$, the compactness of G, the continuity of v^* and lemma 2.5.3. The same lemma also implies that we can find a neighbourhood $U' \subset U$ of 0 in ker L such that $\Gamma(s)(U') \subset U$ for all $s \in G$. Then let

$$U_1 = \bigcup_{s\in G} \Gamma(s)(U') \subset U \quad , \quad V_1 = \bigcup_{s\in G} \Gamma(s)(V) \ .$$

It is clear that $\Gamma(s)(U_1) = U_1$ and $\Gamma(s)(V_1) = V_1$ for each $s \in G$. Moreover, $U_1 \times V_1 \subset \bigcup_{s\in G} \Gamma(s)(U\times V) \subset \Omega$. Let us show that for $(u_1,v_1,\lambda) \in U_1\times V_1\times\omega_0$ the equation (1.4a) is satisfied if and only if $v_1 = v^*(u_1,\lambda)$.

Let $(u_1,v_1,\lambda) \in U_1\times V_1\times\omega_0$ be a solution of (1.4a). Then $v_1 = \Gamma(s)v$ for some $s \in G$ and some $v \in V$. It follows from the equivariance of M and Q that also $(\Gamma^{-1}(s)u_1,v,\lambda) \in U\times V\times\omega_0$ is a solution of (1.4a). Then lemma 1.3 implies that $v = v^*(\Gamma^{-1}(s)u_1,\lambda)$ and $v_1 = \Gamma(s)v^*(\Gamma^{-1}(s)u_1,\lambda)$. Since $\Gamma^{-1}(s)u_1 \in U$ and $\lambda \in \omega_0$, it follows that $v_1 \in V$. Then the result follows from lemma 1.3. Moreover, (8) follows from the foregoing and the fact that $(\Gamma(s)u_1,\Gamma(s)v^*(u_1,\lambda),\lambda) \in U_1\times V_1\times\omega_0$ is a solution of (1.4a) for each $s \in G$, and each $(u_1,\lambda) \in U_1\times\omega_0$. □

From now on we will assume implicitly that the neighbourhoods U and V satisfy (7).

3.3.6. Theorem. Assume (H). Then the mappings $x^* : U\times\omega_0 \to X$ and $F : U\times\omega_0 \to R(Q)$, defined by (1.11) and (1.12), are equivariant, i.e., we have for all $s \in G$ and all $(u,\lambda) \in U\times\omega_0$:

$$x^*(\Gamma(s)u,\lambda) = \Gamma(s)x^*(u,\lambda) \tag{9}$$

and

$$F(\Gamma(s)u,\lambda) = \tilde{\Gamma}(s)F(u,\lambda) \ . \tag{10}$$

P r o o f. (9) follows from (8), while (10) follows from (8), (2.7) and (5). □

3.3.7. Conclusion. As a result of the foregoing we may conclude that the bifurcation equation

$$F(u,\lambda) = 0 \tag{11}$$

obtained from (1) by the Liapunov-Schmidt reduction inherites the symmetry properties of the original equation (1), provided the projection operators used for the reduction are equivariant with respect to the symmetry operators.

3.3.8. Bibliographical note. The Liapunov-Schmidt reduction for symmetric equations as given in this section is a generalization of the treatment in Vanderbauwhede [225], where attention was restricted to finite groups. Somewhat similar results were obtained by Loginov and Trenogin [148], who introduced the commutation relations (4) and (5) as a supplementary hypothesis, and by Sattinger [195], who used particular projections P and Q, defined in terms of dual bases. The possibility to choose P and Q such that (4) and (5) are satisfied was recognized by Sattinger in [197], where also the formula (2.5.11) is given.

3.4. SYMMETRIC SOLUTIONS

3.4.1. In this section we show that there is a one-to-one relation between the solutions of

$$M(x,\lambda) = 0 \tag{1}$$

which are invariant under the symmetry-operators $\{\Gamma(s) \mid s \in G\}$, and the symmetry-invariant solutions of the bifurcation equation

$$F(u,\lambda) = 0 \ . \tag{2}$$

We assume the hypotheses (H) of the preceding section, and use also the notation of that section.

Let X_0 and Z_0 be the subspaces of X, respectively Z, given by (2.11) and (2.12). By theorem 2.5.13 we can define projections P_0 and Q_0 on X_0 and Z_0, respectively, by :

$$P_0 x = \int_G \Gamma(s)x\,dm(s) \qquad , \qquad \forall x \in X \tag{3}$$

and

$$Q_0 z = \int_G \tilde{\Gamma}(s)z\,dm(s) \qquad , \qquad \forall z \in Z \ . \tag{4}$$

3.4.2. Lemma. We have

$$PP_0 = P_0 P \tag{5}$$

and

$$QQ_0 = Q_0Q \ . \tag{6}$$

P r o o f. Let $x \in X$. Then :

$$PP_0x = P\int_G \Gamma(s)xdm(s) = \int_G P\Gamma(s)xdm(s) = \int_G \Gamma(s)Pxdm(s)$$
$$= P_0Px \ .$$

An analogous proof holds for (6). □

3.4.3. Lemma. Let $u \in U \cap X_0$ and $\lambda \in \omega_0$. Then :

$$x^*(u,\lambda) = P_0x^*(u,\lambda) \tag{7}$$

and

$$F(u,\lambda) = Q_0F(u,\lambda) \ . \tag{8}$$

P r o o f. We have, for $u \in U \cap X_0$, $\lambda \in \omega_0$ and $s \in G$:

$$x^*(u,\lambda) = x^*(\Gamma(s)u,\lambda) = \Gamma(s)x^*(u,\lambda)$$

and

$$F(u,\lambda) = F(\Gamma(s)u,\lambda) = \tilde{\Gamma}(s)F(u,\lambda) \ .$$

This implies (7) and (8). □

3.4.4. Theorem. Assume (H). Let U, V, ω_0, x^* and F be as in lemma 3.5 and theorem 3.6. Then the following statements are equivalent for each $x \in U \times V$, $u \in U$ and $\lambda \in \omega_0$:

(i) $x \in X_0$, $Px = u$ and $M(x,\lambda) = 0$;

(ii) $u \in X_0$, $x = x^*(u,\lambda)$ and

$$Q_0F(u,\lambda) = 0 \ . \tag{9}$$

P r o o f. Let (i) be satisfied. Then $P_0u = P_0Px = PP_0x = Px = u$; so $u \in U \cap X_0$. Further, we have from theorem 1.6 that $x = x^*(u,\lambda)$ and $F(u,\lambda) = 0$. This implies (ii).
Conversely, let (ii) be satisfied. Then, using the preceding lemma :

$$x = x^*(u,\lambda) = P_0x^*(u,\lambda) = P_0x$$

and

$$F(u,\lambda) = Q_0F(u,\lambda) = 0 .$$

By theorem 1.6 this implies (i). □

3.4.5. Corollary. Under the hypotheses of theorem 4, let $x \in U \times V$, $\lambda \in \omega_0$, $u = Px$ and $M(x,\lambda) = 0$. Then $x \in X_0$ if and only if $u \in X_0$. □

3.4.6. Conclusion. It follows from theorem 4 and corollary 5 that, as long as we restrict to a sufficiently small neighbourhood of the origin, there is a one-to-one relationship between the solutions $x \in X_0$ of (1), and the solutions $u \in \ker L \cap X_0$ of (2). When one wants to construct the bifurcation equations for such solutions two approaches are possible :

(i) For $x \in X_0$ we can replace (1) by

$$Q_0M(x,\lambda) = 0 \tag{10}$$

(see remark 2.5(ii)). The left-hand side of (10) is a map from $(\Omega \cap X_0)\times\omega$ into Z_0. Then we can apply the Liapunov-Schmidt method to this reduced equation, using the projections $P_1 = P|_{X_0}$ and $Q|_{Z_0}$. By lemma 2 these projections have the desired properties.

(ii) We can also construct the bifurcation equation (2) for the equation (1); when restricting attention to $u \in U \cap X_0$ in (2), we can replace (2) by :

$$Q_0F(u,\lambda) = 0 , \tag{11}$$

where the left-hand side of (11) is considered as a map from $(U \cap X_0)$ into $R(Q) \cap Z_0$.

By the theory above both approaches give exactly the same bifurcation equations for the reduced problem.

3.3.7. Bibliographical note. Stokes [210] has given a somewhat different approach to the problem of the reduction of the bifurcation equations for solutions in certain invariant subspaces. A discussion of the relation between both methods is given in [226]. The same paper also describes a set of hypotheses which are intermediary between (H) and the basic hypotheses of [210]; starting from these intermediary hypotheses one can recover most of the results of this chapter.

3.5. APPLICATION : REVERSIBLE SYSTEMS

In this section we show how some results of Hale [77] for reversible systems (also called systems with "property E") can be derived from the results of the preceding sections.

3.5.1. The problem. Consider the ordinary differential equation

$$\dot{x} = A(t)x + f(t,x,\lambda) \tag{1}$$

where

(i) $x \in \mathbb{R}^n$

(ii) $A : \mathbb{R} \to L(\mathbb{R}^n)$ is continuous and 2π-periodic;

(iii) $f : \mathbb{R} \times \mathbb{R}^n \times \omega \to \mathbb{R}^n$ is continuous, and C^1 in (x,λ); here ω is a neighbourhood of 0 in some Banach space Λ ;

(iv) $f(t,x,\lambda)$ is 2π-periodic in t;

(v) $f(t,x,0) = 0 \quad , \quad \forall (t,x) \in \mathbb{R} \times \mathbb{R}^n$.

We will be looking for 2π-periodic solutions of (1).

3.5.2. The functional formulation. Let us introduce the Banach spaces (of 2π-periodic functions) X and Z, and the linear operator $L : X \to Z$ as in section 2.2. Defining the nonlinear operator $N : X \times \omega \to Z$ by

$$N(x,\lambda)(.) = f(.,x(.),\lambda) \tag{2}$$

we can reformulate the problem as follows :

Find $(x,\lambda) \in X\times\omega$ such that

$$Lx = N(x,\lambda) \ . \tag{3}$$

It is easy to see that N is once continuously Fréchet-differentiable, and

$$N(x,0) = 0 \quad , \quad D_x N(x,0) = 0 \quad , \quad \forall x \in X \ . \tag{4}$$

We use the results of section 2.2. Let P and Q be the projections defined by (2.2.19) and (2.2.20). Also, let $K : R(L) \to \ker P$ be the pseudo-inverse of L; i.e., for each $z \in R(L)$, Kz is the unique 2π-periodic solution of

$$\dot{x} = A(t)x + z(t) \tag{5}$$

also satisfying $Px = 0$. The operator $K(I-Q)$ is then a bounded linear operator from Z into X.

Let us define the following isomorphisms between $\mathbb{R}^p$ ($p = \dim \ker L$) and ker L, respectively R(Q) :

$$\begin{aligned} \Phi : \mathbb{R}^p &\to \ker L \quad , \quad a \mapsto \Phi a = \sum_{i=1}^{p} a_i\phi_i \\ \Psi : \mathbb{R}^p &\to R(Q) \quad , \quad a \mapsto \Psi a = \sum_{i=1}^{p} a_i\psi_i \ . \end{aligned} \tag{6}$$

Using an adapted version of the Liapunov-Schmidt method, we can prove the following result.

3.5.3. <u>Theorem</u>. For each $R > 0$, there exists a $\lambda_0 = \lambda_0(R) > 0$ such that, for each $(a,\lambda) \in \mathbb{R}^p\times\Lambda$ with $\|a\| \leq R$ and $\|\lambda\| \leq \lambda_0(R)$, the equation

$$x = \Phi a + K(I-Q)N(x,\lambda) \tag{7}$$

has a unique solution satisfying $\|(I-P)x\| \leq R$. This solution $x^*(a,\lambda)$ is a C^1-function of (a,λ) and $x^*(a,0) = \Phi a$.

Moreover, if for $\|a\| \leq R$ and $\|\lambda\| \leq \lambda_0(R)$ we define $G(a,\lambda) \in \mathbb{R}^p$ by

$$G(a,\lambda) = \Psi^{-1}[QN(x^*(a,\lambda),\lambda)] \tag{8}$$

then, for each $(a,x,\lambda) \in \mathbb{R}^p\times X\times\Lambda$ satisfying $\|a\| \leq R$, $\|(I-P)x\| \leq R$ and $\|\lambda\| \leq \lambda_0(R)$, the following are equivalent :

(i) $Px = \Phi a$, and $Lx = N(x,\lambda)$;

(ii) $x = x^*(a,\lambda)$ and

$$G(a,\lambda) = 0 . \tag{9}$$

P r o o f. As in the Liapunov-Schmidt reduction one proves that equation (3) is equivalent to

$$v = K(I-Q)N(u+v,\lambda) \tag{10.a}$$

$$0 = QN(u+v,\lambda) . \tag{10.b}$$

Fix some $R > 0$, and let $A(R) = \{v\in X_{I-P} \mid \|v\| \leq R\}$. Also, for each $a \in \mathbb{R}^p$ and $\lambda \in \omega$ consider the map :

$$F(a,\lambda) : A(R) \to X_{I-P} \quad , \quad v \mapsto F(a,\lambda)v = K(I-Q)N(\Phi a+v,\lambda) . \tag{11}$$

Choose $\lambda_0(R) > 0$ sufficiently small, such that :

$$\sup\{\|f(t,x,\lambda)\| \mid t\in\mathbb{R},\ \|x\| \leq (1+\|\Phi\|)R,\ \|\lambda\| \leq \lambda_0(R)\}$$
$$\leq R\|K(I-Q)\|^{-1}$$

and

$$\sup\{\|\frac{\partial f}{\partial x}(t,x,\ )\| \mid t\in\mathbb{R},\ \|x\| \leq (1+\|\Phi\|)R,\ \|\lambda\| \leq \lambda_0(R)\}$$
$$\leq \frac{1}{2}\|K(I-Q)\|^{-1} .$$

This is possible because of (4). Then it is easy to see that for each $(a,\lambda) \in \mathbb{R}^p\times\omega$ satisfying $\|a\| \leq R$ and $\|\lambda\| \leq \lambda_0(R)$ we have :

$$F(a,\lambda) : A(R) \to A(R)$$

while $F(a,\lambda)$ is a contraction, with a contraction constant (= 1/2) independent of (a,λ) in the domain considered (i.e. we have a *uniform* contraction). It follows that the map $F(a,\lambda)$ has a unique fixed point $v^*(a,\lambda) \in A(R)$, for each (a,λ) such that $\|a\| \leq R$, $\|\lambda\| \leq \lambda_0(R)$. It is then clear that $x^*(a,\lambda) = \Phi a + v^*(a,\lambda)$ will satisfy the requirements of the first part of the theorem. The second part follows then as in the Liapunov-Schmidt approach, reminding that Ψ is an isomorphism between $\mathbb{R}^p$ and $R(Q)$. □

For each λ, the bifurcation equations (8) form a set of p scalar equations in the p unknowns a_i $(i = 1,\ldots,p)$. Also

$$G(a,0) = 0 \quad , \quad \forall a \in \mathbb{R}^p \tag{11}$$

while $G(a,\lambda)$ is a C^1-function of its arguments (where defined).

Now we introduce some symmetry in the problem.

3.5.4. Definition. Let $g : \mathbb{R} \times \mathbb{R}^n \times \omega \to \mathbb{R}^n$, and consider the first order differential equation

$$\dot{x} = g(t,x,\lambda) \ . \tag{12}$$

Then we say that (1) is *reversible* if there exists a symmetric $S \in L(\mathbb{R}^n)$ such that $S^2 = I$ and

$$Sg(-t,Sx,\lambda) = -g(t,x,\lambda) \quad , \quad \forall (t,x,\lambda) \in \mathbb{R} \times \mathbb{R}^n \times \omega \ . \tag{13}$$

One also says that system (12) has the "*property E with respect to S*" (see Hale [77]).

3.5.5. Reversible periodic systems. Assume that equation (1) is reversible :

$$SA(-t) = -A(t)S \quad , \quad \forall t \in \mathbb{R} \tag{14}$$

and

$$Sf(-t,Sx,\lambda) = -f(t,x,\lambda) \quad , \quad \forall t \in \mathbb{R}^n,\ \forall x \in \mathbb{R}^n,\ \forall \lambda \in \omega \ . \tag{15}$$

In order to introduce the formalism of the preceding sections, consider a two-element group $G = \{e,i\}$, where e is the identity, and $i^2 = e$. We define representations Γ and $\tilde{\Gamma}$ of G by :

$$\Gamma(e)x = x \quad , \quad \Gamma(i)x(t) = Sx(-t) \quad , \quad \forall x \in X,\ \forall t \in \mathbb{R} \ ; \tag{16}$$

$$\tilde{\Gamma}(e)z = z \quad , \quad \tilde{\Gamma}(i)z(t) = -Sz(-t) \quad , \quad \forall z \in Z,\ \forall t \in \mathbb{R} \ . \tag{17}$$

It is easily verified from (14) and (15) that

$$\tilde{\Gamma}(s)L = L\Gamma(s) \quad , \quad \forall s \in G$$

and

$$N(\Gamma(s)x,\lambda) = \tilde{\Gamma}(s)N(x,\lambda) \quad , \quad \forall x \in X,\ \forall \lambda \in \omega,\ \forall s \in G \ .$$

That means : (3) is equivariant with respect to $(G,\Gamma,\tilde{\Gamma})$. Remark that Γ and $\tilde{\Gamma}$ do not coincide on X.

According to the example 2.5.12 the projections P and Q satisfy the commutation relations (3.4) and (3.5). So we are in a position to apply the results of the preceding sections.

3.5.6. Since $\Gamma(i)(\ker L) = \ker L$ and $\tilde{\Gamma}(i)(R(Q)) = R(Q)$ there exist non-singular $(p\times p)$-matrices B and $\tilde{B}$ such that :

$$S\phi_j(-t) = \sum_{k=1}^{p} B_{kj}\phi_k(t) \quad , \quad j = 1,\ldots,p \ , \ \forall t \in \mathbb{R} \tag{18}$$

and

$$-S\psi_j(-t) = \sum_{k=1}^{p} \tilde{B}_{kj}\psi_k(t) \quad , \quad j = 1,\ldots,p \ , \ \forall t \in \mathbb{R} \ . \tag{19}$$

This implies

$$\Gamma(i)\Phi a = \Phi B a \quad , \quad \tilde{\Gamma}(i)\Psi a = \Psi\tilde{B}a \quad , \quad \forall a \in \mathbb{R}^p \ . \tag{20}$$

3.5.7. Theorem. Suppose that equation (1) is reversible. Let $x^*(a,\lambda)$ and $G(a,\lambda)$ be the functions given by theorem 3. Then, if $\|\lambda\| \leqslant \lambda_0(\|a\|)$:

$$x^*(Ba,\lambda)(t) = Sx^*(a,\lambda)(-t) \quad , \quad \forall t \in \mathbb{R} \tag{21}$$

and

$$G(Ba,\lambda) = \tilde{B}G(a,\lambda) \ . \quad \square \tag{22}$$

3.5.8. Theorem. Under the conditions of theorem 7 let $a \in \mathbb{R}^p$ be such that

$$a = Ba \ . \tag{23}$$

Then :

$$x^*(a,\lambda)(t) = Sx^*(a,\lambda)(-t) \quad , \quad \forall t \in \mathbb{R} \tag{24}$$

and

$$G(a,\lambda) = \tilde{B}G(a,\lambda) \ . \quad \square \tag{25}$$

Theorems 7 and 8 contain precisely the results given by Hale in [77], p.267-269.

The following result (Meire and Vanderbauwhede [163]) may help to determine the elements of ker L. Consider the linear 2π-periodic equation

$$\dot{x} = A(t)x \tag{26}$$

and assume that (26) is reversible, i.e. we have (14) for some S. Let $T(t) = T(t,0)$ be the transition matrix of (26), and $C = T(2\pi)$ the corresponding monodromy matrix. We have :

$$T(t+2\pi) = T(t)C \quad , \quad \forall t \in \mathbb{R} \ . \tag{27}$$

3.5.9. Theorem. Let A(t) satisfy (14). Then :

$$T(-t) = ST(t)S \quad , \quad \forall t \in \mathbb{R} \tag{28}$$

and

$$C = ST^{-1}(\pi)ST(\pi) \; . \tag{29}$$

P r o o f. It is immediate to see that $T_1(t) = ST(-t)S$ will be a fundamental matrix for (26). Since $T_1(0) = I$, we have $T_1(t) = T(t)$ for all $t \in \mathbb{R}$. This proves (28).

Putting $t = -\pi$ in (27) and using (28) gives (29). □

3.5.10. Corollary. Let A(t) in (26) satisfy

$$A(-t) = -A(t) \qquad , \qquad \forall t \in \mathbb{R} \; . \tag{30}$$

Then all solutions of (26) are even and 2π-periodic.

P r o o f. We can take $S = I$ in theorem 9. Then $C = I$. The result follows from (28), (29), and the fact that all solutions of (26) have the form $x(t) = T(t)x_0$ for some $x_0 \in \mathbb{R}^n$. □

3.5.11. Example. Suppose that the right-hand side of (1) is odd in t; i.e., (30) holds while also

$$f(-t,x,\lambda) = -f(t,x,\lambda) \; . \tag{31}$$

Then equation (1) is reversible, using $S = I$. It follows from corollary 10 that $\dim \ker L = n$, while the elements of $\ker L$ are even functions of t. So $B = I$. Similarly, the elements of $\ker L^*$ are even functions of t, and $\tilde{B} = -I$. We conclude from theorem 8 that :

$$x^*(a,\lambda)(-t) = x^*(a,\lambda)(t) \qquad , \qquad \forall t \in \mathbb{R} \tag{32}$$

and

$$G(a,\lambda) = -G(a,\lambda) = 0 \; . \tag{33}$$

So, for each $(a,\lambda) \in \mathbb{R}^n \times \omega$ satisfying $\|\lambda\| \leqslant \lambda_0(\|a\|)$, $x^*(a,\lambda)(t)$ is an even function of t, and is a 2π-periodic solution of (1). Or : for $\|\lambda\|$ sufficiently small, all solutions of (1) are even and 2π-periodic.

The foregoing forms a local version of a global result which can easily be obtained using the approach of "autosynartetic solutions" of differential equations, as developed by D.C. Lewis ([258],[259]). In fact, the argument of the proof of theorem 9 is a special case of the arguments used by Lewis. The result is that all solutions of

$$\dot{x} = f(t,x) \tag{34}$$

are even and 2π-periodic, as soon as f is of class C^1, odd and 2π-periodic in t.

As another application of theorem 8 we derive the frequency response curves for a class of oscillation equations with a small nonlinearity and a small forcing term.

3.5.12. Example. Consider the following scalar equation :

$$\frac{d^2x}{d\tau^2} + x + g(x) = h(\omega\tau) \ , \tag{35}$$

where $g : \mathbb{R} \to \mathbb{R}$ is a C^1-function, $h : \mathbb{R} \to \mathbb{R}$ is 2π-periodic and even, while ω is a real number near 1. We also assume that g and h are small, in a sense to be made precise. Making the time rescale $\omega\tau = t$, we can rewrite (35) as a first order system :

$$\begin{aligned} \dot{x} &= y \\ \dot{y} &= -x - (\omega^{-2}-1)x + \omega^{-2}[-g(x)+h(t)] \ . \end{aligned} \tag{36}$$

We are looking for 2π-periodic solutions of (36); these will correspond to $2\pi/\omega$-periodic solutions of (35). Equation (36) has the form (1), with :

$$\begin{aligned} A &= \begin{pmatrix} 0 & 1 \\ -1 & 0 \end{pmatrix} , \quad f_1(t,x,y;\omega,g,h) = 0 \ , \\ f_2(t,x,y;\omega,g,h) &= -(\omega^{-2}-1)x + \omega^{-2}[-g(x)+h(t)] \ . \end{aligned} \tag{37}$$

This means that we consider $\lambda = (\omega,g,h)$ as the parameter; we let $\omega \in \mathbb{R}^+$ and

$h \in \{z : \mathbb{R} \to \mathbb{R} \mid z$ is continuous, even and 2π-periodic$\}$, with the usual supremum-norm. The space of the functions g is given a Banach space structure as follows. We choose in advance a value of $R > 0$ as appearing in theorem 3. All solutions in the bounded set described by theorem 3 will satisfy $|x(t)| \leq 2R$, $\forall t \in \mathbb{R}$. We take

$$g \in \{\tilde{g} : [-2R, 2R] \to \mathbb{R} \mid \tilde{g} \text{ is of class } C^1\} ;$$

this last space is equiped with the C^1-norm.

Let us now look for the symmetries of (36), and apply theorem 8.

3.5.13. The system (36) is reversible, using

$$S = \begin{pmatrix} 1 & 0 \\ 0 & -1 \end{pmatrix}$$

We can take :

$$\phi_1(t) = \psi_1(t) = \begin{pmatrix} \cos t \\ -\sin t \end{pmatrix} \quad , \quad \phi_2(t) = \psi_2 t = \begin{pmatrix} \sin t \\ \cos t \end{pmatrix} .$$

It follows that

$$B = \begin{pmatrix} 1 & 0 \\ 0 & -1 \end{pmatrix} \quad , \quad \tilde{B} = \begin{pmatrix} -1 & 0 \\ 0 & 1 \end{pmatrix} .$$

By theorem 8 one of the two bifurcation equations will be identically satisfied if we take $a = (r,0)^T$, for some $|r| \leq R$. The remaining bifurcation equation takes, after multiplication by ω^2, the form :

$$(\omega^2-1)r + \frac{1}{\pi}\int_0^{2\pi} [-g(r\cos t + v_1^*(r,0,\omega,g,h)(t)) + h(t)]\cos t \, dt = 0 . \qquad (38)$$

Let (ω,g,h) be given, with $|\omega-1|$, $\|g\|$ and $\|h\|$ sufficiently small, in accordance with theorem 3. Then to each solution r of (38), with $|r| \leq R$, there corresponds a 2π-periodic solution of (36); by theorem 8, this solution will also be an even function of t. Usually (38) is approximated by :

$$(\omega^2-1)r + \frac{1}{\pi}\int_0^{2\pi} [-g(r\cos t) + h(t)] \cos t \, dt = 0 . \qquad (39)$$

For given g and h (sufficiently small), this is a relation between the frequency ω and the "amplitude" r of the corresponding solution of (36). The curves in (ω,r)-plane given by (39) (or (38)) are referred to as frequency response curves. For a discussion of this curves, for example in the case of the Duffing equation, see Hale [77].

3.5.14. In [160], Mawhin defined an extension of the symmetry property E, as follows. Let S be a constant symmetric matrix satisfying $S^2 = I$, and let ε and τ be real numbers such that $\varepsilon^2 = 1$ and $(1+\varepsilon)\tau = 2\pi m$, for some $m \in \mathbb{Z}$. Assume that the function g in (12) is 2π-periodic with respect to t. Then we say that equation (12) has the property E with respect to (S,ε,τ) when :

$$Sg(\varepsilon t+\tau,Sx,\lambda) = \varepsilon g(t,x,\lambda) \quad , \quad \forall t \in \mathbb{R},\ \forall x \in \mathbb{R}^n,\ \forall \lambda \in \omega \ . \tag{40}$$

For $\varepsilon = -1$ and $\tau = 0$ this reduces to definition 4.

We can repeat the foregoing approach for this case; (16) and (17) are replaced by :

$$\Gamma(i)x(t) = Sx(\varepsilon t+\tau) \tag{41}$$

and

$$\tilde{\Gamma}(i)z(t) = \varepsilon Sz(\varepsilon t+\tau) \ . \tag{42}$$

Similar changes have to be made in the definition of B and $\tilde{B}$, and in (21) and (24).

3.6. FURTHER RESULTS AND APPLICATIONS

When further hypotheses are imposed on the restrictions of the symmetry operators to ker L and R(Q), then more specific conclusions can be obtained from the results of sections 3 and 4. The next chapters contain several such results; here we give a few of the most direct examples, together with a number of applications.

3.6.1. Theorem. Suppose that (H) is satisfied, while

$$\ker L \subset X_0 \ . \tag{1}$$

Then we can find a neighbourhood W of 0 in X, and a neighbourhood ω_0 of 0 in Λ, such that all solutions $(x,\lambda) \in W\times\omega_0$ of the equation

$$M(x,\lambda) = 0 \qquad (2)$$

satisfy :

$$\Gamma(s)x = x \quad , \quad \forall s \in G ; \qquad (3)$$

that is : $x \in X_0$.

P r o o f. This follows from theorem 1.6 and theorem 4.4. □

As a consequence all sufficiently small solutions of (2) can be found by solving the reduced bifurcation equation (4.11). Using property 2.4(ii) we have the following corollary which formulates a practical way to apply theorem 1.

3.6.2. Corollary. Assume (H), and let

$$G_L = \{s\in G \mid \Gamma(s)u = u, \forall u\in \ker L\} . \qquad (4)$$

Then there exist a neighbourhood W of 0 in X, and a neighbourhood ω_0 of 0 in Λ, such that all solutions $(x,\lambda) \in W\times\omega_0$ of equation (2) satisfy

$$\Gamma(s)x = x \quad , \quad \forall s \in G_L . \quad □ \qquad (5)$$

3.6.3. Theorem. Assume (H) and :

$$Z_0 \subset R(L) , \qquad (6)$$

i.e.

$$Z_0 \cap R(Q) = \{0\} . \qquad (7)$$

Let U be the neighbourhood of the origin in ker L given by theorem 4.4. Then $(x^*(u,\lambda),\lambda)$ is a solution of (2), for each $(u,\lambda) \in (u\cap X_0) \times \omega_0$; also, each of these solutions belongs to X_0.

P r o o f. It follows from (7) that, for $z \in R(Q)$, $Q_0 z = 0$. So the bifurcation equation (4.11) is trivially satisfied. □

3.6.4. Example. The foregoing theorems are abstractions of the results in example 5.11. Indeed, in this example the restriction of $\Gamma(s)$ to ker L is the identity, for each $s \in G$; also, the restriction of $\tilde{\Gamma}(s)$ to $R(Q)$ equals minus the identity. For this example the conditions (1) and (6) are satisfied.

3.6.5. In the next theorem we will make a more detailed hypothesis on the symmetry properties of the elements of ker L. In order to formulate the hypothesis, consider a closed subgroup G' of G, together with the corresponding invariant subspace

$$X_0' = \{x \in X \mid \Gamma(s)x = x, \ \forall s \in G'\} \tag{8}$$

and projection operator

$$P_0'x = \int_{G'} \Gamma(s)x\,dm'(s) \ ; \tag{9}$$

(m' is the Haar measure of G'). We will assume that $\ker L \cap X_0'$ generates ker L by the action of the symmetry operators; more precisely :

(R) For each $u \in \ker L$ one can find $s \in G$ such that :

$$\Gamma(s)u \in \ker L \cap X_0' \ . \tag{10}$$

Hypotheses similar to (R) will play an important role at several places in the following chapters, and more in particular in the main result of chapter 4.

3.6.6. Theorem. Assume (H) and (R). Let U, V and ω_0 be the neighbourhoods given by lemma 3.5. Let $(x,\lambda) \in (U\times V)\times\omega_0$ be a solution of (2). Then there exists some $s \in G$ such that

$$\Gamma(s)x \in X_0' \ . \tag{11}$$

P r o o f. Let $u = Px$; then $x = x^*(u,\lambda)$ and $F(u,\lambda) = 0$. By (R) we can find $s \in G$ such that $\Gamma(s)u \in \ker L \cap X_0'$. Then :

$$\Gamma(s)x = \Gamma(s)x^*(u,\lambda) = x^*(\Gamma(s)u,\lambda) ,$$

and, for each $t \in G'$:

$$\Gamma(t)\Gamma(s)x = x^*(\Gamma(t)\Gamma(s)u,\lambda) = x^*(\Gamma(s)u,\lambda) = \Gamma(s)x .$$

This proves the theorem. □

3.6.7. Corollary. Under the hypotheses of theorem 6, all solutions $(x,\lambda) \in (U\times V)\times\omega_0$ of (2) have the form

$$x = \Gamma(s)x_0 ,$$

with $s \in G$, and with (x_0,λ) a solution of (2) such that $x_0 \in X_0'$. Conversely, of (x_0,λ) is such a solution, then $(\Gamma(s)x_0,\lambda)$ is also a solution of (2), for each $s \in G$. □

This corollary allows us to restrict attention to solutions of (2) belonging to the subspace X_0'. By theorem 4.4 this gives a reduction of the bifurcation equation.

3.6.8. Example : subharmonic solutions. Consider again the 2π-periodic equation

$$\dot{x} = A(t)x + f(t,x,\lambda) ; \tag{12}$$

making the same assumptions as in subsection 3.5.1, we will this time however look for *subharmonic* solutions of (12). These are solutions having a least period which is an entire multiple of the basic period 2π of (12). Using the adapted Liapunov-Schmidt reduction for (12), (as in section 5) and theorem 1, we obtain the following negative result :

Let $m \in \mathbb{N}\setminus\{0\}$ and suppose that each $2\pi m$-periodic solution of

$$\dot{x} = A(t)x \tag{13}$$

is in fact 2π-periodic. Then each $2\pi m$-periodic solution of (12) with sufficiently small, will also be 2π-periodic.

For the proof, we replace 2π by $2\pi m$ in the treatment of section 5. For example, we consider the spaces $X^{(m)}$ and $Z^{(m)}$ of continuously differentiable, respectively continuous, $2\pi m$-periodic functions. Consider also the group $G = \mathbb{Z}$ (mod m) and its representation

$$(\Gamma(p)z)(t) = z(t+2\pi p) \quad , \quad \forall t \in \mathbb{R} \ , \ \forall z \in Z^{(m)} \ , \qquad p = 0,1,\ldots,m-1 \ . \tag{14}$$

Defining $L^{(m)} : X^{(m)} \to Z^{(m)}$ formally as before, the hypothesis shows that $\Gamma(p)u = u$ for each $u \in \ker L^{(m)}$ and each $p = 0,1,\ldots,m-1$. The result follows from theorem 1. The region where the result is valid is the same as the one given by theorem 5.3 for the validity of the Liapunov-Schmidt reduction.

3.6.9. Example : conservative oscillation equation. Consider the problem of finding 2π-periodic solutions of the autonomous scalar equation :

$$\ddot{x} + x = g(x,\lambda) \tag{15}$$

where $g : \mathbb{R}\times\Lambda \to \mathbb{R}$ is a C^1-function, satisfying $g(0,0) = \frac{\partial g}{\partial x}(0,0) = 0$. Of course there is no reason whe we should only consider 2π-periodic solutions of (15). However, if we want to find periodic solutions of (15) having a period near to 2, then by a time rescale (as in example 5.12) the problem can be reduced to that of finding 2π-periodic solutions of another equation, which is of the form (15) if we include the (unknown) rescaling parameter in the new parameter λ.

Let

$$Z = \{z : \mathbb{R}\to\mathbb{R} \mid z \text{ is continuous and } 2\pi\text{-periodic}\} \ ;$$

$$X = \{x\in Z \mid x \text{ is of class } C^2\} \ ,$$

$$L : X \to Z \quad , \quad x \mapsto Lx = \ddot{x} + x$$

and

$$N : X\times\Lambda \to Z \quad , \quad (x,\lambda) \mapsto N(x,\lambda)(t) = g(x(t),\lambda) \ .$$

Our problem can be written in the form

$$Lx = N(x,\lambda) \ . \tag{16}$$

We have :

$$\ker L = \text{span}\{\cos(.),\sin(.)\} \tag{17}$$

and

$$R(L) = \{z \in Z \mid \int_0^{2\pi} z(t)\cos t \, dt = \int_0^{2\pi} z(t)\sin t \, dt = 0\} \ . \tag{18}$$

Let $G = O(2)$, and define a representation of G over Z as follows :

$$(\Gamma(\alpha)z)(t) = z(t+\alpha) \quad , \quad \forall\alpha \in \mathbb{R}, \forall z \in Z, \forall t \in \mathbb{R} \ , \tag{19}$$

and

$$(\Gamma_\sigma z)(t) = z(-t) \quad , \quad \forall z \in Z, \forall t \in \mathbb{R} \ . \tag{20}$$

(See subsection 2.6.13 for the notations $\Gamma(\alpha)$ and Γ_σ). The restrictions of $\Gamma(\alpha)$ and Γ_σ to X define a representation of $O(2)$ over X, which we also denote as Γ. The mappings L and N are equivariant with respect to (G,Γ); the hypothesis (R) is satisfied for $G' = \{\phi(0) = I,\sigma\}$, and Z_0' is the space of continuous, 2π-periodic and even functions. An application of theorem 6 and corollary 7 gives the following conclusion :

Each solution $(x,\lambda) \in X\times\Lambda$ of (15), with $\|x\|$ and $\|\lambda\|$ sufficiently small, becomes an even function of t after an appropriate phase shift; i.e.

$$x(-t+\alpha) = x(t+\alpha) \quad , \quad \forall t \in \mathbb{R} \ ,$$

for some $\alpha \in \mathbb{R}$. When solving (15) for small 2π-periodic solutions it is sufficient to consider even 2π-periodic solutions. The bifurcation equations for (16) form a two-dimensional problem for each fixed λ; by our result, this reduces to a one-dimensional problem.

3.6.10. Example : Dirichlet boundary value problem. Let Ω be a bounded $C^{2,\alpha}$-domain in $\mathbb{R}^n$ ($n \geq 2$, $\alpha \in]0,1[$), and consider the following Dirichlet problem :

$$\begin{aligned} \Delta u + \mu_j u &= f(x,u,\lambda) \quad \text{in } \Omega , \\ u &= 0 \quad \text{in } \partial\Omega , \end{aligned} \tag{21}$$

where μ_j is an eigenvalue for the problem

$$\begin{aligned} \Delta u + \mu u &= 0 \quad \text{in } \Omega , \\ u &= 0 \quad \text{in } \partial\Omega . \end{aligned} \tag{22}$$

We suppose that $f : \bar{\Omega} \times \mathbb{R} \times \Lambda \to \mathbb{R}$ is of class C^1, and satisfies

$$f(x,0,0) = 0 \quad , \quad \frac{\partial f}{\partial u}(x,0,0) = 0 \quad , \quad \forall x \in \bar{\Omega} . \tag{23}$$

Let $X = C_0^{2,\alpha}(\bar{\Omega}) = \{u \in C^{2,\alpha}(\bar{\Omega}) \mid u(x) = 0, \forall x \in \partial\Omega\}$ and $Z = C^{\alpha}(\bar{\Omega})$; define

$$\begin{aligned} L_j &: X \to Z \quad , \quad u \mapsto L_j u = \Delta u + \mu_j u ; \\ N &: X \times \Lambda \to Z \quad , \quad (u,\lambda) \mapsto N(u,\lambda)(x) = f(x,u(x),\lambda) . \end{aligned}$$

Our problem takes the form :

$$L_j u = N(u,\lambda) . \tag{24}$$

We know from the results of section 2.3 that L_j is a Fredholm operator with zero index.

Let G be the group of transformations $S : \mathbb{R}^n \to \mathbb{R}^n$, which are compositions of translations, rotations and reflexions, and which leave the domain Ω invariant : $S(\Omega) = \Omega$. Define a representation of G over $C^0(\bar{\Omega})$ (and so also over X and Z) as follows :

$$(\Gamma(S)u)(x) = u(S^{-1}x) \quad , \quad \forall u \in C^0(\bar{\Omega}), \ \forall x \in \bar{\Omega}, \ \forall S \in G . \tag{25}$$

Assume also that

$$f(Sx,u,\lambda) = f(x,u,\lambda) \quad , \quad \forall(x,u,\lambda) \in \bar{\Omega}\times\mathbb{R}\times\Lambda, \ \forall S \in G \ . \tag{26}$$

This condition is in particular satisfied when $f(x,u,\lambda) = f(u,\lambda)$ does not depend explicitly on x. Application of corollary 2 gives the following result :

Let f in (21) satisfy (23) and (26), and let

$$G_j = \{S\in G \mid \Gamma(S)u = u, \ \forall u\in \ker L_j\} \ .$$

Then each solution (u,λ) of (21), with $\|u\|_{2,\alpha}$ and $\|\lambda\|$ sufficiently small, satisfies

$$u(Sx) = u(x) \quad , \quad \forall x \in \bar{\Omega} \ , \ \forall S \in G_j \ . \tag{27}$$

A more precise result can be given for bifurcation from the lowest eigenvalue μ_1 of (22).

3.6.11. <u>Theorem</u>. Consider the boundary value problem

$$\begin{aligned} \Delta u + \mu_1 u &= f(x,u,\lambda) \quad && \text{in } \Omega \\ u &= 0 && \text{in } \partial\Omega \end{aligned} \tag{28}$$

where f satisfies (23) and (26), and μ_1 is the lowest eigenvalue for (22). Then each solution of (28) with $\|u\|_{2,\alpha}$ and $\|\lambda\|$ sufficiently small satisfies :

$$u(Sx) = u(x) \quad , \quad \forall x \in \bar{\Omega} \ , \ \forall S \in G \ . \tag{29}$$

P r o o f. From the foregoing it follows that it is sufficient to show that $G_1 = G$. By theorem 2.3.31 we have $\dim \ker L_1 = 1$; let $\ker L_1 = \{\alpha u_1 \mid \alpha\in\mathbb{R}\}$. By the same theorem, u_1 can be chosen such that $u_1(x) > 0$, $\forall x \in \Omega$.

Since $\Gamma(S)(\ker L_1) = \ker L_1$, for each $S \in G$, we have

$$\Gamma(S)u_1 = \alpha(S)u_1 \quad , \quad \forall s \in G,$$

for some $\alpha(S) \in \mathbb{R}\setminus\{0\}$. Suppose $|\alpha(S)| < 1$; since G is compact (Ω is bounded) we may suppose that the sequence $S,S^2,S^3,\ldots$ converges to some $S_0 \in G$ (in the operator topology). But then

$$\Gamma(S_0)u_1 = \lim_{n\to\infty} \Gamma(S^n)u_1 = \lim_{n\to\infty} (\alpha(S))^n u_1 = 0 ,$$

contradicting the fact that $\Gamma(S_0)$ has an inverse. In case $|\alpha(S)| > 1$, then $|\alpha(S^{-1})| = |\alpha(S)|^{-1} < 1$, again leading to a contradiction. So $|\alpha(S)| = 1$. Finally, also $\alpha(S) = -1$ is impossible, since

$$(\Gamma(S)u_1)(x) = u_1(S^{-1}x) > 0 \quad , \quad \forall x \in \Omega .$$

We conclude that $\Gamma(S)u_1 = u_1$, for each $S \in G$. This proves the theorem. □

3.6.12. Example. As a particular example of problem (21), let $\Omega = B = \{x\in\mathbb{R}^2 \mid \|x\|^2 < 1\}$ be the unit ball in $\mathbb{R}^2$. The eigenvalues can easily be found by using polar coordinates and a Fourier expansion in the angle variable. One finds that each eigenvalue μ must satisfy :

$$J_k(\sqrt{\mu}) = 0 \tag{30}$$

for some $k \in \mathbb{N}$, $J_k(x)$ being the k-th order Bessel function. For each $k \in \mathbb{N}$, equation (30) has a sequence of solutions $\{\mu_{km} \mid m\in\mathbb{N}\}$, tending to $+\infty$ as $m\to\infty$. The eigenvalues μ_{0m} $(m\in\mathbb{N})$ are simple, and correspond to radially symmetric eigenfunctions $J_0(\sqrt{\mu_{0m}}r)$; μ_{00} is the lowest eigenvalue. For $k \geqslant 1$, μ_{km} is a double eigenvalue, the corresponding eigenfunctions being linear combinations of

$$J_k(\sqrt{\mu_{km}}r)\cos k\phi \quad \text{and} \quad J_k(\sqrt{\mu_{km}}r)\sin k\phi .$$

Assume now that f is radially symmetric; that means, we consider the Dirichlet problem :

$$\begin{aligned} &\Delta u + \mu_{km}u = f(r,u,\lambda) && \text{in } B \\ &u = 0 && \text{in } \partial B \end{aligned} \tag{31}$$

with

$$f(r,0,0) = \frac{\partial f}{\partial u}(r,0,0) \quad , \quad \forall r\in[0,1] . \tag{32}$$

The symmetry group leaving B invariant is the group O(2). Next theorem follows then from corollaries 2 and 7.

3.6.13. <u>Theorem</u>. Under the foregoing conditions, let (u,λ) be a solution of (31), with $\|u\|_{2,\alpha}$ and $\|\lambda\|$ sufficiently small. Then :

(i) $u = u(r)$, in case $k = 0$;

(ii) $u(\phi(\frac{2\pi}{k}p)x) = u(x)$, $\forall x \in \bar{B}$, $p = 0,1,\ldots,k-1$, in case $k \geqslant 1$.

Also, for each such solution there is a $\beta \in \mathbb{R}$ such that the function $v(x) = u(\phi(\beta)x)$ is symmetric around the x_1-axis :

$$v(\sigma x) = v(x) \quad , \quad \text{or} \quad v(r,-\phi) = v(r,\phi) \; . \qquad \square \tag{33}$$

So, when solving (31) for small solutions, one can restrict attention to solutions satisfying (33). In a later chapter we will discuss a 3-dimensional version of problem (31).

3.6.14. <u>Symmetry of the von Kårmån equations</u>. Another class of problems for which similar results can be proved are the buckling problems for plates considered in section 2.4, and described by the von Kårmån equations. We want first to investigate the symmetry properties of the operators appearing in the abstract formulation of the problem. The approach is the same for each of the different cases considered in section 2.4 (clamped plate, simply supported rectangular plate and simply supported cylindrical plate); so we will not make the distinction here.

Let $\Omega \subset \mathbb{R}^2$ be the basic domain of the plate problem, and let $S : \mathbb{R}^2 \to \mathbb{R}^2$ be any distance preserving transformation leaving Ω invariant : $S(\Omega) = \Omega$. Since the elements of H, the basic Hilbert space, are continuous functions, we can define the representation :

$$(\Gamma(S)u)(x) = u(S^{-1}x) \quad , \quad \forall u \in H, \; \forall x \in \bar{\Omega} \; . \tag{34}$$

Let $u,v,\phi \in H$ be sufficiently smooth, and $w = B(u,v)$, then :

$$\int_\Omega [u,v]\phi = \int_\Omega (\Delta w)(\Delta\phi) \; .$$

This implies :

$$\int_\Omega [\Gamma(S)u,\Gamma(S)v]\phi = \int_\Omega [u,v](S^{-1}x)\phi(x)dx$$

$$= \int_\Omega [u,v](x)\phi(Sx)dx$$

$$= \int_\Omega \Delta w(x).\Delta\phi(Sx)dx$$

$$= \int_\Omega \Delta w(S^{-1}x).\Delta\phi(x)dx$$

$$= (\Gamma(S)w,\phi) \ .$$

We conclude that

$$B(\Gamma(S)u,\Gamma(S)v) = \Gamma(S)B(u,v) \quad , \quad \forall u,v \in H \ . \tag{35}$$

It follows also that

$$C(\Gamma(S)w) = \Gamma(S)C(w) \quad , \quad \forall w \in H \ . \tag{36}$$

If the function F_0 appearing in the definition (2.4.13) of the operator A satisfies :

$$F_0(Sx) = F_0(x) \quad , \quad \forall x \in \Omega \tag{37}$$

(that means : the external forces have the same symmetry as the domain Ω), then also :

$$A\Gamma(S) = \Gamma(S)A \ . \tag{38}$$

Finally, also the operator Q appearing in the equation for a cylindrical plate satisfies :

$$Q(\Gamma(S)w) = \Gamma(S)Q(w) \quad , \quad \forall w \in H \ . \tag{39}$$

3.6.15. The clamped plate. Consider the buckling problem for a clamped plate, which is also subject to a normal load; this load is supposed to be proportional to a small parameter ν. The equation takes the form :

$$(I - \lambda A)w + C(w) = \nu p \tag{40}$$

where p is a fixed element of $H = W_0^{2,2}(\Omega)$. We let G be the group of transformations S considered in the preceding subsection; suppose that

$$\Gamma(S)p = p \quad , \quad \forall S \in G . \tag{41}$$

Then the solutions of (40) have symmetry properties similar to those of the solutions of the Dirichlet problem considered in example 10. More precisely :

Let λ_j be a characteristic value of A, and $L_j = I-\lambda_j A$. Let

$$G_j = \{S \in G \mid u(Sx) = u(x), \forall x \in \Omega, \forall u \in \ker L_j\} .$$

Then for $|\lambda-\lambda_j|$ and $|\nu|$ sufficiently small, each sufficiently small solution w of (40) satisfies

$$w(Sx) = w(x) \quad , \quad \forall x \in \Omega \, , \, \forall S \in G_j . \tag{42}$$

For example, in the case of a circular plate we have essentially the same result as in theorem 13. In particular, when solving (40) for a circular plate, we can restrict attention to solutions which are symmetric around the x_1-axis.

3.6.16. The rectangular plate. In case $p = 0$ (or $\nu = 0$) in (40), this equation has an additional symmetry : all operators commute with $-I$; if (w,λ) is a solution, then the same is true for $(-w,\lambda)$. This will double the number of symmetry operators, as compared with the group G considered in the preceding subsection.

As an example, consider the simply supported rectangular plate. Denoting the elements of the group G by their representation over H, we have for this case :

$$G = \{I,\Gamma_x,\Gamma_y,\Gamma_{xy},-I,-\Gamma_x,-\Gamma_y,-\Gamma_{xy}\} , \qquad (43)$$

where :

$$\begin{aligned} \Gamma_x w(x,y) &= w(-x,y) \\ \Gamma_y w(x,y) &= w(x,-y) \qquad (44) \\ \Gamma_{xy} &= \Gamma_x \Gamma_y . \end{aligned}$$

The characteristic values λ_{mn} $(m,n \in \mathbb{N}\backslash\{0\})$ are given by (2.4.51); they correspond to the eigenfunctions ϕ_{mn} given by (2.4.52). Denoting by G_{mn} the subgroup of G which leaves ϕ_{mn} invariant, we have :

$$\begin{aligned} G_{mn} &= \{I,-\Gamma_x,-\Gamma_y,\Gamma_{xy}\} && \text{if m = even and n = even ;} \\ &= \{I,-\Gamma_x,\Gamma_y,-\Gamma_{xy}\} && \text{if m = even and n = odd ;} \\ &= \{I,\Gamma_x,-\Gamma_y,-\Gamma_{xy}\} && \text{if m = odd and n = even ;} \\ &= \{I,\Gamma_x,\Gamma_y,\Gamma_{xy}\} && \text{if m = odd and n = odd .} \end{aligned}$$

Solutions (w,λ) with $\|w\|_H$ and $|\lambda-\lambda_{mn}|$ sufficiently small will be invariant under the operators of G_{mn}, under the condition that λ_{mn} is a simple characteristic value.

3.6.17. <u>The cylindrical plate</u>. In the case of a simply supported cylindrical plate the equation is no longer invariant under the operator $-I$, even when the normal load is zero; this is a consequence of the appearance of the quadratic operator Q in the equation.

Suppose that the normal load has the symmetry of the cylinder (see (41)); then we can take

$$G = \{\Gamma(\alpha),\Gamma(\alpha)\Gamma_x,\Gamma(\alpha)\Gamma_y,\Gamma(\alpha)\Gamma_{xy} \mid \alpha\in\mathbb{R}\} , \qquad (45)$$

where Γ_x, Γ_y and Γ_{xy} are as before, and

$$(\Gamma(\alpha)w)(x,y) = w(x,y+\alpha) \quad , \quad \forall\alpha \in \mathbb{R} . \tag{46}$$

The subgroups G_{mn} corresponding to the characteristic values λ_{mn} given in subsection 2.4.28 are :

$$G_{m,0} = G \quad \text{if} \quad m = \text{odd} ;$$

$$G_{m,0} = \{\Gamma(\alpha), \Gamma(\alpha)\Gamma_x \mid \alpha \in \mathbb{R}\} \quad , \quad \text{if } m = \text{even} ;$$

$$G_{m,n} = \{\Gamma(\tfrac{2\pi}{n}p), \Gamma(\tfrac{2\pi}{n}p)\Gamma_x \mid p = 0,1,\ldots,n-1\} \ , \text{ if } m = \text{odd and } n \geq 1 ;$$

$$G_{m,n} = \{\Gamma(\tfrac{2\pi}{n}p) \mid p = 0,1,\ldots,n-1\} \quad , \quad \text{if } m = \text{even and } n \geq 1 .$$

Solutions (w,λ,ν) in a neighbourhood of $(0,\lambda_{mn},0)$ will be invariant under the operators of G_{mn}. Also, when $n > 1$, one can restrict attention to solutions satisfying $\Gamma_y w = w$; this follows from corollary 7.

3.6.18. Bibliographical note. For other applications of symmetry arguments in bifurcation theory (e.g. for the Benard problem), one can see the recent lecture notes by Sattinger [265].

4 Perturbations of symmetric nonlinear equations

4.1. INTRODUCTION

Consider the problem of finding 2π-periodic solutions of the ordinary differential equation :

$$\ddot{x} + x = g(x,\lambda) + \mu p(t) , \tag{1}$$

where $g(0,0) = \frac{\partial g}{\partial x}(0,0) = 0$, $p : \mathbb{R} \to \mathbb{R}$ is continuous and 2π-periodic, and μ is a small parameter. In many particular examples of such problem the function $p(t)$ is also even : $p(-t) = p(t)$. Under such condition one usually restricts attention to 2π-periodic solutions which are also even in t. (See e.g. Hale [77], Hale and Rodrigues [88]). There is a very practical reason for this restriction : using the results of section 3.4 it is easily seen that the bifurcation equations for this problem reduce from two scalar equations in the general case to only one scalar equation if one restricts to even solutions.

In case $\mu = 0$ the restriction to even solutions can be justified by corollary 3.6.7 : each solution of (1) with $\mu = 0$ becomes an even function of t after an appropriate phase shift. When $\mu \neq 0$ the corollary is no longer applicable; indeed, equation (1) is for $\mu \neq 0$ not invariant under arbitrary phase shifts. Although the restriction to even functions gives us information on *some* solutions, the question remains whether or not there are any other solutions. In this chapter we describe conditions which ensure that all sufficiently small solutions of (1) are even when $\mu \neq 0$.

In the particular case of the Duffing equation, where

$$g(x,\lambda) = \lambda x - x^3 , \tag{2}$$

such result was proved by Hale and Rodrigues [88]. The main property of (1) used in their proof can best be expressed by the hypothesis (R) used in section 3.6. Let $G = O(2)$, with a representation over the space of continuous 2π-periodic functions as given by (3.6.19) and (3.6.20); let $G' = \{\phi(0),\sigma\}$.

For $\mu = 0$ the equation (1) is equivariant with respect to (G,Γ); for $\mu \neq 0$ the equation is only equivariant with respect to the subgroup G'. This means that the perturbation destroys the symmetry of the unperturbed equation. For $\mu = 0$ (1) reduces to the equation studied in subsection 3.6.9; it follows that (1) satisfies the hypothesis (R) of section 3.6. The difference with the situation of section 3.6 is that (1) is no longer equivariant with respect to $G = O(2)$ when $\mu \neq 0$.

In section 2 we prove some abstract theorems which generalize the result of Hale and Rodrigues to a large class of nonlinear equations having symmetry properties similar to those of (1). Our presentation is an adaption of some earlier joint work of H.M. Rodrigues and the author [182]. In section 3 we discuss the particular case where the unperturbed equation has an O(2)-symmetry; we also consider applications to periodic solutions of forced conservative oscillatory equations such as (1), elliptic boundary value problems in the unit circle, and some buckling problems having rotational symmetry. In section 4 we briefly discuss the situation when the unperturbed equation has an O(3)-symmetry.

4.2. THE ABSTRACT RESULTS

4.2.1. The hypotheses. Let X, Z, Λ and Σ be real Banach spaces, Ω a neighbourhood of the origin in $X\times\Lambda\times\Sigma$ and $M : \Omega \to Z$; we will consider the equation

$$M(x,\lambda,\sigma) = 0 . \tag{1}$$

Remark that we have replaced the parameter space Λ used in the previous chapter by a product space $\Lambda\times\Sigma$. When $\sigma = 0$ we call (1) the *unperturbed equation*, while the *perturbed equation* corresponds to $\sigma \neq 0$. We make the following assumptions about M.

(H1) M is of class C^2, $M(0,0,0) = 0$ and $L = D_xM(0,0,0)$ satisfies the hypothesis (C) of section 3.1.

(H2) There is a compact topological group G, a closed subgroup G_0, and representations $\Gamma : G \to L(X)$ and $\tilde{\Gamma} : G \to L(Z)$, such that M is equivariant with respect to $(G,\Gamma,\tilde{\Gamma})$ when $\sigma = 0$, and with respect to $(G_0,\Gamma,\tilde{\Gamma})$ when $\sigma \neq 0$:

$$M(\Gamma(s)x,\lambda,0) = \tilde{\Gamma}(s)M(x,\lambda,0) \quad , \quad \forall s \in G, \forall (x,\lambda,0) \in \Omega \tag{2}$$

and

$$M(\Gamma(s)x,\lambda,\sigma) = \tilde{\Gamma}(s)M(x,\lambda,\sigma) \quad , \quad \forall s \in G_0, \forall (x,\lambda,\sigma) \in \Omega \ . \tag{3}$$

As in chapter 3 it follows from (H1) and (H2) that L is equivariant with respect to $(G,\Gamma,\tilde{\Gamma})$, and that we can find equivariant projections $P \in L(X)$ and $Q \in L(Z)$ such that $R(P) = \ker L$ and $\ker Q = R(L)$. We will use the following subspaces of X and Z :

$$\begin{aligned} X_I &= \{x \in X \mid \Gamma(s)x = x, \forall s \in G\} \ , \ X_0 = \{x \in X \mid \Gamma(s)x = x, \forall s \in G_0\} \ , \\ Z_I &= \{z \in Z \mid \tilde{\Gamma}(s)z = z, \forall s \in G\} \ , \ Z_0 = \{z \in Z \mid \tilde{\Gamma}(s)z = z, \forall s \in G_0\} \ , \end{aligned} \tag{4}$$

together with the corresponding projections :

$$\begin{aligned} P_I x &= \int_G \Gamma(s)x\,dm(s) \ , \ P_0 z = \int_{G_0} \Gamma(s)x\,dm_0(s) \ , \quad \forall x \in X \ , \\ Q_I z &= \int_G \tilde{\Gamma}(s)z\,dm(s) \ , \ Q_0 z = \int_{G_0} \tilde{\Gamma}(s)z\,dm_0(s) \ , \quad \forall z \in Z \ . \end{aligned} \tag{5}$$

In (5) m and m_0 are the Haar measures of G and G_0 respectively; one has

$$R(P_I) = X_I \ , \ R(P_0) = X_0 \ , \ R(Q_I) = Z_I \ , \ R(Q_0) = Z_0 \ . \tag{6}$$

We will also use the mappings $\gamma : G \to \mathbb{R}$ and $\tilde{\gamma} : G \to \mathbb{R}$ defined by :

$$\gamma(s) = \sup\{\|(I-P_0)\Gamma(s^{-1})u\| \mid u \in \ker L \cap X_0, \|u\| = 1\}$$

and

$$\tilde{\gamma}(s) = \inf\{\|(I-Q_0)\tilde{\Gamma}(s)w\| \mid w \in R(Q) \cap Z_0, \|w\| = 1\} \ . \tag{7}$$

Now we can formulate the remaining hypotheses :

(H3) For each $u \in \ker L$ there is some $s \in G$ such that $\Gamma(s)u \in \ker L \cap X_0$; i.e. :

$$\ker L = \{\Gamma(s)u_0 \mid u_0 \in \ker L \cap X_0,\ s \in G\} . \tag{8}$$

(H4) There is a $\beta > 0$ such that

$$\beta\gamma(s) \leqslant \tilde{\gamma}(s) \quad , \quad \forall s \in G . \tag{9}$$

Hypothesis (H3) is in fact the hypothesis (R) used in section 3.6; (H4) is a technical hypothesis, which can be verified as soon as ker L and R(Q) are determined. At the end of this section we will briefly return on these hypotheses.

4.2.2. Theorem. Assume (H1), (H2) and (H3). Then there is a neighbourhood W of (0,0) in $X \times \Lambda$ such that, for each $(x,\lambda) \in W$,

$$M(x,\lambda,0) = 0 \tag{10}$$

implies the existence of some $s \in G$ such that

$$\Gamma(s)x \in X_0 ,$$

i.e. :

$$\Gamma(t)\Gamma(s)x = \Gamma(s)x , \qquad \forall t \in G_0 . \tag{11}$$

P r o o f. This is just a restatement of theorem 3.6.6. □

The next theorem forms the main result of this chapter; it describes conditions which ensure that the solutions of the perturbed equation (1) have the same symmetry as the equation itself.

4.2.3. Theorem. Assume (H1), (H2), (H3) and (H4). Let $S \subset \Sigma\backslash\{0\}$ be a compact subset such that :

$$\inf_{\sigma \in S} \|\sigma\|^{-1} \|QD_\sigma M(0,0,0)\sigma\| \equiv \delta(S) > 0 . \tag{12}$$

Let

$$C_S = \{\mu\sigma \mid \mu \in \mathbb{R},\ \sigma \in S\} \tag{13}$$

be the cone generated by S.

Then one can find a neighbourhood W of (0,0) in $X\times\Lambda$, and a neighbourhood ω of 0 in Σ, such that for each solution $(x,\lambda,\sigma) \in W\times(\omega\cap C_S)$ of (1), with $\sigma \neq 0$, we have $x \in X_0$, i.e. :

$$\Gamma(s)x = x \quad , \quad \forall s \in G_0 . \tag{14}$$

P r o o f. Let (x,λ,σ) be a solution of (1), sufficiently near to the origin, such that the Liapunov-Schmidt reduction holds. Let $u = Px$; then we have :

$$x = x^*(u,\lambda,\sigma) \quad \text{and} \quad F(u,\lambda,\sigma) = 0 . \tag{15}$$

Applying the results of chapter 3 separately for the cases $\sigma = 0$ and $\sigma \neq 0$, we see that x^* and F have the following symmetry properties :

$$x^*(\Gamma(s)u,\lambda,\sigma) = \Gamma(s)x^*(u,\lambda,\sigma) , \qquad \forall s \in G_0 \text{ if } \sigma \neq 0, \forall s \in G \text{ if } \sigma = 0 ; \tag{16}$$

$$F(\Gamma(s)u,\lambda,\sigma) = \tilde{\Gamma}(s)F(u,\lambda,\sigma) , \qquad \forall s \in G_0 \text{ if } \sigma \neq 0, \forall s \in G \text{ if } \sigma = 0 . \tag{17}$$

By (H3) we can find some $s \in G$ and some $u_0 \in \ker L\cap X_0$ such that $u = \Gamma(\bar{s})u_0$, where, for simplicity of notation, we put $\bar{s} = s^{-1}$ for each $s \in G$. Then (15) implies that

$$G(s,u_0,\lambda,\sigma) \equiv (I-Q_0)\tilde{\Gamma}(s)F(\Gamma(\bar{s})u_0,\lambda,\sigma) = 0 . \tag{18}$$

The mapping G is defined for all $s \in G$ and for (u_0,λ,σ) in a sufficiently small neighbourhood of the origin in $(\ker L\cap X_0)\times\Lambda\times\Sigma$. It follows from (17) that

$$G(s,u_0,\lambda,0) = 0 \quad , \quad \forall s \in G , \ \forall(u_0,\lambda) \tag{19}$$

and

$$G(s,u_0,\lambda,\sigma) = 0 \quad , \quad \forall s \in G_0, \ \forall(u_0,\lambda,\sigma) . \tag{20}$$

Let us analyse G in somewhat more detail. Using (19) we can write :

$$G(s,u_0,\lambda,\sigma) = (I-Q_0)\tilde{\Gamma}(s)\int_0^1 D_\sigma F(\Gamma(\bar{s})u_0,\lambda,t\sigma).\sigma dt$$

$$= (I-Q_0)\tilde{\Gamma}(s)\int_0^1 D_\sigma F(P_0\Gamma(\bar{s})u_0,\lambda,t\sigma).\sigma dt + (I-Q_0)\tilde{\Gamma}(s)\int_0^1 dt\int_0^1 dt'$$

$$D_u D_\sigma F(P_0\Gamma(\bar{s})u_0 + t'(I-P_0)\Gamma(\bar{s})u_0,\lambda,t\sigma).(\sigma,(I-P_0)\Gamma(\bar{s})u_0)$$

$$= G_1(s,u_0,\lambda,\sigma) + G_2(s,u_0,\lambda,\sigma) .$$

We have :

$$\|G_2(s,u_0,\lambda,\sigma)\| \leqslant \gamma(s)\|\sigma\|\|u_0\|C_2(s,u_0,\lambda,\sigma) \tag{21}$$

where $C_2(s,u_0,\lambda,\sigma)$ remains bounded for $s \in G$ and (u_0,λ,σ) in a neighbourhood of the origin; indeed G is compact and $D_u D_\sigma F(u,\lambda,\sigma)$ is continuous.

Since $F(P_0 u,\lambda,\sigma) \in R(Q) \cap Z_0$ for each (u,λ,σ), it follows that $D_\sigma F(P_0 u,\lambda,\sigma).\sigma' \in R(Q) \cap Z_0$. Then the definition of $\tilde{\gamma}(s)$ implies that :

$$\|G_1(s,u_0,\lambda,\sigma)\| \geqslant \tilde{\gamma}(s)\|\sigma\|C_1(s,u_0,\lambda,\sigma) , \tag{22}$$

where, for $\sigma \neq 0$:

$$C_1(s,u_0,\lambda,\sigma) = \|\int_0^1 D_\sigma F(\Gamma(\bar{s})u_0,\lambda,t\sigma).\frac{\sigma}{\|\sigma\|}dt\| . \tag{23}$$

Since $D_\sigma F(u,\lambda,\sigma)$ is a continuous function of its arguments, and

$$D_\sigma F(0,0,0) = QD_\sigma M(0,0,0) ,$$

we have by (12) and the compactness of S :

$$C_1(s,u_0,\lambda,\sigma) \geqslant \tfrac{1}{2}\delta(S) > 0 \tag{24}$$

for all $s \in G$ and all (u_0,λ,σ) sufficiently near to the origin, with $\sigma \neq 0$ and $\sigma \in C_S$.

Now suppose the theorem is false. Then we can find a sequence $\{(x_n,\lambda_n,\sigma_n)|$

$n \in \mathbb{N}\}$, converging to $(0,0,0)$ as $n \to \infty$, each term being a solution of (1), and such that for each $n \in \mathbb{N}$ we have $\sigma_n \in C_S \setminus \{0\}$ and $x_n \notin X_0$. Let $u_n = Px_n$; then $x_n = x^*(u_n,\lambda_n,\sigma_n)$ for n large enough. Since $u_n \in X_0$ implies $x_n \in X_0$, by (16), we have $u_n \notin X_0$. Let $s_n \in G$ and $u_{0,n} \in \ker L \cap X_0$ be such that $u_n = \Gamma(\bar{s}_n)u_{0,n}$; then it follows from the definition of $\gamma(s)$ and from $(I-P_0)u_n \neq 0$ that $\gamma(s_n) \neq 0$ for all $n \in \mathbb{N}$. Moreover, our foregoing analysis shows that

$$(\gamma(s_n)\|\sigma_n\|)^{-1}G(s_n,u_{0,n},\lambda_n,\sigma_n) = 0 \quad , \quad \forall n \in \mathbb{N} . \tag{25}$$

Taking the limit for $n \to \infty$ and using the estimates (21), (22) and the hypothesis (H4), we conclude that :

$$\lim_{n\to\infty} C_1(s_n,u_{0,n},\lambda_n,\sigma_n) = 0 . \tag{26}$$

This, however, contradicts (24). □

4.2.4. <u>Corollary</u>. Assume (H1)-(H4). Suppose that $\dim \Sigma < \infty$ and that the operator $QD_\sigma M(0,0,0) : \Sigma \to R(Q)$ is injective. Then there is a neighbourhood W of (0,0) in $X \times \Lambda$ and a neighbourhood ω of 0 in Σ such that $(x,\lambda,\sigma) \in W \times (\omega \setminus \{0\})$ and $M(x,\lambda,\sigma) = 0$ imply $x \in X_0$.
P r o o f. We apply theorem 3 with $S = \{\sigma \in \Sigma \mid \|\sigma\| = 1\}$. S is compact since $\dim \Sigma < \infty$; together with the injectivity of $QD_\sigma M(0,0,0)$ this also implies the condition (12) of the theorem. As for the conclusion, remark that $C_S = \Sigma$. □

4.2.5. <u>Corollary</u>.Let Ω be a neighbourhood of the origin in $X \times \Lambda$, and $\sigma \in \mathbb{R}$. Consider the equation

$$M(x,\lambda) = \sigma M_0(x,\lambda) . \tag{27}$$

Assume :

(i) $M : \Omega \to Z$ is of class C^2, $M(0,0,0) = 0$, $L = D_x M(0,0)$ satisfies (C) and M is equivariant with respect to some $(G,\Gamma,\tilde{\Gamma})$, with G compact;

(ii) $M_0 : \Omega \to Z$ is of class C^2 and equivariant with respect to some closed subgroup G_0 of G;

(iii) (H3) and (H4) are satisfied;

(iv) $QM_0(0,0) \neq 0$.

Then there is a neighbourhood W of (0,0) in $X \times \Lambda$ and a $\sigma_0 > 0$ such that for each solution of (27) with $(x,\lambda) \in W$ and $0 < |\sigma| < \sigma_0$ we have $x \in X_0$.

P r o o f. The hypotheses of theorem 3 are easily verified by taking $\Sigma = \mathbb{R}$ and $S = \{1\}$. □

4.2.6. Corollary. Let Ω be a neighbourhood of the origin in $X \times \Lambda$, $M : \Omega \to Z$ and $p \in Z$. Consider the equation

$$M(x,\lambda) = p \ . \tag{28}$$

Assume :

(i) M is of class C^1, $M(0,0) = 0$, $L = D_x M(0,0)$ satisfies (C) and M is equivariant with respect to some $(G,\Gamma,\tilde{\Gamma})$ (G compact);

(ii) (H3) and (H4) are satisfied for some closed subgroup G_0 of G;

(iii) $p_0 \in Z_0$ and $Qp_0 \neq 0$.

Then there is a neighbourhood W of the origin in $X \times \Lambda$, a neighbourhood ω of p_0 in $Z_0 \setminus \{0\}$ and a number $\sigma_0 > 0$ such that for each $p \in C_\omega = \{\mu p \mid p \in \omega, \sigma \in \mathbb{R}\}$ with $0 < \|p\| \leq \sigma_0$, each solution $(x,\lambda) \in W$ of (28) is such that $x \in X_0$.

P r o o f. Let $\Sigma = Z_0$ and ω a neighbourhood of p_0 in Z_0 on which $\|Qp\|$ remains bounded away from zero; such neighbourhood exists because of condition (iii). If $\dim \Sigma < \infty$ we can immediately apply theorem 3, with $S = \omega$. In the general case one has to reconsider the last part of the proof of theorem 3; from the expression for $C_1(s,u_0,\lambda,\sigma)$ it is easily seen that (26) gives a contradiction. Also, a careful examination of the proof shows that it is sufficient for M to be of class C^1. □

4.2.7. Corollary. Under the conditions of corollary 6, consider the equation

$$M(x,\lambda) = \sigma p_0 \ , \tag{29}$$

with $\sigma \in \mathbb{R}$. Then there is a neighbourhood W of the origin in $X \times \Lambda$ and a

$\sigma_0 > 0$ such that, for $0 < |\sigma| \leqslant \sigma_0$, each solution $(x,\lambda) \in W$ of (29) is such that $x \in X_0$. □

4.2.8. Remark. Let the hypotheses (H3) and (H4) be satisfied for some closed subgroup G_0 of G, and let $s_1 \in G$. Consider the subgroup :

$$G_1 = \{s_1.s.\bar{s}_1 \mid s \in G_0\} . \tag{30}$$

We claim that (H3) and (H4) are also satisfied for the subgroup G_1.

Indeed, we have :

$$X_1 = \{x \in X \mid \Gamma(s)x = x,\ \forall s \in G_1\} = \{\Gamma(s_1)x \mid x \in X_0\} . \tag{31}$$

This shows that $x \in X_1$ if and only if $x = \Gamma(s_1)P_0\Gamma(\bar{s}_1)x$. So the corresponding projection on X_1 is given by $P_1 = \Gamma(s_1)P_0\Gamma(\bar{s}_1)$. Now we have :

$$\begin{aligned}\ker L &= \{\Gamma(s)u_0 \mid s \in G,\ u_0 \in \ker L \cap X_0\} \\ &= \{\Gamma(s.\bar{s}_1)\Gamma(s_1)u_0 \mid s \in G,\ u_0 \in \ker L \cap X_0\} \\ &= \{\Gamma(s)u_1 \mid s \in G,\ u_1 \in \ker L \cap X_1\} .\end{aligned}$$

This proves (H3) for the subgroup G_1.

As for (H4), we have for each $s \in G$:

$$\begin{aligned}\gamma_1(s) &= \sup\{\|u\|^{-1}\|(I-P_1)\Gamma(\bar{s})u\| \mid u \in \ker L \cap X_1,\ u \neq 0\} \\ &= \sup\{\|\Gamma(s_1)u\|^{-1}\|\Gamma(s_1)(I-P_0)\Gamma(\bar{s}_1.\bar{s}.s_1)u\| \mid u \in \ker L \cap X_0,\ u \neq 0\} \\ &\leqslant \|\Gamma(s_1)\|\|\Gamma(\bar{s}_1)\|\gamma(\bar{s}_1.s.s_1) ,\end{aligned}$$

and similarly :

$$\tilde{\gamma}_1(s) \geqslant (\|\tilde{\Gamma}(s_1)\|.\|\tilde{\Gamma}(\bar{s}_1)\|)^{-1}\tilde{\gamma}(\bar{s}_1.s.s_1) .$$

It follows that $\beta_1\gamma_1(s) \leqslant \tilde{\gamma}_1(s)$ for each $s \in G$, with

$$\beta_1 = \beta[\|\Gamma(s_1)\|\|\Gamma(\bar{s}_1)\|\|\tilde{\Gamma}(s_1)\|\|\tilde{\Gamma}(\bar{s}_1)\|]^{-1} .$$

Using this remark one can easily prove some variants of the foregoing results. For example, one obtains the following modification of corollary 7.

4.2.9. Corollary. In the hypotheses of corollary 7, replace the condition $p_0 \in Z_0$ by the condition $\Gamma(s_1)p_0 \in Z_0$, for some $s_1 \in G$. Then the solution of (29) considered in the conclusion of corollary 7 will satisfy :

$$\Gamma(s_1)x \in X_0 \ . \quad \square \tag{31}$$

4.2.10. We conclude this section with a remark on the hypotheses (H1)-(H4). The following situation frequently appears in applications : X is continuously imbedded and dense in Z, while $Z = \ker L \oplus R(L)$, and $\dim \ker L < \infty$. Let Q be a projection in Z onto ker L, and such that $\ker Q = R(L)$; then we can take $P = Q|_X$. Assume also that $\Gamma(s) = \tilde{\Gamma}(s)|_X$, for each $s \in G$; we will denote both representations by Γ.

The claim is that under such conditions the hypotheses (H1)-(H4) imply that the restriction of Γ to ker L is irreducible over ker L, except when $\ker L \subset X_0$. This last case is uninteresting as far as theorem 3 is concerned, since the conclusion of theorem 3 is in that case an immediate consequence of the results of chapter 3.

In order to prove the claim, suppose that $\ker L = U_1 \oplus U_2$, with U_1 and U_2 nontrivial subspaces, such that

$$\Gamma(s)(U_1) \subset U_1 \quad , \quad \Gamma(s)(U_2) \subset U_2 \quad , \quad \forall s \in G \ .$$

By (H3) we may assume that $U_i \cap X_0 \neq \{0\}$, for $i = 1,2$. Excluding the case $\ker L \subset X_0$, we may also suppose that e.g. $U_2 \setminus X_0 \neq \phi$. Take $u_1 \in U_1 \cap X_0$, $u_1 \neq 0$, and $u_2 \in U_2 \setminus X_0$. By (H3) there is some $s \in G$ such that $\Gamma(s)(u_1+u_2) \in X_0$, i.e. such that

$$\Gamma(s)u_1 \in U_1 \cap X_0 \quad \text{and} \quad \Gamma(s)u_2 \in U_2 \cap X_0 \ .$$

Since $u_1 \in X_0$ and $\Gamma(s)u_1 \in X_1$ it follows that $\tilde{\gamma}(s) = 0$. Also, $\Gamma(s)u_2 \in X_0$ and

$$(I-Q_0)\Gamma(\bar{s})\Gamma(s)u_2 = (I-Q_0)u_2 \neq 0$$

imply that $\gamma(s) > 0$. This however contradicts (H4).

4.3. PERTURBATIONS OF EQUATIONS WITH O(2)-SYMMETRY

In this section we apply the preceding results to the particular case where the unperturbed equation is equivariant with respect to the symmetry group $G = O(2)$. We will denote the elements of O(2) by $\phi(\alpha)$ and $\tau\circ\phi(\alpha)$, where $\alpha \in \mathbb{R}$ and :

$$\phi(\alpha) = \begin{pmatrix} \cos\alpha & \sin\alpha \\ -\sin\alpha & \cos\alpha \end{pmatrix} \quad , \quad \tau = \begin{pmatrix} 1 & 0 \\ 0 & -1 \end{pmatrix} . \tag{1}$$

If $\Gamma : O(2) \to L(\mathbb{R}^n)$ is a representation, then we denote $\Gamma(\alpha) = \Gamma(\phi(\alpha))$ and $\Gamma_\tau(\alpha) = \Gamma_\tau.\Gamma(\alpha) = \Gamma(\tau)\Gamma(\phi(\alpha))$. We refer to section 2.6 for the irreducible representations of O(2).

4.3.1. The hypotheses. We will consider equations of the form :

$$M(x,\lambda,\sigma) = 0 \tag{2}$$

with hypotheses on M similar to the ones described in subsection 4.2.10. More precisely, we assume :

(a) $M : X\times\Lambda\times\Sigma \to Z$ is of class C^2 and $M(0,0,0) = 0$, where X is continuously imbedded and dense in Z;

(b) if $L = D_x M(0,0,0)$, then $\dim \ker L < \infty$ and $Z = \ker L \oplus R(L)$;

(c) there is a representation $\Gamma : O(2) \to L(Z)$ such that

$$M(\Gamma(R)x,\lambda,0) = \Gamma(R)M(x,\lambda,0) \quad , \quad \forall R \in O(2),\ \forall (x,\lambda) \in X\times\Lambda ; \tag{3}$$

(d) the restriction of Γ to ker L is irreducible.

By the theory of section 2.6 the hypothesis (d) implies that dim ker L = 1 or 2. We will consider separately the three possibilities given by theorem 2.6.19.

4.3.2. Theorem. Assume (a)-(d), dim ker L = 1 and $\Gamma_\tau u = u$ for each $u \in \ker L$. Suppose also that

$$M(\Gamma_\tau x,\lambda,\sigma) = \Gamma_\tau M(x,\lambda,\sigma) \quad , \quad \forall(x,\lambda,\sigma) . \tag{4}$$

Then each solution (x,λ,σ) of (2), sufficiently near to the origin, satisfies :

$$\Gamma_\tau x = x . \tag{5}$$

P r o o f. This follows from theorem 3.6.1, using the basic group $G_0 = \{\phi(0),\tau\}$. □

4.3.3. <u>Theorem</u>. Assume (a)-(d), dim ker L = 1 and $\Gamma_\tau u = -u$ for each $u \in \ker L$. Suppose (4) and

$$M(-x,\lambda,\sigma) = -M(x,\lambda,\sigma) \quad , \quad \forall(x,\lambda,\sigma) . \tag{6}$$

Then each solution (x,λ,σ) of (2), sufficiently near to the origin, satisfies :

$$\Gamma_\tau x = -x . \tag{7}$$

P r o o f. Define $\tilde{\Gamma} : O(2) \to L(Z)$ by

$$\tilde{\Gamma}(\alpha) = \Gamma(\alpha) \quad , \quad \tilde{\Gamma}_\tau(\alpha) = -\Gamma_\tau(\alpha) \quad , \quad \forall\alpha \in \mathbb{R} . \tag{8}$$

It is immediate that $\tilde{\Gamma}$ defines a new representation of O(2), while (3), (4) and (6) imply that the conditions of theorem 2 are satisfied, if we replace Γ by $\tilde{\Gamma}$. The result follows then from theorem 2. □

4.3.4. <u>Lemma</u>. Assume (a)-(d) and dim ker L = 2. Let $G_0 = \{\phi(0),\tau\}$. Then the hypotheses (H3) and (H4) of section 2 are satisfied.

P r o o f. By theorem 2.6.19 there is a basis $\{u_1,u_2\}$ of ker L such that :

$$\begin{aligned} &\Gamma(\alpha)u_1 = \cos k\alpha . u_1 - \sin k\alpha . u_2 , \\ &\Gamma(\alpha)u_2 = \sin k\alpha . u_1 + \cos k\alpha . u_2 , \qquad \forall\alpha \in \mathbb{R} \\ &\Gamma_\tau u_1 = u_1 \quad , \quad \Gamma_\tau u_2 = -u_2 , \end{aligned} \tag{9}$$

for some $k \in \mathbb{N}\setminus\{0\}$. It follows that $\ker L \cap X_0 = \mathrm{span}\{u_1\}$. Also, for each $a,b \in \mathbb{R}$ we can find some $\rho \geq 0$ and some $\alpha \in \mathbb{R}$ such that

$$a = \rho \cos k\alpha \quad , \quad b = -\rho \sin k\alpha \ .$$

Then we have :

$$au_1 + bu_2 = \Gamma(\alpha)(\rho u_1) \ ,$$

which proves (H3).

As for (H4), we have $P_0 = Q_0 = \frac{1}{2}(I+\Gamma_\tau)$. It follows that :

$$(I-P_0)\Gamma(\alpha)u_1 = -\sin k\alpha . u_2 \quad , \quad (I-P_0)\Gamma_\tau(\alpha)u_1 = \sin k\alpha . u_2 \ .$$

This shows that :

$$\gamma(\alpha) = \gamma_\tau(\alpha) = \tilde{\gamma}(\alpha) = \tilde{\gamma}_\tau(\alpha) = |\sin k\alpha| \quad , \quad \forall \alpha \in \mathbb{R} \ ,$$

and that the hypothesis (H4) is satisfied. □

Using remark 2.9 we see that (H3) and (H4) will also be satisfied if we take $G_0 = \{\phi(0), \tau\circ\phi(2\alpha_0)\}$, for some fixed $\alpha_0 \in \mathbb{R}$. Then X_0 consists of those $x \in X$ such that

$$\Gamma_\tau(\alpha_0)x = \Gamma(\alpha_0)x \ . \tag{10}$$

4.3.5. <u>Theorem</u>. Assume (a)-(d), $\dim \ker L = 2$ and :

(i) M is equivariant with respect to the subgroup $G_0 = \{\phi(0), \tau\}$, i.e. (4) holds;

(ii) for a compact $S \subset \Sigma\setminus\{0\}$ we have :

$$\delta(S) \equiv \inf_{\sigma \in S} \|\sigma\|^{-1} \|QD_\sigma M(0,0,0)\sigma\| > 0 \ . \tag{11}$$

Then there is a neighbourhood W of $(0,0)$ in $X\times\Lambda$, and a neighbourhood ω of 0 in Σ, such that each solution $(x,\lambda,\sigma) \in W \times (C_S \cap \omega)$ of (2), with $\sigma \neq 0$, satisfies (5).

P r o o f. This is an immediate consequence of theorem 2.3. □

4.3.6. Corollary. If in theorem 2 and theorem 5 the condition (4) is replaced by

$$M(\Gamma_\tau(2\alpha_0)x,\lambda,\sigma) = \Gamma_\tau(2\alpha_0)M(x,\lambda,\sigma) \quad , \qquad \forall(x,\lambda,\sigma) \in X\times\Lambda\times\Sigma \tag{12}$$

for some $\alpha_0 \in \mathbb{R}$, then the theorems remain valid, if we replace the conclusion (5) by (10).

P r o o f. This follows from the remark before theorem 5, using $G_0 = \{\phi(0),\tau\circ\phi(2\alpha_0)\}$. □

4.3.7. Application : Periodic perturbations of autonomous oscillation equations. Consider the problem of finding 2π-periodic solutions of the scalar equation

$$\ddot{x} + x = g(x,\lambda) + h(t,x,\lambda,\sigma) \ . \tag{13}$$

We assume the following :

(i) $g : \mathbb{R}\times\Lambda \to \mathbb{R}$ is of class C^2, with

$$g(0,0) = 0 \quad , \quad \frac{\partial g}{\partial x}(0,0) = 0 \ .$$

(ii) $h : \mathbb{R}\times\mathbb{R}\times\Lambda\times\Sigma \to \mathbb{R}$ is of class C^2, 2π-periodic in t, and

$$h(t,x,\lambda,0) = 0 \quad , \quad \forall(t,x,\lambda) \ .$$

Using the formalism of subsection 3.6.9, this problem can be brought in the form (2). Using the representation

$$(\Gamma(\alpha)x)(t) = x(t+\alpha) \quad , \quad (\Gamma_\tau(\alpha)x)(t) = x(-t+\alpha) \ , \tag{14}$$

the hypotheses (a)-(d) are easily verified. We have

$$\ker L = \operatorname{span}\{cos(.),sin(.)\} \ , \tag{15}$$

which transforms under Γ according to (9), with $k = 1$. For the projection Q we can take :

$$(Qz)(t) = \frac{1}{\pi}\cos t\int_0^{2\pi} z(s)\cos s\,ds + \frac{1}{\pi}\sin t\int_0^{2\pi} z(s)\sin s\,ds \ , \qquad \forall z \in Z \ . \tag{16}$$

Application of theorem 5 gives the following results.

4.3.8. Theorem. Suppose that

$$h(-t,x,\lambda,\sigma) = h(t,x,\lambda,\sigma) \ . \tag{17}$$

Let S be a compact subset of $\Sigma\setminus\{0\}$ such that

$$\delta(S) = \inf_{\sigma\in S} \|\sigma\|^{-1}\left|\int_0^{2\pi} \cos t D_\sigma h(t,0,0,0)\sigma dt\right| > 0 \ . \tag{18}$$

Then for $\|\lambda\|$ and $\|\sigma\|$ sufficiently small, $\sigma \in C_S$ and $\sigma \neq 0$, each sufficiently small 2π-periodic solution of (13) will be an even function of t :

$$x(-t) = x(t) \quad , \quad \forall t \in \mathbb{R} \ . \tag{19}$$

P r o o f. (17) and (19) correspond to the conditions (i) and (ii) of theorem 5. □

4.3.9. Particular case. Let $\Sigma = \mathbb{R}$ and

$$h(t,x,\lambda,\sigma) = \sigma p(t) \ , \tag{20}$$

where $p(t)$ is continuous, 2π-periodic and even. If

$$\int_0^{2\pi} \cos t\ p(t)dt \neq 0 \ , \tag{21}$$

then, for $\|\lambda\|$ and $|\sigma|$ sufficiently small, and for $\sigma \neq 0$, each sufficiently small 2π-periodic solution of

$$\ddot{x} + x = g(x,\lambda) + \sigma p(t) \tag{22}$$

will be an even function of t. For the case of the Duffing equation this result was proved by Hale and Rodrigues in [88]. Remark that for this case it is sufficient for $g(x,\lambda)$ to be of class C^1 (see corollary 2.6).

4.3.10. Theorem. Suppose that

$$g(-x,\lambda) = -g(x,\lambda) \quad , \quad \forall (x,\lambda) \tag{23}$$

and

$$h(-t,-x,\lambda,\sigma) = -h(t,x,\lambda,\sigma) \; , \quad \forall (t,x,\lambda,\sigma) \; . \tag{24}$$

Let S be a compact subset of $\Sigma\setminus\{0\}$ such that

$$\delta(S) = \inf_{\sigma\in S} \|\sigma\|^{-1} \left| \int_0^{2\pi} \sin t D_\sigma h(t,0,0,0)\sigma dt \right| > 0 \; . \tag{25}$$

Then, for $\|\lambda\|$ and $\|\sigma\|$ sufficiently small, with $\sigma \in C_S$ and $\sigma \neq 0$, each sufficiently small 2π-periodic solution of (13) will be odd in t :

$$x(-t) = -x(t) \quad , \quad \forall t \in \mathbb{R} \; . \tag{26}$$

P r o o f. This result is proved in a similar way as theorem 8, this time using the following representation of O(2) over Z, the space of continuous 2π-periodic functions :

$$(\Gamma(\alpha)z)(t) = z(t+\alpha) \quad , \quad (\Gamma_\tau(\alpha)z)(t) = -z(-t+\alpha) \; . \quad \Box \tag{27}$$

4.3.11. Application : an elliptic boundary value problem.
Let $B = \{x\in\mathbb{R}^2 \mid \|x\| < 1\}$ be the unit sphere in $\mathbb{R}^2$. ($\|.\|$ is the Euclidean norm). We will denote by (r,θ) the polar coordinates of a point $x \in \bar{B}$; correspondingly, the value of a function u at the point x will be denoted by u(x) or $u(r,\theta)$.

Let $f : \bar{B}\times\mathbb{R}\times\Lambda \to \mathbb{R}$ and $h : \bar{B}\times\mathbb{R}\times\Lambda \to \mathbb{R}$ be C^2-functions; we will assume that f does not depend on θ :

$$f(x,u,\lambda) = f(r,u,\lambda) \; , \tag{28}$$

and that

$$f(x,0,0) = \frac{\partial f}{\partial u}(x,0,0) = 0 \quad , \quad \forall x \in \bar{B} . \tag{29}$$

Consider the following boundary value problem :

$$\begin{aligned} \Delta u + \mu_{km} u &= f(x,u,\lambda) + \sigma h(x,u,\lambda) \quad , \quad x \in B \\ u(x) &= 0 \quad , \quad x \in \partial B . \end{aligned} \tag{30}$$

Here $\sigma \in \mathbb{R}$ is a small parameter, and μ_{km} is an eigenvalue for the Dirichlet problem for the Laplacian in B :

$$\begin{aligned} \Delta u + \mu u &= 0 \quad , \quad x \in B , \\ u(x) &= 0 \quad , \quad x \in \partial B . \end{aligned} \tag{31}$$

We have seen in subsection 3.6.10 how the problem (30) can be brought into the abstract form (2). We have $X = C_0^{2,\alpha}(\bar{B})$ and $Z = C^{0,\alpha}(\bar{B})$ for some $\alpha \in]0,1[$. The operator $L : X \to Z$ defined by

$$Lu = \Delta u + \mu_{km} u \quad , \quad \forall u \in C_0^{2,\alpha}(\bar{B}) \tag{32}$$

is a Fredholm operator with zero index (see section 2.3).

The representation Γ of O(2) over $C^0(\bar{B})$, as introduced in section 3.6, takes the form :

$$(\Gamma(\alpha)u)(r,\theta) = u(r,\theta+\alpha) \quad , \quad (\Gamma_\tau(\alpha)u)(r,\theta) = u(r,-\theta+\alpha) . \tag{33}$$

The discussion of the solutions of (31), as given in section 2.3 and subsection 3.6.12, shows that the hypothesis (a)-(d) of the foregoing general theory are satisfied. We will denote the eigenfunctions, given in 3.6.12, as follows :

$$\begin{aligned} \chi_{0,m}(r,\theta) &= J_0(\sqrt{\mu_{0m}}\, r) \quad , \quad k = 0 \\ \chi_{k,m}(r,\theta) &= J_k(\sqrt{\mu_{km}}\, r)\cos k\theta \quad , \quad k \geq 1 \\ \zeta_{k,m}(r,\theta) &= J_k(\sqrt{\mu_{km}}\, r)\sin k\theta \quad , \quad k \geq 1 . \end{aligned} \tag{34}$$

For the projection Q we take :

$$(Qu)(x) = c_{km}\chi_{km}(x) \int_B u(y)\chi_{km}(y)dy + c_{km}\zeta_{km}(x) \int_B u(y)\zeta_{km}(y)dy \; . \tag{35}$$

Here c_{km} is a normalisation constant, such that Q is indeed a projection; if $k = 0$, then the second term in (35) does not appear.

4.3.12. Theorem. Let $k = 0$ in (30). Assume that

$$h(r,-\theta,u,\lambda) = h(r,\theta,u,\lambda) \quad , \quad \forall(r,\theta,u,\lambda) \; . \tag{36}$$

Then each solution of (30), with $\|u\|_{2,\alpha}$, $\|\lambda\|$ and $|\sigma|$ sufficiently small, satisfies :

$$u(r,-\theta) = u(r,\theta) \quad , \quad \forall(r,\theta) \; . \tag{37}$$

P r o o f. If $k = 0$, then $\dim \ker L = 1$, and theorem 2 applies. □

4.3.13. Theorem. Let $k \geq 1$ in (30). Assume (36) and

$$\int_B h(x,0,0)\chi_{km}(x)dx \neq 0 \; . \tag{38}$$

Then each solution of (30), with $\sigma \neq 0$ and $\|u\|_{2,\alpha}$, $\|\lambda\|$ and $|\sigma|$ sufficiently small, satisfies (37).

P r o o f. By application of theorem 5. □

4.3.14. Corollary. If in theorem 12 or theorem 13 the condition (36) is replaced by

$$h(r,\theta_0-\theta,u,\lambda) = h(r,\theta_0+\theta,u,\lambda) \quad , \quad \forall(r,\theta,u,\lambda) \tag{39}$$

for some $\theta_0 \in \mathbb{R}$, then the theorems remain valid, if in the conclusion (37) is replaced by

$$u(r,\theta_0-\theta) = u(r,\theta_0+\theta) \quad , \qquad \forall(r,\theta) \ . \tag{40}$$

In the case of theorem 13, the condition (38) must also be replaced by

$$\left(\int_B h(x,0,0)\chi_{km}(x)dx, \int_B h(x,0,0)\zeta_{km}(x)dx\right) \neq (0,0) \ . \tag{41}$$

P r o o f. By application of corollary 6. □

4.3.15. Theorem. Let $k \geqslant 1$ in (30), and assume that

$$\text{(i)} \quad f(r,-u,\lambda) = -f(r,u,\lambda) \quad , \qquad \forall(r,u,\lambda) \ ; \tag{42}$$

$$\text{(ii)} \quad h(r,-\theta,-u,\lambda) = -h(r,\theta,u,\lambda) \quad , \qquad \forall(r,\theta,u,\lambda) \ ; \tag{43}$$

$$\text{(iii)} \quad \int_B h(x,0,0)\zeta_{km}(x)dx \neq 0 \ . \tag{44}$$

Then each solution of (30), with $\|u\|_{2,\alpha}$, $\|\lambda\|$ and $|\sigma|$ sufficiently small, and $\sigma \neq 0$, will be such that

$$u(r,-\theta) = -u(r,\theta) \quad , \qquad \forall(r,\theta) \ . \tag{45}$$

P r o o f. The proof is similar to that of the foregoing theorems, by using this time the following representation of O(2) over $C^0(\bar{B})$:

$$(\Gamma(\alpha)u)(r,\theta) = u(r,\theta+\alpha) \quad , \quad (\Gamma_\tau(\alpha)u) = -u(r,-\theta+\alpha) \ . \qquad \square \tag{46}$$

4.3.16. Remark. Theorem 15 has a corollary analogous to the corollary 14 of theorem 13. Also, one can combine several of the foregoing results, on condition that the function h has enough symmetry. The main point to be careful about is that the symmetry imposed on h does not contradict the condition (41). We give a few examples.

4.3.17. Corollary. Let k be even in (30). Suppose that :

$$\text{(i)} \quad h(-x_1,x_2,u,\lambda) = h(x_1,-x_2,u,\lambda) = h(x_1,x_2,u,\lambda) \ , \tag{47}$$

for all values of the arguments ;

(ii) (38) holds, in case $k \neq 0$.

Then each solution of (30), with $\|u\|_{2,\alpha}$, $\|\lambda\|$ and $|\sigma|$ sufficiently small, and with $\sigma \neq 0$, will be such that :

$$u(-x_1,x_2) = u(x_1,-x_2) = u(x_1,x_2) \quad , \quad \forall(x_1,x_2) \in \bar{B} . \tag{48}$$

P r o o f. By combination of theorem 12 (if k = 0) or theorem 13 (if $k \neq 0$) with corollary 14, taking $\theta_0 = \pi/2$. One could also use theorem 2.3 directly, by taking $G_0 = \{\phi(0),\phi(\pi),\tau\circ\phi(0), \tau\circ\phi(\pi)\}$. If k is odd, then the condition (47) on h implies that both integrals in (41) vanish; so the conclusion only holds for k even. □

4.3.18. Corollary. Let k be odd in (30). Assume (42), (44) and

$$h(-x_1,x_2,u,\lambda) = h(x_1,x_2,u,\lambda) = -h(x_1,-x_2,u,\lambda) . \tag{49}$$

Then each solution of (30), with $\|u\|_{2,\alpha}$, $\|\lambda\|$ and $|\sigma|$ sufficiently small, and with $\sigma \neq 0$, will be such that

$$u(-x_1,x_2) = u(x_1,x_2) = -u(x_1,-x_2) \quad , \quad \forall(x_1,x_2) \in \bar{B} . \tag{50}$$

P r o o f. By combination of theorem 15 with corollary 14, in which one takes $\theta_0 = \pi/2$. Again, under the symmetry condition (49) for h, (41) can only be satisfied if k is odd. □

4.3.19. Remark. For $k \geq 1$, the elements of ker L remain invariant under the symmetry operators :

$$\{\Gamma(\tfrac{2\pi}{k}j) \mid j = 0,1,\dots,k-1\} .$$

It follows from the theory of chapter 3 that, if

$$h(r,\theta+\tfrac{2\pi}{k}j,u,\lambda) = h(r,\theta,u,\lambda) \quad , \quad j = 0,1,\dots,k-1 , \tag{51}$$

then also :

$$u(r,\theta+\tfrac{2\pi}{k}j) = u(r,\theta) \quad , \quad j = 0,1,\dots,k-1 , \tag{52}$$

for sufficiently small solutions of (30). In case $k = 0$ one can use any subgroup of $O(2)$ to get a similar result.

4.3.20. Application : the circular and cylindrical plate. The buckling problems for a clamped circular plate and for a simply supported cylindrical plate have, in the absence of normals loads, also an $O(2)$-symmetry; in the case of the cylindrical plate the symmetry group is even larger than $O(2)$. A physically interesting perturbation is obtained by the introduction of a normal load. The problem dealt with in this chapter reduces for these examples to the question under what conditions the symmetry of the normal load determines the symmetry of the corresponding solution. Our abstract results can be applied in a way similar as for the problem (30). Here we will simply state a few of the results which one can obtain.

Using the appropriate Hilbert space H (see section 2.4) the buckling problem for the clamped circular plate is described by the equation :

$$(I-\lambda A)w + C(w) = \nu p \; ; \qquad (53)$$

here ν is a small parameter, controlling the amplitude of the normal load. For the simply supported cylindrical plate we have the equation :

$$(I-\lambda A+\alpha^2 A^2)w + \alpha Q(w) + C(w) = \nu p \, . \qquad (54)$$

For problem (53) we denote by λ_{mn} ($m = 1,2,\ldots,\ n = 0,1,2,\ldots$) the characteristic values of A; for $n = 0$, λ_{m0} corresponds to a radially symmetric eigenfunction :

$$\phi_{m0}(r,\theta) = a_m(r) \; ; \qquad (55)$$

for $n \geq 1$, λ_{mn} corresponds to the eigenfunctions :

$$\phi_{mn}(r,\theta) = a_{mn}(r)\cos n\theta \qquad (56)$$

and

$$\psi_{mn}(r,\theta) = a_{mn}(r)\sin n\theta \, . \qquad (57)$$

In the case of the cylindrical plate λ_{mn} will denote the characteristic values given in subsection 2.4.28, with corresponding eigenfunctions ϕ_{mn} and ψ_{mn}, as given in (2.8.71)-(2.8.73). If we assume that all characteristic values are distinct, then the hypotheses (a)-(d) of this section can easily be verified, using the symmetry operators introduced in section 3.6, and using the fact that A is self-adjoint.

4.3.21. Theorem. Assume that

$$p(r,\theta_0-\theta) = p(r,\theta_0+\theta) \quad , \quad \forall (r,\theta) \tag{58}$$

for some $\theta_0 \in \mathbb{R}$. Let $m \in \mathbb{N}\backslash\{0\}$ and $n \in \mathbb{N}$; in case $n \geq 1$, assume that

$$((p,\phi_{mn}),(p,\psi_{mn})) \neq (0,0) \ . \tag{59}$$

Then each solution (w,λ,ν) of (53), with $\|w\|$, $|\lambda-\lambda_{mn}|$ and $|\nu|$ sufficiently small, and with $\nu \neq 0$ in case $n \geq 1$, will be such that

$$w(r,\theta_0-\theta) = w(r,\theta_0+\theta) \quad , \quad \forall (r,\theta) \ . \quad \square \tag{60}$$

4.3.22. Theorem. If the condition (58) of theorem 21 is replaced by

$$p(r,\theta_0-\theta) = -p(r,\theta_0+\theta) \quad , \quad \forall (r,\theta) \ , \tag{61}$$

then in the conclusion (60) should be replaced by

$$w(r,\theta_0-\theta) = -w(r,\theta_0+\theta) \quad , \quad \forall (r,\theta) \ . \quad \square \tag{62}$$

4.3.23. Theorem. Consider (54) and assume that

$$p(x,y_0-y) = p(x,y_0+y) \quad , \quad \forall (x,y) \ , \tag{63}$$

for some $y_0 \in \mathbb{R}$. Let $m \in \mathbb{N}\backslash\{0\}$ and $n \in \mathbb{N}$. In case $n \geq 1$, let (59) be satisfied. Then each solution of (54), with $\|w\|$, $|\lambda-\lambda_{mn}|$ and $|\nu|$ sufficiently small, and with $\nu \neq 0$ if $n \geq 1$, will be such that

$$w(x,y_0-y) = w(x,y_0+y) \quad , \quad \forall (x,y) \ . \quad \square \tag{64}$$

4.3.24. Remark. For the cylindrical plate there is no analogue of theorem 22; the reason for this is the presence of the quadratic term Q(w) in (54).

4.4. AXISYMMETRIC PERTURBATIONS OF A PROBLEM WITH O(3)-SYMMETRY

In this section we briefly discuss an application of theorem 2.3 and its corollaries to perturbations of a boundary value problem with O(3)-symmetry. The results which we can obtain are far less general than for perturbations of problems with O(2)-symmetries. This is mainly due to the fact that the irreducible representations of O(3) (and, more generally of O(n) with $n \geq 3$) are much more complicated than those of O(2). The main difficulty is in finding a representation of O(3) and a subgroup G_0 for which (H3) and (H4) are satisfied. We will only give one particular example.

4.4.1. The problem. In $B = \{x \in \mathbb{R}^3 \mid \|x\| < 1\}$ we consider the following boundary value problem :

$$\begin{aligned} &\Delta u + \mu_{\ell j} u = f(x,u,\lambda) + \sigma h(x,u,\lambda) \quad , \quad && x \in B \\ &u(x) = 0 \ , && x \in \partial B \end{aligned} \quad . \tag{1}$$

Here $f : \bar{B} \times \mathbb{R} \times \Lambda \to \mathbb{R}$ and $h : \bar{B} \times \mathbb{R} \times \Lambda \to \mathbb{R}$ are of class C^2, and such that

$$f(Rx,u,\lambda) = f(x,u,\lambda) \quad , \quad \forall R \in O(3) \tag{2}$$

and

$$f(x,0,0) = \frac{\partial f}{\partial u}(x,0,0) = 0 \quad , \quad \forall x \in \bar{B} \ ; \tag{3}$$

further $\sigma \in \mathbb{R}$ is a small parameter, while $\mu_{\ell j}$ is an eigenvalue for the Dirichlet problem :

$$\begin{aligned} &\Delta u + \mu u = 0 \quad && \text{in } B \\ &u = 0 && \text{on } \partial B \end{aligned} \quad . \tag{4}$$

(See further on for the notation $\mu_{\ell j}$).

We bring this problem in the form (2.1) by defining $X = C_0^{2,\alpha}(\bar{B})$, $Z = C^{0,\alpha}(\bar{B})$ and

$$M(u,\lambda,\sigma)(x) = \Delta u(x) + \mu_{\ell j}u(x) - f(x,u(x),\lambda) - \sigma h(x,u(x),\lambda) \ , \qquad \forall x \in \bar{B} \ . \tag{5}$$

The operator $L = D_u M(0,0,0)$ is given by :

$$Lu(x) = \Delta u(x) + \mu_{\ell j}u(x) \quad , \quad \forall u \in X \ , \ \forall x \in \bar{B} \ . \tag{6}$$

We know from the theory of section 2.3 that L is a Fredholm with index zero.

On $C^0(\bar{B})$ we can define the following representation of $O(3)$:

$$\Gamma(R)u(x) = u(R^{-1}x) \qquad , \quad \forall x \in \bar{B} \ , \ \forall R \in O(3) \ . \tag{7}$$

Then it is clear from (2) that $M(u,\lambda,0)$ is equivariant with respect to $(O(3),\Gamma)$.

4.4.2. The eigenvalue problem (4). In order to obtain some more information on ker L, let us consider the eigenvalue problem (4). It appears that because of the spherical symmetry of this problem, it can be handled in a suitable way by using the spherical harmonics introduced in subsection 2.6.30. (See e.g. Courant and Hilbert [48]). The space U_ℓ of spherical harmonics of order ℓ has dimension $2\ell+1$; let $\{Y_{\ell m} \mid m = 1,2,\dots,2\ell+1\}$ be a basis of U_ℓ, which is orthonormal with respect to the $L_2(S^2)$-inner product. One can show that $\{Y_{\ell m} \mid \ell \in \mathbb{N}, m = 1,2,\dots,2\ell+1\}$ forms a complete orthonormal subset of $L_2(S^2)$. Then we can solve (4) by expanding u in spherical harmonics; one writes

$$u(r,\theta) = \sum_{\ell \in \mathbb{N}} \sum_{m=1}^{2\ell+1} \eta_{\ell m}(r) Y_{\ell m}(\theta) \tag{8}$$

(where (r,θ) are polar coordinates for $x \in \bar{B}$) and brings (8) into (4). It follows that $\eta_{\ell m}(r)$ must be a solution of

$$D_\ell(\mu)\eta = 0 \ , \ \eta(r) \text{ regular at } r = 0 \text{ and } \eta(1) = 0 \ , \tag{9}$$

where $D_\ell(\mu)$ is a linear second order ordinary differential operator, which is singular at $r = 0$, and depends on ℓ and μ, but not on m. For fixed ℓ problem (9) has nontrivial solutions if and only if μ belongs to an infinite sequence $\{\mu_{\ell j} \mid j \in \mathbb{N}\}$ of eigenvalues; if $\mu = \mu_{\ell j}$ then (9) has a one-dimensional space of solutions, spanned by a function $\eta_{\ell j}(r)$. The eigenvalues $\mu_{\ell j}$ are strictly positive, $\mu_{\ell j} \to \infty$ as $j \to \infty$, and $\mu_{\ell j} \neq \mu_{\ell' j'}$ if $\ell \neq \ell'$.

We conclude that (4) has nontrivial solutions if and only if $\mu = \mu_{\ell j}$ for some $(\ell,j) \in \mathbb{N} \times \mathbb{N}$; to each eigenvalue $\mu_{\ell j}$ there corresponds a $(2\ell+1)$-dimensional eigenspace, spanned by the functions

$$\eta_{\ell j}(r) Y_{\ell m}(\theta) \quad , \quad m = 1,2,\ldots,2\ell+1 .$$

Under rotations the eigenspace transforms according to the irreducible representation $\Gamma^{(\ell)}$ of $O(3)$, introduced in subsection 2.6.30. So the index ℓ in $\mu_{\ell j}$ refers to the dimension $2\ell+1$ of the corresponding eigenspace and to the way the eigenvectors transform under rotations.

4.4.3. The case $\ell = 0$. First suppose that $\ell = 0$ in (1). This means that $\dim \ker L = 1$ and that the elements of $\ker L$ are spherically symmetric. Using the theory of chapter 3 one then immediately proves the following : if G_0 is any closed subgroup of $O(3)$, and

$$h(Rx,u,\lambda) = h(x,u,\lambda) \quad , \quad \forall R \in G_0 , \tag{9}$$

then each sufficiently small solution of (1) will satisfy

$$u(Rx) = u(x) \quad , \quad \forall R \in G_0 . \tag{10}$$

In particular an analogue of theorem 3.12 holds.

For a particular subgroup, namely $G_0 = \{I,-I\}$, this result can be extended for all $\ell \in \mathbb{N}$. Indeed, for general $\ell \in \mathbb{N}$ one has :

$$\Gamma_- u = \Gamma(-I)u = (-1)^{\ell} u \quad , \quad \forall u \in \ker L , \tag{11}$$

(see remark 2.6.32). Then the theory of chapter 3 gives us the following result.

4.4.4. Theorem. Suppose that ℓ is even in (1), and that

$$h(-x,u,\lambda) = h(x,u,\lambda) \quad , \quad \forall (x,u,\lambda) . \tag{12}$$

Then each sufficiently small solution (u,λ,σ) of (1) will be even :

$$u(-x) = u(x) \quad , \quad \forall x \in \bar{B} . \tag{13}$$

In case ℓ is odd,

$$f(x,-u,\lambda) = -f(x,u,\lambda) \quad , \quad \forall (x,u,\lambda) \tag{14}$$

and

$$h(-x,-u,\lambda) = -h(x,u,\lambda) \quad , \quad \forall (x,u,\lambda) , \tag{15}$$

then each sufficiently small solution will be odd :

$$u(-x) = -u(x) \quad , \quad \forall x \in \bar{B} . \quad \square \tag{16}$$

4.4.5. The case $\ell = 1$. Let now $\ell = 1$ in (1). It is immediately seen from the definition that to each spherical harmonic $Y_1 \in U_1$ there corresponds a unique $a \in \mathbb{R}^3$ such that

$$Y_1(\theta) = a.\theta \quad , \quad \forall \theta \in S^2 . \tag{17}$$

Then :

$$(\Gamma(R)Y_1)(\theta) = (Ra).\theta \quad , \quad \forall R \in O(3) , \tag{18}$$

and for each $Y_1 \in U_1$ we can find some $R \in O(3)$ such that

$$(\Gamma(R)Y_1)(\theta) = ce_3.\theta \quad , \quad \forall \theta \in S^2 , \tag{19}$$

where $c \in \mathbb{R}$ and e_3 is the unit vector along the x_3-axis. It follows that the hypothesis (H3) of section 2 is satisfied if we take

$$G_0 = \{R \in O(3) \mid Re_3 = e_3\} . \tag{20}$$

As for (H4), we are in a situation as described in subsection 2.10. As a norm for elements in $\ker L = R(Q)$ we can take $\|a\|$, the Euclidean norm of the vector $a \in \mathbb{R}^3$ which determines Y_1 via (17). The restriction of Q_0 to ker L = R(Q) corresponds to projection of a onto the x_3-axis. Using this correspondence it is easy to see that $\gamma(R) = \tilde{\gamma}(R)$ for all $R \in O(3)$. So also (H4) is satisfied, and we can apply the results of section 2.

4.4.6. Theorem. Let $\ell = 1$ in (1). Suppose that

$$h(Rx,u,\lambda) = h(x,u,\lambda) \qquad , \qquad \forall R \in G_0 , \tag{21}$$

and

$$\int_0^1 r^2 dr \int_{S^2} h(r,\theta,0,0)\eta_{1j}(r)\theta_3 d\theta \neq 0 . \tag{22}$$

Then, for $\|\lambda\|$ and $|\sigma|$ sufficiently small and $\sigma \neq 0$, each sufficiently small solution of (1) will satisfy :

$$u(Rx) = u(x) \qquad , \qquad \forall R \in G_0 . \tag{23}$$

P r o o f. The result follows from corollary 2.5. The condition (22) implies $Qh(.,0,0) \neq 0$. Remark that, because of (21), the integral in (22) becomes zero if we replace θ_3 by θ_1 or θ_2. □

4.4.7. Remark. The vector e_n in the definition of G_0 can of course be replaced by any $e \in S^2$. We conclude that in general an axisymmetric perturbation will lead to axisymmetric solutions with the same axis, if $\ell = 1$.

5 Generic bifurcation and symmetry

5.1. INTRODUCTION

In this chapter we start the study of the bifurcation equation and the bifurcation set for the important case that dim ker L = codim R(L) = 1. It appears that the behaviour of the bifurcation set is mainly determined by the degree of the first nonvanishing parameter-independent term in the Taylor expansion of the bifurcation function. We will discuss the cases where the dominant term is either quadratic or cubic. Our presentation combines elements from Chow, Hale and Mallet-Paret [39] with the approach used in [229].

We will show that in both cases considered in this chapter it is possible to describe the bifurcation set as a finite union of submanifolds in the parameter space; each of these submanifolds has a finite codimension. For each value of the parameter in the same connected component of the complement of the bifurcation set, the bifurcation problem has the same number of solutions. This number of solutions can only change when λ crosses the bifurcation set, i.e. when a bifurcation takes place.

In section 2 we discuss the case where the dominant term is quadratic, while the case of a cubic dominant term is studied in sections 3 and 4. It appears that under a certain "generic condition" the bifurcation set for this second case is cusp-shaped. In section 5 we analyse the situation when the equivariance of the equation under a symmetry group prevents the generic condition from being satisfied. An example of such a nongeneric situation can be obtained by considering the buckling problem for rectangular plates, subjected to symmetric normal loads; in section 6 we study how different types of symmetry for the normal load affect the corresponding bifurcation set.

A number of results in this chapter show a close relationship to some results from singularity theory. We will briefly discuss this relationship. In this connection let us remark that we obtain our results by "classical" methods (rescaling techniques and the implicit function theorem), while singularity theory uses as one of its most important tools the less classical and more difficult preparation theorem of Malgrange and Mather (see e.g. [26], [74]).

5.2. BIFURCATION PROBLEMS WITH A QUADRATIC DOMINANT TERM

5.2.1. The problem. Let X, Z and Λ be real Banach spaces, $\Omega \subset X$ and $\omega \subset \Lambda$ open subsets, and $M : \Omega \times \omega \to Z$ a C^r-function, with $r \geq 1$. We want to study the solution set of the equation

$$M(x,\lambda) = 0 \tag{1}$$

under the following hypothesis :

(H1) For some $(x_0,\lambda_0) \in \Omega \times \omega$, we have $M(x_0,\lambda_0) = 0$, while $L = D_xM(x_0,\lambda_0)$ is a Fredholm operator, with dim ker L = codim R(L) = 1.

More in particular, since simple examples (see chapter 1) show that (1) may have a different number of solutions for different values of the parameter, we would like to solve the following problem : determine a neighbourhood Ω_1 of x_0 in X, a neighbourhood ω_1 of λ_0 in Λ, and a partition of ω_1, such that for λ varying within each of the subregions determined by the partition, the number of solutions $x \in \Omega_1$ of (1) remains constant. This will also determine the bifurcation set β for (1); this is the set of all $\lambda \in \omega_1$ which are a bifurcation point for (1) at a solution $x \in \Omega_1$ (see the definition in chapter 1).

Without loss of generality, we may assume that $(0,0) \in \Omega \times \omega$ and $(x_0,\lambda_0) = (0,0)$.

5.2.2. The generic problem. A particular problem of the form just described is the following. Let X, Z and Ω be as before; let $C^r(\Omega;Z)$ denote the Banach space of all r-times continuously differentiable functions $m : \Omega \to Z$ satisfying :

$$|m|_r = \sup\{\|m(x)\| + \|Dm(x)\| + \ldots + \|D^rm(x)\| \mid x \in \Omega\} < \infty .$$

Let $x_0 \in \Omega$, $m_0 \in C^r(\Omega;Z)$ and $m_0(x_0) = 0$. Suppose also that $L = Dm_0(x_0)$ is a Fredholm operator with dim ker L = codim R(L) = 1. We want to determine, for each $m \in C^r(\Omega;Z)$ with $|m-m_0|_r$ sufficiently small, the number of solutions of the equation

$$m(x) = 0 \tag{2}$$

belonging to a sufficiently small neighbourhood of x_0. This problem has the form (1) by taking $\Lambda = C^r(\Omega;Z)$ and

$$M(x,m) = m(x) \quad , \quad \forall x \in \Omega \ , \ \forall m \in C^r(\Omega;Z) \ . \tag{3}$$

It is clear that this "generic problem" contains each of the problems (1); once we have solved the generic problem, then it is sufficient to restrict attention to $\tau = \{M(.,\lambda) \mid \lambda \in \omega\} \subset C^r(\Omega;Z)$ to find the solution for (1). We will nevertheless give the analysis for the equation (1).

5.2.3. Reduction of the equation. We can apply the Liapunov-Schmidt reduction to equation (1). Let $P \in L(X)$ and $Q \in L(Z)$ be projections, such that $\ker L = R(P)$ and $\ker Q = R(L)$. Let $\ker L = \text{span}\{u_0\}$ and $R(Q) = \text{span}\{z_0\}$. Define the linear functional $Q_0 \in Z^*$ by :

$$Qz = Q_0(z).z_0 \quad , \quad \forall z \in Z \ . \tag{4}$$

Let $v^*(u,\lambda)$ be the unique solution of the auxiliary equation (3.1.4a). It is a C^r-function, defined for (u,λ) in a neighbourhood of the origin in $\ker L\times\Lambda$, and taking values in $\ker P$. Also $v^*(0,0) = 0$ and $D_u v^*(0,0) = 0$. Equation (1) reduces then to the following bifurcation equation :

$$F(\rho,\lambda) \equiv Q_0 M(\rho u_0 + v^*(\rho u_0,\lambda),\lambda) = 0 \ . \tag{5}$$

$F(\rho,\lambda)$ is a real valued C^r-function, defined for (ρ,λ) in a neighbourhood of the origin in $\mathbb{R}\times\Lambda$, and satisfying :

$$F(0,0) = 0 \quad , \quad D_\rho F(0,0) = 0 \ . \tag{6}$$

5.2.4. Now we supplement (H1) with the following hypothesis :

(H2) $r \geq 2$, and $D_x^2 M(x_0,\lambda_0).(u_0,u_0) \notin R(L)$.

Here, as before, u_0 is a vector generating $\ker L$. By (H2) we are allowed to take

$$z_0 = QD_x^2M(x_0,\lambda_0).(u_0,u_0) \tag{7}$$

in (4). Assuming again $(x_0,\lambda_0) = (0,0)$ we find then :

$$D_\rho^2F(0,0) = Q_0D_x^2M(0,0).(u_0,u_0) = Q_0z_0 = 1\ . \tag{8}$$

Remark that the hypotheses (H1) and (H2) are in fact hypotheses on the function $m_0 : \Omega \to Z$ defined by $m_0(x) = M(x,\lambda_0)$, $\forall x \in \Omega$.

5.2.5. Theorem. Assume (H1) and (H2). Then there is a neighbourhood $\Omega_1\times\omega_1$ of (x_0,λ_0) in $X\times\Lambda$ and a C^1-functional $\alpha : \omega_1 \to \mathbb{R}$ such that for $\lambda \in \omega_1$ the number of solutions of (1) in Ω_1 equals :

(i) zero , if $\alpha(\lambda) > 0$;
(ii) one , if $\alpha(\lambda) = 0$;
(iii) two , if $\alpha(\lambda) < 0$.

P r o o f. By (8) and the continuity of $D_\rho^2F(\rho,\lambda)$ we can find $\delta > 0$ and a neighbourhood ω_1 of the origin in Λ such that :

$$D_\rho^2F(\rho,\lambda) > 0 \quad , \quad \forall\rho \in [-\delta,\delta] \quad , \quad \forall\lambda \in \omega_1\ ,$$

$$D_\rho F(\delta,\lambda) > 0 \quad , \quad D_\rho F(-\delta,\lambda) < 0\ , \quad \forall\lambda \in \omega_1\ ,$$

and

$$F(\delta,\lambda) > 0 \quad , \quad F(-\delta,\lambda) > 0 \quad , \quad \forall\lambda \in \omega_1\ .$$

Let $\lambda \in \omega_1$. Then $D_\rho F(\rho,\lambda)$ is a strictly increasing function of $\rho \in [-\delta,\delta]$, and has exactly one zero in the interior of this interval; let $\rho_0(\lambda)$ be this zero. The function $\lambda \mapsto \rho_0(\lambda)$ is continuously differentiable, as follows from an application of the implicit function theorem on the equation

$$D_\rho F(\rho,\lambda) = 0\ . \tag{9}$$

Also $\rho_0(0) = 0$.

It follows that the function $\rho \mapsto F(\rho,\lambda)$ has a strict minimum at the point

$\rho = \rho_0(\lambda)$. Let $\alpha(\lambda)$ be the corresponding minimal value :

$$\alpha(\lambda) = F(\rho_0(\lambda),\lambda) \qquad , \qquad \forall\lambda \in \omega_1 . \tag{10}$$

The equation (5) has no solution $\rho \in [-\delta,\delta]$ if $\alpha(\lambda) > 0$. There is exactly one such solution if $\alpha(\lambda) = 0$, namely $\rho = \rho_0(\lambda)$. Finally, (5) has two different solutions $\rho \in [-\delta,\delta]$ in case $\alpha(\lambda) < 0$. This proves the theorem. □

5.2.6. Theorem. Suppose (H1) and (H2). Then there is a $\delta > 0$, a neighbourhood ω_1 of λ_0 in Λ, and a constant $C > 0$ such that each solution $(\rho,\lambda) \in [-\delta,\delta] \times \omega_1$ of (5) satisfies :

$$|\rho-\rho_0(\lambda)| \leq C|\alpha(\lambda)|^{1/2} . \tag{11}$$

P r o o f. If the theorem is not true, then there is a sequence $\{(\rho_n,\lambda_n) \mid n\in\mathbb{N}\}$, converging to $(0,0)$, such that $F(\rho_n,\lambda_n) = 0$, $\rho_n \neq \rho_0(\lambda_n)$, while also $\alpha(\lambda_n).(\rho_n-\rho_0(\lambda_n))^{-2}$ converges to zero. By the definition of $\rho_0(\lambda)$ and $\alpha(\lambda)$ we have :

$$F(\rho,\lambda) = \alpha(\lambda) + \frac{1}{2}(\rho-\rho_0(\lambda))^2 D_\rho^2 F(\rho_0(\lambda),\lambda) + o(|\rho-\rho_0(\lambda)|^2) .$$

Dividing $F(\rho_n,\lambda_n) = 0$ by $(\rho_n-\rho_0(\lambda_n))^2$ and taking the limit for $n \to \infty$, we find :

$$\frac{1}{2}D_\rho^2 F(0,0) = 0 .$$

This, however, contradicts (8). □

5.2.7. Corollary. Suppose (H1) and (H2). Suppose also that there is a $\tilde{\lambda} \in \Lambda$ such that

$$D_\lambda M(x_0,\lambda_0).\tilde{\lambda} \notin R(L) . \tag{12}$$

Then $\lambda \in \omega_1$ is a bifurcation point for (1) corresponding to a solution $x \in \Omega_1$, if and only if $\alpha(\lambda) = 0$. The bifurcation set

$$\{\lambda \in \omega_1 \mid \alpha(\lambda) = 0\} \tag{13}$$

contains λ_0, and is a submanifold of Λ, with codimension equal to 1.

P r o o f. To see that (13) is a submanifold of codimension 1, we remark that

$$D_\lambda \alpha(0).\tilde{\lambda} = Q_0 D_\lambda M(0,0).\tilde{\lambda} \quad , \quad \forall \lambda \in \Lambda .$$

Because of (12) we can find some $\tilde{\lambda} \in \Lambda$ for which this expression is different from zero. So $\alpha : \omega_1 \to \mathbb{R}$ is a submersion in a neighbourhood of λ_0; this proves the last part of the theorem (see Lang [141]).

It is clear that bifurcation points of (1) corresponds to bifurcation points of (5), at least if we restrict to a small neighbourhood of the origin. When $\alpha(\lambda) > 0$ then (5) has no solution, and λ cannot be a bifurcation point. When $\alpha(\lambda) < 0$ and $F(\rho,\lambda) = 0$, then it follows from the proof of theorem 5 that $D_\rho F(\rho,\lambda) \neq 0$; an application of the implicit function theorem shows that λ is not a bifurcation point.

Let finally $\lambda \in \omega_1$ be such that $\alpha(\lambda) = 0$. It follows from (12) that each neighbourhood of λ contains parameter values λ' for which $\alpha(\lambda') < 0$; for such λ' the equation $F(\rho,\lambda') = 0$ has two different solutions. When $\lambda' \to \lambda$ in the region $\{\lambda' \in \omega_1 \mid \alpha(\lambda') < 0\}$, then the two corresponding solutions will converge to $\rho_0(\lambda)$, as follows from theorem 6; also $\rho_0(\lambda)$ is the unique solution of $F(\rho,\lambda) = 0$. This shows that λ is a bifurcation point. □

5.2.8. Remarks. The submanifold (13) divides the neighbourhood in two connected components $\omega_1^{(+)} = \{\lambda \in \omega_1 \mid \alpha(\lambda) > 0\}$ and $\omega_1^{(-)} = \{\lambda \in \omega_1 \mid \alpha(\lambda) < 0\}$. For $\lambda \in \omega_1^{(+)}$, equation (1) has no solutions in Ω_1, for $\lambda \in \omega_2^{(-)}$ there are two solutions in Ω_1.

The condition (12) is satisfied for the problem (2); indeed, we have :

$$D_m M(x_0,m_0).\tilde{m} = \tilde{m}(x_0) \quad , \quad \forall \tilde{m} \in C^r(\Omega;Z) . \tag{14}$$

This does not belong to R(L) for an appropriate $\tilde{m} \in C^r(\Omega;Z)$. A reformulation of the foregoing for this generic problem gives precisely the results contained in Chow, Hale and Mallet-Paret [39].

More generally, the first part of the proof of corollary 7 shows that the condition (12) means that the inclusion map $\lambda \mapsto M(.,\lambda)$ from Λ into $C^r(\Omega,Z)$ is transversal to the submanifold $B = \{m \in C^r(\Omega;Z) \mid \alpha(m) = 0\}$ in the point $m_0 = M(.,\lambda_0)$; B is the bifurcation set for the generic problem (2). For this reason we can call (12) a transversality condition. This condition will "generically" (i.e. almost always) be satisfied, as soon as $\dim \Lambda \geq 1$. So we may also refer to (12) as a generic condition.

The hypotheses (H1) and (H2) imply that the bifurcation function $F(\rho,\lambda)$ is such that $F_0(\rho) = F(\rho,0) = \rho^2$ + higher order terms. We may consider $F(\rho,\lambda)$ as an unfolding of the function F_0. Since $D_\lambda F(0,0).\tilde{\lambda} = Q_0 D_\lambda M(0,0).\tilde{\lambda}$, it follows from singularity theory (see e.g. Bröcker [26], Golubitsky and Guillemin [72], Martinet [261]) that (12) is precisely the necessary and sufficient condition for $F(\rho,\lambda)$ to be a *universal unfolding* of the function $F_0(\rho)$.

For $\lambda \in \omega_1$ we define :

$$x_0(\lambda) = \rho_0(\lambda)u_0 + v^*(\rho_0(\lambda)u_0,\lambda) . \tag{15}$$

The following theorems show that when $\alpha(\lambda) = 0$, i.e. when λ is a bifurcation point, then the hypotheses (H1) and (H2) are also satisfied at the point $(x_0(\lambda),\lambda)$.

5.2.9. Theorem. Suppose (H1) is satisfied at the origin. Then there is a neighbourhood $\Omega_1 \times \omega_1$ of the origin in $X \times \Lambda$ and a $\delta > 0$ such that the only points $(x,\lambda) \in \Omega_1 \times \omega_1$ at which (H1) is satisfied have the form $(\rho u_0 + v^*(\rho u_0,\lambda),\lambda)$ for some $(\rho,\lambda) \in]-\delta,\delta[\times \omega_1$ such that :

$$F(\rho,\lambda) = 0 \quad \text{and} \quad D_\rho F(\rho,\lambda) = 0 . \tag{16}$$

For all other $(x,\lambda) \in \Omega_1 \times \omega_1$ we have either $M(x,\lambda) \neq 0$ or $\dim \ker D_x M(x,\lambda) = \operatorname{codim} R(D_x M(x,\lambda)) = 0$.

P r o o f. Since (H1) includes the condition $M(x,\lambda) = 0$, it is clear that (x,λ) should have the given form, with (ρ,λ) a solution of $F(\rho,\lambda) = 0$; this is a consequence of the Liapunov-Schmidt reduction.

For each (ρ,λ) near the origin we define

$$\tilde{L}(\rho,\lambda) = D_x M(\rho u_0 + v^*(\rho u_0,\lambda),\lambda) \ ;$$

we have $\tilde{L}(0,0) = L$, and we want to determine $\ker \tilde{L}(\rho,\lambda)$ and $R(\tilde{L}(\rho,\lambda))$. To do so, we consider the equation

$$\tilde{L}(\rho,\lambda)\bar{x} = z \ , \tag{17}$$

with $z \in Z$ given. Let $\bar{x} = \bar{\rho}u_0 + \bar{v}$, with $\bar{\rho} \in \mathbb{R}$ and $\bar{v} \in \ker P$; then (17) splits into two equations :

$$(I-Q)\tilde{L}(\rho,\lambda)(\bar{\rho}u_0 + \bar{v}) = (I-Q)z \tag{18}$$

and

$$Q_0\tilde{L}(\rho,\lambda)(\bar{\rho}u_0 + \bar{v}) = Q_0 z \ . \tag{19}$$

Moreover, differentiating the equation defining $v^*(\rho u_0,\lambda)$, we find :

$$(I-Q)\tilde{L}(\rho,\lambda).(u_0 + D_u v^*(\rho u_0,\lambda)u_0) = 0 \ . \tag{20}$$

Multiplying this relation by $\bar{\rho}$, and subtracting from (18), we obtain :

$$(I-Q)\tilde{L}(\rho,\lambda).(\bar{v} - \bar{\rho}D_u v^*(\rho u_0,\lambda)u_0) = (I-Q)z \ . \tag{21}$$

Since $(I-Q)\tilde{L}(0,0) = L$ is an isomorphism between $\ker P$ and $R(L) = \ker Q$, the same holds for $(I-Q)\tilde{L}(\rho,\lambda)$; denote by $\tilde{K}(\rho,\lambda)$ the inverse of this isomorphism. Then (21) gives :

$$\bar{v} = \bar{\rho}D_u v^*(\rho u_0,\lambda)u_0 + \tilde{K}(\rho,\lambda)(I-Q)z \ . \tag{22}$$

Substitution of (22) into (19) gives :

$$\bar{\rho}D_\rho F(\rho,\lambda) = Q_0[I - \tilde{L}(\rho,\lambda)\tilde{K}(\rho,\lambda)(I-Q)]z \ . \tag{23}$$

This equation has a unique solution $\bar{\rho}$ for each $z \in Z$ if and only if $D_\rho F(\rho,\lambda) \neq 0$; then $\ker \tilde{L}(\rho,\lambda) = \{0\}$ and $R(\tilde{L}(\rho,\lambda)) = Z$. If $D_\rho F(\rho,\lambda) = 0$, then

$$\ker \tilde{L}(\rho,\lambda) = \mathrm{span}\{u_0 + D_u v^*(\rho u_0,\lambda)u_0\} \ ; \tag{24}$$

to see this, it is sufficient to put $z = 0$ in the preceding calculations. Under the same condition (23) has a solution if and only if the right-hand side of (23) is zero. For $(\rho,\lambda) = (0,0)$, this right-hand side reduces to $Q_0(z)$; since Q_0 is a nontrivial functional, we conclude that codim $R(\tilde{L}(\rho,\lambda))$ $= 1$ if $D_\rho F(\rho,\lambda) = 0$. This proves the theorem. □

5.2.10. Theorem. Suppose (H1) and (H2) are satisfied at the point $(0,0) \in X \times \Lambda$. Then the set of points $(x,\lambda) \in \Omega_1 \times \omega_1$ at which (H1) is satisfied is given by

$$\{(x_0(\lambda),\lambda) \mid \lambda \in \omega_1,\ \alpha(\lambda) = 0\} . \tag{25}$$

At these points, also (H2) is satisfied.

P r o o f. The first part follows immediately from theorem 9 and the definition of $\alpha(\lambda)$. Let $L_1(\lambda) = \tilde{L}(\rho_0(\lambda),\lambda)$, $K_1(\lambda) = \tilde{K}(\rho_0(\lambda),\lambda)$,

$$u_1(\lambda) = u_0 + D_u v^*(\rho_0(\lambda)u_0,\lambda)u_0$$

(see (24)), and

$$z_1(\lambda) = D_x^2 M(x_0(\lambda),\lambda)(u_1(\lambda),u_1(\lambda)) .$$

If $\alpha(\lambda) = 0$, then $z_1(\lambda) \notin R(L_1(\lambda))$ if and only if :

$$Q_0[I - L_1(\lambda)K_1(\lambda)(I-Q)]\, z_1(\lambda) \neq 0 . \tag{26}$$

The left-hand side of this inequality is continuous in λ, and reduces for $\lambda = 0$ to $Q_0 D_x^2 M(0,0)(u_0,u_0)$, which is different from zero by assumption. We conclude that (26) will be satisfied for all λ in a sufficiently small neighbourhood of the origin in Λ. In particular, (H2) will be satisfied at the points of (25). □

5.3. BIFURCATION PROBLEMS WITH A CUBIC DOMINANT TERM

In this section we discuss the number of solutions of the bifurcation equation (2.5) in the case where $D_\rho^2 F(0,0) = 0$ and $D_\rho^3 F(0,0) \neq 0$. This case has already been treated in the basic paper of Chow, Hale and Mallet-Paret [39], and also in Vanderbauwhede [229] . In this last paper a rescaling technique was used. The presentation here keeps somewhat the middle between these contributions. Our results allow in a number of cases an easy calculation of the approximate form of the bifurcation set. The approach seems to be particularly useful when many parameters appear in the problem; we will illustrate this by an example in section 6.

5.3.1. The hypothesis. In this section we replace the hypotheses (H2) of the preceding section by :

(H3) $r \geqslant 3$, and the bifurcation function $F(\rho,\lambda)$ given by (2.5) satisfies :

$$D^2_\rho F(0,0) = 0 \tag{1}$$

and

$$D^3_\rho F(0,0) \neq 0 . \tag{2}$$

The condition (1) can be reformulated as :

$$D^2_x M(0,0).(u_0,u_0) \in R(L) \tag{3}$$

(see the previous section). As for (2), we obtain from a direct calculation the following expression for $D^3_\rho F(0,0)$:

$$\begin{aligned} D^3_\rho F(0,0) &= Q_0 D^3_x M(0,0).(u_0,u_0,u_0) \\ &\quad - 3Q_0 D^2_x M(0,0).(u_0,K(I-Q)D^2_x M(0,0).(u_0,u_0)) . \end{aligned} \tag{4}$$

We remark that again (H3) is a condition on the function $m_0(x) = M(x,\lambda_0)$, and does not involve parameter values $\lambda \neq \lambda_0$. Let

$$a_0 = \frac{1}{3!} D^3_\rho F(0,0) . \tag{5}$$

We may suppose that $a_0 > 0$; in case $a_0 < 0$, it suffices to replace z_0 by $-z_0$ in the definition of Q_0. The next lemma's describe the solutions of the bifurcation equation

$$F(\rho,\lambda) = 0 \ . \tag{6}$$

5.3.2. Lemma. Suppose (H1) and (H3). Then there is a $\delta > 0$, a neighbourhood ω_1 of the origin in Λ, and a C^1-map $\gamma_1 : \omega_1 \to \mathbb{R}$ such that for each $\lambda \in \omega_1$ satisfying $\gamma_1(\lambda) \geq 0$ the equation (1) has a unique solution $\rho \in]-\delta,\delta[$. Moreover, $\gamma_1(0) = 0$.

P r o o f. Since $F(\rho,0) = a_0\rho^3 + o(\rho^3)$ it follows from continuity arguments that we can find $\delta > 0$ and a neighbourhood ω_1 of the origin in Λ such that

$$D_\rho^3 F(\rho,\lambda) > 0 \quad , \quad \forall\rho \in [-\delta,\delta] \quad , \quad \forall\lambda \in \omega_1 \ , \tag{7}$$

$$D_\rho^2 F(\delta,\lambda) > 0 \quad , \quad D_\rho^2 F(-\delta,\lambda) < 0 \quad , \quad \forall\lambda \in \omega_1 \ , \tag{8}$$

$$D_\rho F(\delta,\lambda) > 0 \quad , \quad D_\rho F(-\delta,\lambda) > 0 \quad , \quad \forall\lambda \in \omega_1 \ , \tag{9}$$

and

$$F(\delta,\lambda) > 0 \quad , \quad F(-\delta,\lambda) < 0 \quad , \quad \forall\lambda \in \omega_1 \ . \tag{10}$$

From (7) and (8) it follows that, for each $\lambda \in \omega_1$, the equation

$$D_\rho^2 F(\rho,\lambda) = 0 \tag{11}$$

has a unique solution $\rho = \rho_0(\lambda) \in]-\delta,\delta[$. The implicit function theorem shows that $\rho_0 : \omega_1 \to]-\delta,\delta[$ is continuously differentiable.

Let now

$$\gamma_1(\lambda) = D_\rho F(\rho_0(\lambda),\lambda) \quad , \quad \forall\lambda \in \omega_1 \ . \tag{12}$$

Application of theorem 2.5 on the equation

$$D_\rho F(\rho,\lambda) = 0 \tag{13}$$

shows that (13) has no solution $\rho \in]-\delta,\delta[$ when $\gamma_1(\lambda) > 0$, just one solution $\rho = \rho_0(\lambda)$ when $\gamma_1(\lambda) = 0$, and two solutions in case $\gamma_1(\lambda) < 0$. Fix some

$\lambda \in \omega_1$. If $\gamma_1(\lambda) > 0$, then the map $\rho \mapsto F(\rho,\lambda)$ is strictly increasing; because of (10) the equation (13) has exactly one solution in $]-\delta,\delta[$. If $\gamma_1(\lambda) = 0$ then $F(\rho,\lambda)$ has a strictly positive derivative, except at the point $\rho = \rho_0(\lambda)$. This implies that $F(\rho,\lambda)$ is strictly increasing, and again has a unique solution $\rho \in]-\delta,\delta[$. This proves the lemma. □

5.3.3. Lemma. Using the notation of lemma 2, define, for $\lambda \in \omega_1$:

$$\gamma_0(\lambda) = F(\rho_0(\lambda),\lambda) . \tag{14}$$

If $\lambda \in \omega_1$ and $\gamma_1(\lambda) \geq 0$, then λ can only be a bifurcation point for (6) at a solution $\rho \in]-\delta,\delta[$ if $\gamma_0(\lambda) = \gamma_1(\lambda) = 0$.

P r o o f. It follows from the implicit function theorem that $\lambda \in \omega_1$ can only be a bifurcation point for (6) at a solution ρ if $F(\rho,\lambda) = 0$ and $D_\rho F(\rho,\lambda) = 0$. We see from the proof of lemma 2 that the equation $D_\rho F(\rho,\lambda) = 0$ has no solution if $\gamma_1(\lambda) > 0$. If $\gamma_1(\lambda) = 0$ then the equation has exactly one solution, namely $\rho = \rho_0(\lambda)$. This will also be a solution of $F(\rho,\lambda) = 0$ if and only if $\gamma_0(\lambda) = 0$. This proves the lemma. □

Remark that for all $\lambda \in \omega_1$ satisfying $\gamma_0(\lambda) = \gamma_1(\lambda) = 0$ we have

$$F(\rho_0(\lambda),\lambda) = D_\rho F(\rho_0(\lambda),\lambda) = D_\rho^2 F(\rho_0(\lambda),\lambda) = 0 . \tag{15}$$

5.3.4. Lemma. Suppose (H1) and (H3). For $\lambda \in \omega_1$, let

$$a(\lambda) = \frac{1}{3!} D_\rho^3 F(\rho_0(\lambda),\lambda) . \tag{16}$$

Then there exist functions $\zeta_+(\lambda,\eta)$ and $\zeta_-(\lambda,\eta)$, defined and continuous for $\lambda \in \omega_1$ and $|\eta|$ sufficiently small, and continuously differentiable for $\eta \neq 0$, such that :

(i) $\zeta_\pm(\lambda,0) = \pm(3a(\lambda))^{-1/2}$; (17)

(ii) if $\gamma_1(\lambda) \leq 0$, then the solutions of (13) in $]-\delta,\delta[$ are given by :

$$\rho = \rho_\pm(\lambda) = \rho_0(\lambda) + \zeta_\pm(\lambda,\eta(\lambda))\eta(\lambda) , \tag{18}$$

where

$$\eta(\lambda) = (-\gamma_1(\lambda))^{1/2} . \tag{19}$$

P r o o f. First, let us study the equation :

$$H(\tilde{\rho},\eta,\lambda) \equiv D_\rho F(\rho_0(\lambda) + \tilde{\rho},\lambda) - \gamma_1(\lambda) - \eta^2 = 0 . \tag{20}$$

By the argument of theorem 5.2.6 it is easily shown that solutions of (20) in a neighbourhood of the origin will satisfy $|\tilde{\rho}| \leqslant C|\eta|$ for some constant $C > 0$. Therefore we put $\tilde{\rho} = \eta\zeta$ in (20), which gives us the equation :

$$H_1(\eta,\zeta,\lambda) \equiv D_\rho F(\rho_0(\lambda) + \eta\zeta,\lambda) - \gamma_1(\lambda) - \eta^2 = 0 . \tag{21}$$

The function H_1 is of class C^1 in all variables, and of class C^2 in η and ζ; moreover $H_1(0,\zeta,\lambda) = D_\eta H(0,\zeta,\lambda) = 0$ for all (ζ,λ). So we can write :

$$H_1(\eta,\zeta,\lambda) = \eta^2 H_2(\eta,\zeta,\lambda) , \tag{22}$$

where

$$H_2(\eta,\zeta,\lambda) = \int_0^1 sds \int_0^1 ds' D_\rho^3 F(\rho_0(\lambda) + ss'\eta\zeta,\lambda)\zeta^2 - 1 . \tag{23}$$

It follows from (22) that in the domain $\eta \neq 0$ the function H_2 has the same smoothness properties as the function H_1. Further, H_2 is continuous everywhere, and

$$H_2(0,\zeta,\lambda) = 3a(\lambda)\zeta^2 - 1 , \tag{24}$$

which shows that $H_2(0,\zeta,\lambda)$ is continuously differentiable in ζ. Finally

$$\lim_{\eta\to 0} \eta^{-2} D_\zeta H_1(\eta,\zeta,\lambda) = 6a(\lambda)\zeta ,$$

so that we can conclude that H_2 has a continuous partial derivative in the variable ζ.

Now $H_2(0,\pm(3a(\lambda))^{-1/2},\lambda) = 0$ and $D_\zeta H_2(0,\pm(3a(\lambda))^{-1/2},\lambda) = \mp 2(3a(\lambda))^{1/2} \neq 0$.

An application of the implicit function theorem gives us the existence of functions $\zeta_\pm(\lambda,\eta)$ having the smoothness properties given in the statement of the lemma, satisfying (17) and such that for (η,λ) near the origin, $\zeta = \zeta_\pm(\lambda,\eta)$ are the only solutions of $H_2(\eta,\zeta,\lambda) = 0$ in $[-C,C]$. Consequently the only solutions of (20) in a neighbourhood of the origin are given by $\tilde{\rho} = \eta\zeta_\pm(\lambda,\eta)$. Part (ii) of the statement of the lemma then follows from the observation that (20) reduces to (13) if we put $\tilde{\rho} = \rho-\rho_0(\lambda)$ and $\eta = (-\gamma_1(\lambda))^{1/2}$. □

<u>Remark</u>. It is possible to define the functions $\rho_\pm(\lambda)$, as given by (18), for all λ in a neighbourhood of the origin by taking $\eta(\lambda) = |\gamma_1(\lambda)|^{1/2}$. The functions obtained in this way are continuous, and of class C^1 in $\{\lambda \in \omega_1 \mid \gamma_1(\lambda) \neq 0\}$. Moreover, $\rho_\pm(\lambda)$ are of class C^2 in $\omega_1^- = \{\lambda \in \omega_1 \mid \gamma_1(\lambda) < 0\}$, since for $\lambda \in \omega_1^-$, $\rho = \rho_\pm(\lambda)$ solve (13), while F is of class C^3 and $D_\rho^2 F(\rho_\pm(\lambda),\lambda) \neq 0$.

If we define $\gamma(\lambda)$ for $\lambda \in \omega_1$ by :

$$\gamma(\lambda) = F(\rho_-(\lambda),\lambda).F(\rho_+(\lambda),\lambda) \ , \tag{25}$$

then $\gamma(\lambda)$ has the same smoothness properties as the functions $\rho_\pm(\lambda)$, in particular, $\gamma(\lambda)$ is of class C^2 in ω_1^-.

5.3.5. <u>Lemma</u>. Assume (H1) and (H3), and define $\gamma(\lambda)$ by (25). Then we have the following for each $\lambda \in \omega_1^-$:

(i) if $\gamma(\lambda) > 0$, then (6) has exactly one simple solution $\rho \in]-\delta,\delta[$;

(ii) if $\gamma(\lambda) = 0$, then (6) has one simple and one double solution in $]-\delta,\delta[$;

(iii) if $\gamma(\lambda) < 0$, then (6) has three simple solutions in $]-\delta,\delta[$.

P r o o f. If $\gamma_1(\lambda) < 0$, then it is easily verified that $F(\rho,\lambda)$ has a local maximum at $\rho = \rho_-(\lambda)$ and a local minimum at $\rho = \rho_+(\lambda)$. It follows from (10) that (6) has one simple solution if the corresponding values of $F(\rho,\lambda)$ have the same sign, and three simple solutions when these values have an opposite sign. When one of these values is zero, then there is one simple and one

double solution. The result follows since $\gamma(\lambda)$ is precisely the product of these maximum and minimum values of $F(\rho,\lambda)$. □

5.3.6. There is another way of characterizing the set $\{\lambda \in \omega_1 \mid \gamma_1(\lambda) \leq 0$ and $\gamma(\lambda) = 0\}$. It follows from the definition of $\gamma_0(\lambda)$ and $\gamma_1(\lambda)$ that

$$F(\rho_0(\lambda)+\tilde{\rho},\lambda) = \gamma_0(\lambda) + \gamma_1(\lambda)\tilde{\rho} + \tilde{\rho}^3 R(\tilde{\rho},\lambda) \ , \tag{26}$$

where $R(\tilde{\rho},\lambda)$ is a continuous function which can be defined by :

$$R(\tilde{\rho},\lambda) = \int_0^1 s^2 ds \int_0^1 s'ds' \int_0^1 ds'' D_\rho^3 F(\rho_0(\lambda) + ss's''\tilde{\rho},\lambda) \ . \tag{27}$$

It is then easily seen that, if $\gamma_1(\lambda) \leq 0$, then

$$F(\rho_\pm(\lambda),\lambda) = \gamma_0(\lambda) - \sigma_\pm(\lambda,\eta(\lambda))(\eta(\lambda))^3 \ , \tag{28}$$

where $\eta(\lambda) = (-\gamma_1(\lambda))^{1/2}$ and

$$\sigma_\pm(\lambda,\eta) = \zeta_\pm(\lambda,\eta) - (\zeta_\pm(\lambda,\eta))^3 R(\eta\zeta_\pm(\lambda,\eta),\lambda) \ . \tag{29}$$

The functions $\sigma_\pm(\lambda,\eta)$ are continuous, and :

$$\sigma_\pm(\lambda,0) = \pm 2(27a(\lambda))^{-1/2} \ . \tag{30}$$

The following theorem summarizes our results up to now.

5.3.7. Theorem. Assume (H1) and (H3). Then there exist a $\delta > 0$, a neighbourhood ω_1 of the origin in Λ, C^1-functionals $\gamma_0 : \omega_1 \to \mathbb{R}$ and $\gamma_1 : \omega_1 \to \mathbb{R}$, and continuous functionals $\sigma_+(\lambda,\eta)$ and $\sigma_-(\lambda,\eta)$, defined in a neighbourhood of the origin and satisfying (30), such that for $\lambda \in \omega_1$ the following holds :

(i) if $\gamma_1(\lambda) > 0$, then (6) has one simple solution $\rho \in]-\delta,\delta[$;

(ii) if $\gamma_1(\lambda) \leq 0$, and if $\eta(\lambda) = (-\gamma_1(\lambda))^{1/2}$, then (6) has the following solutions in $]-\delta,\delta[$:

(a) one simple solution if

$$\gamma_0(\lambda) > \sigma_+(\lambda,\eta(\lambda))(\eta(\lambda))^3 \text{ or } \gamma_0(\lambda) < \sigma_-(\lambda,\eta(\lambda))(\eta(\lambda))^3 ;$$

(b) three simple solutions if

$$\sigma_-(\lambda,\eta(\lambda))(\eta(\lambda))^3 < \gamma_0(\lambda) < \sigma_-(\lambda,\eta(\lambda))(\eta(\lambda))^3 ;$$

(c) one simple and one double solution if $\gamma_1(\lambda) < 0$ and :

$$\gamma_0(\lambda) = \sigma_+(\lambda,\eta(\lambda))(\eta(\lambda))^3 \text{ or } \gamma_0(\lambda) = \sigma_-(\lambda,\eta(\lambda))(\eta(\lambda))^3 ;$$

(d) one triple solution if $\gamma_0(\lambda) = \gamma_1(\lambda) = 0$. □

5.3.8. Theorem. Assume (H1) and (H3). Then $\delta > 0$ and the neighbourhood ω_1, appearing in the statement of theorem 7, can be chosen sufficiently small such that each solution $(\rho,\lambda) \in]-\delta,\delta[\times \omega_1$ of (6) will satisfy :

$$|\rho-\rho_0(\lambda)| \leq C(|\gamma_0(\lambda)|^{1/3} + |\gamma_1(\lambda)|^{1/2}) \tag{31}$$

for some constant C.

P r o o f. If not, then we can find a sequence of solutions $\{(\rho_n,\lambda_n) \mid n\in\mathbb{N}\}$ of (6), such that $\rho_n \neq \rho_0(\lambda_n)$ for each $n \in \mathbb{N}$, and

$$(\rho_n,\lambda_n) \to (0,0) \quad , \quad \frac{|\gamma_0(\lambda_n)|^{1/3}}{|\rho_n-\rho_0(\lambda_n)|} \to 0 \ , \ \frac{|\gamma_1(\lambda_n)|^{1/2}}{|\rho_n-\rho_0(\lambda_n)|} \to 0$$

as $n \to \infty$. Using the expression (26) for $F(\rho,\lambda)$, and dividing $F(\rho_n,\lambda_n) = 0$ by $|\rho_n-\rho_0(\lambda_n)|^3$, we find in the limit for $n \to \infty$:

$$R(0,0) = 0 . \tag{32}$$

However, $R(0,0) = a_0$, and so (32) contradicts (H3). □

We conclude this section by giving an alternative formulation of the hypothesis (H3). Since the definition of the function $F(\rho,\lambda)$ contains the projections P and Q, and the vectors u_0 and z_0, we can ask whether (H3) will still be satisfied for another choice of these projections and vectors. Next theorem shows that this is indeed the case.

5.3.9. Theorem. Suppose that $m_0(x) = M(x,\lambda_0)$ satisfies (H1). Let P, Q, u_0 and z_0 be as before, and define :

$$F_0(\rho) = F(\rho,\lambda_0) = Q_0M(x_0 + \rho u_0 + v^*(\rho u_0,\lambda_0),\lambda_0) \ . \tag{33}$$

Suppose that $r \geqslant k \geqslant 2$. Then the following statements are equivalent :

(i) $D_\rho^2 F_0(0) = D_\rho^3 F_0(0) = \ldots = D_\rho^{k-1} F_0(0) = 0$,
and $D_\rho^k F_0(0) \neq 0$;

(ii) there is a C^k-map $x^* :]-\delta,\delta[\to X$, with $x^*(0) = x_0$ and $D_\rho x^*(0) \neq 0$, such that

$$m_0(x^*(\rho)) = \alpha(\rho) z_1 \quad , \qquad \forall\rho \in]-\delta,\delta[\tag{34}$$

for some $z_1 \notin R(D_xM_0(x_0))$ and some function $\alpha(\rho)$ satisfying :

$$\alpha(0) = D_\rho\alpha(0) = \ldots = D_\rho^{k-1}\alpha(0) = 0 \quad , \quad D_\rho^k\alpha(0) \neq 0 \ . \tag{35}$$

P r o o f. We remark first that (H1) implies that $F_0(0) = D_\rho F_0(0) = 0$. Assume (i), and let

$$x^*(\rho) = x_0 + \rho u_0 + v^*(\rho u_0,\lambda_0) \ .$$

Then (32) and (33) are satisfied, if we take $z_1 = z_0$ and $\alpha(\rho) = F_0(\rho)$. Conversely, suppose (ii) is satisfied. Let $P(x^*(\rho)-x_0) = \beta(\rho)u_0$; then $\beta(0) = 0$ and

$$x^*(\rho) = x_0 + \beta(\rho)u_0 + v^*(\beta(\rho)u_0,\lambda_0) + \tilde{v}(\rho) \tag{36}$$

for some function $\tilde{v} :]-\delta,\delta[\to \ker P$. From (35),

$$(I-Q)m_0(x^*(\rho)) = \alpha(\rho)(I-Q)z_1 \ ,$$

and the equation defining $v^*(\beta(\rho)u_0,\lambda_0)$ it follows that :

$$\tilde{v}(0) = 0 \quad , \quad D_\rho\tilde{v}(0) = 0 \quad , \quad \ldots \quad , \quad D_\rho^{k-1}\tilde{v}(0) = 0 \ . \tag{37}$$

Now define $F_1(\rho)$ and $F_2(\rho)$ by :

$$F_1(\rho) = Q_0 m_0(x^*(\rho)) \tag{38}$$

and

$$F_2(\rho) = Q_0 m_0(x_0 + \beta(\rho)x_0 + v^*(\beta(\rho)u_0,\lambda_0)) \ . \tag{39}$$

From (36) and (37) it follows that :

$$F_1(0) = F_2(0) \quad , \quad D_\rho F_1(0) = D_\rho F_2(0),\ldots, D_\rho^k F_1(0) = D_\rho^k F_2(0) \ .$$

Indeed, the only term in $D_\rho^k F_1(0)$ containing $D_\rho^k\tilde{v}(0)$ has the form

$$Q_0 D_x m_0(x_0) . D_\rho^k \tilde{v}(0) \ ,$$

which is zero by the definition of Q_0.

Now $Q_0(z_1) \neq 0$, since $z_1 \notin R(D_x m_0(x_0))$. From $F_1(\rho) = \alpha(\rho)Q_0(z_1)$ and (35) it follows that

$$F_2(0) = D_\rho F_2(0) = \ldots = D_\rho^{k-1} F_2(0) = 0 \quad , \quad D_\rho^k F_2(0) \neq 0 \ . \tag{40}$$

Finally, we compare the definition (39) of $F_2(\rho)$ with the definition of $F_0(\rho)$. We obtain :

$$F_2(0) = F_0(0) \ ;$$
$$D_\rho F_2(0) = D_\rho F_0 . D_\rho\beta(0) \ ,$$
$$D_\rho^2 F_2(0) = D_\rho^2 F_0(0).(D_\rho\beta(0))^2 + D_\rho F_2(0).D_\rho^2\beta(0) \ ,$$
$$\ldots$$
$$D_\rho^k F_2(0) = D_\rho^k F_0(0).(D_\rho\beta(0))^k + \text{terms containing}$$
$$(F_0(0), D_\rho F_0(0),\ldots D_\rho^{k-1} F_0(0)).$$

Since $D_\rho x^*(0) \neq 0$ it follows from (36) that $D_\rho \beta(0) \neq 0$. Then (i) follows from (40). □

In the formulation of the next theorem we use the following notation :

$$B_0 = \{(\rho_0(\lambda),\lambda) \mid \lambda \in \omega_1, \ \gamma_0(\lambda) = \gamma_1(\lambda) = 0\} \ , \tag{41}$$

$$B_+ = \{(\rho_+(\lambda),\lambda) \mid \lambda \in \omega_1, \ \gamma_1(\lambda) < 0, \ \gamma_0(\lambda) = \sigma_+(\lambda,\eta(\lambda))(\eta(\lambda))^3\} \ , \tag{42}$$

and

$$B_- = \{(\rho_-(\lambda),\lambda) \mid \lambda \in \omega_1, \ \gamma_1(\lambda) < 0, \ \gamma_0(\lambda) = \sigma_-(\lambda,\eta(\lambda))(\eta(\lambda))^3\} \ . \tag{43}$$

Remember also the notation :

$$x^*(\rho,\lambda) = \rho u_0 + v^*(\rho u_0,\lambda) \ . \tag{44}$$

5.3.10. Theorem. Suppose that (H1) and (H3) are satisfied at (0,0). Then there is a neighbourhood $\Omega_1 \times \omega_1$ of the origin in $X \times \Lambda$ such that for $(x,\lambda) \in \Omega_1 \times \omega_1$ the hypothesis (H1) is satisfied at (x,λ) if and only if $x = x^*(\rho,\lambda)$ with $(\rho,\lambda) \in B = B_0 \cup B_+ \cup B_-$.
If $(\rho,\lambda) \in B_0$ then also (H3) is satisfied at the point $(x^*(\rho,\lambda),\lambda)$.
If $(\rho,\lambda) \in B_+ \cup B_-$, then (H2) is satisfied at $(x^*(\rho,\lambda),\lambda)$.

P r o o f. The first part follows from theorem 2.9 and the fact that $F(\rho,\lambda) = D_\rho F(\rho,\lambda) = 0$ if and only if $(\rho,\lambda) \in B$. The second part follows from theorem 9, as follows :

Let $(\rho,\lambda) \in B$, and define :

$$\bar{x}(\bar{\rho}) = (\rho + \bar{\rho})u_0 + v^*((\rho + \bar{\rho})u_0,\lambda) \ .$$

Then : $M(\bar{x}(\bar{\rho}),\lambda) = F(\rho + \bar{\rho},\lambda)z_0 = \alpha(\bar{\rho})z_0$. For $(\rho,\lambda) \in B_0$ we have $\alpha(0) = D_{\bar{\rho}}\alpha(0) = D_{\bar{\rho}}^2\alpha(0) = 0$ and $D_{\bar{\rho}}^3\alpha(0) \neq 0$. If $(\rho,\lambda) \in B_+ \cup B_-$, then $\alpha(0) = D_{\bar{\rho}}\alpha(0) = 0$ and $D_{\bar{\rho}}^2\alpha(0) \neq 0$. This proves the theorem. □

5.4. GENERIC AND NON-GENERIC BIFURCATION

5.4.1. Introduction. Consider again the equation

$$M(x,\lambda) = 0 \tag{1}$$

and suppose that the hypotheses (H1) and (H3) from the preceding sections are satisfied. For a sufficiently small neighbourhood $\Omega_1 \times \omega_1$ of the origin in $X \times \Lambda$, we define :

$$\beta = \{\lambda \in \omega_1 \mid \lambda \text{ is a bifurcation point for (1) at a solution } x \in \Omega_1\} . \tag{2}$$

From theorem 3.10 we see that $\beta \subset \beta_0 \cup \beta_+ \cup \beta_-$, where

$$\begin{aligned} \beta_0 &= \{\lambda \in \omega_1 \mid \gamma_0(\lambda) = \gamma_1(\lambda) = 0\} , \\ \beta_+ &= \{\lambda \in \omega_1 \mid \gamma_1(\lambda) < 0,\ \gamma_0(\lambda) = \sigma_+(\lambda,\eta(\lambda))(\eta(\lambda))^3\} , \\ \beta_- &= \{\lambda \in \kappa_1 \mid \gamma_1(\lambda) < 0,\ \gamma_0(\lambda) = \sigma_-(\lambda,\eta(\lambda))(\eta(\lambda))^3\} . \end{aligned} \tag{3}$$

However, we have not necessarily that $\beta = \beta_0 \cup \beta_+ \cup \beta_-$. For example, assume that $M(x,\lambda)$ is such that $\gamma_1(\lambda) = 0$ for all $\lambda \in \omega_1$; then the equation (1) will have exactly one solution in Ω_1 for each $\lambda \in \omega_1$, and the bifurcation set β will be empty.

In order to determine β we have to study the following subsets :

$$M_0 = \{\lambda \in \omega_1 \mid \gamma_0(\lambda) = 0\} \tag{4}$$

and

$$M_1 = \{\lambda \in \omega_1 \mid \gamma_1(\lambda) = 0\} . \tag{5}$$

5.4.2. Theorem. Suppose that (H1) and (H3) are satisfied at (0,0) for the equation (1). Define a map $\Psi : \omega_1 \to \mathbb{R}^2$ by :

$$\Psi(\lambda) = (\gamma_0(\lambda),\gamma_1(\lambda)) \quad , \quad \forall \lambda \in \omega_1 . \tag{6}$$

Suppose that the linear map $D_\lambda\Psi(0) \in L(\Lambda,\mathbb{R}^2)$ is surjective. Then M_0 and M_1 are C^1-submanifolds of Λ, with codimension equal to 1, and transversal at the origin $\lambda = 0$. Moreover, under these conditions we have $\beta = \beta_0 \cup \beta_+ \cup \beta_-$.

P r o o f. The condition implies in particular that the linear functionals $D_\lambda\gamma_0(0) \in L(\Lambda;\mathbb{R})$ and $D_\lambda\gamma_1(0) \in L(\Lambda;\mathbb{R})$ are surjective. This shows that M_0 and M_1 are C^1-submanifolds with codimension equal to 1.

Let now $\tilde{\lambda}_1 \in \Lambda$ and $\tilde{\lambda}_2 \in \Lambda$ be such that $D_\lambda\Psi(0).\tilde{\lambda}_1$ and $D_\lambda\Psi(0).\tilde{\lambda}_2$ are linearly independent. We may suppose that $D_\lambda\gamma_0(0).\tilde{\lambda}_1 \neq 0$. Then :

$$\Lambda = \mathrm{span}\{\tilde{\lambda}_1\} \oplus \ker D_\lambda\gamma_0(0) .$$

This implies that $\tilde{\lambda}_2 = \alpha\tilde{\lambda}_1 + \bar{\lambda}_2$, for some $\alpha \in \mathbb{R}$ and $\bar{\lambda}_2 \in \ker D_\lambda\gamma_0(0)$. The vectors $D_\lambda\Psi(0).\tilde{\lambda}_1$ and $D_\lambda\Psi(0).\bar{\lambda}_2 = (0, D_\lambda\gamma_1(0).\bar{\lambda}_2)$ are linearly independent; so $D_\lambda\gamma_1(0).\bar{\lambda}_2 \neq 0$. It follows that

$$\Lambda = \mathrm{span}\{\bar{\lambda}_2\} \oplus \ker D_\lambda\gamma_1(0) .$$

Since $\bar{\lambda}_2 \in \ker D_\lambda\gamma_0(0)$ we conclude that Λ is spanned by $\ker D_\lambda\gamma_0(0)$ and $\ker D_\lambda\gamma_1(0)$; i.e., M_0 and M_1 are transversal at the point $\lambda = 0$. This also implies that $\beta_0 = M_0 \cap M_1$ is a C^1-submanifold with codimension equal to 2.

Each point $\lambda_0 \in \beta_0$ is a bifurcation point. Indeed, each neighbourhood of λ_0 contains λ at which $\gamma_1(\lambda) < 0$, and, for example, $\gamma_0(\lambda) = 0$; for such λ the equation (1) has three different solutions in Ω_1; it follows from theorem 3.8 that these solutions converge to the unique solution of $M(x,\lambda_0) = 0$ in Ω_1 as $\lambda \to \lambda_0$.

For $\lambda_+ \in \beta_+$ equation (1) has one single and one double solution; let x_+ be this double solution. It follows from theorem 3.10 that the conditions of corollary 2.7 are satisfied at the point (x_+,λ_+). So λ_+ is a bifurcation point for (1); the bifurcation takes place from the double solution. A similar argument shows that $\lambda_- \in \beta_-$ is a bifurcation point. □

The following figures show the bifurcation set, which forms a cusp along β_0, and the corresponding solutions of the bifurcation equation $F(\rho,\lambda) = 0$. The bifurcation set β divides ω_1 into two subregions; for λ in region I, the equation (1) has a unique solution in Ω_1, for λ belonging to region II there are three such solutions.

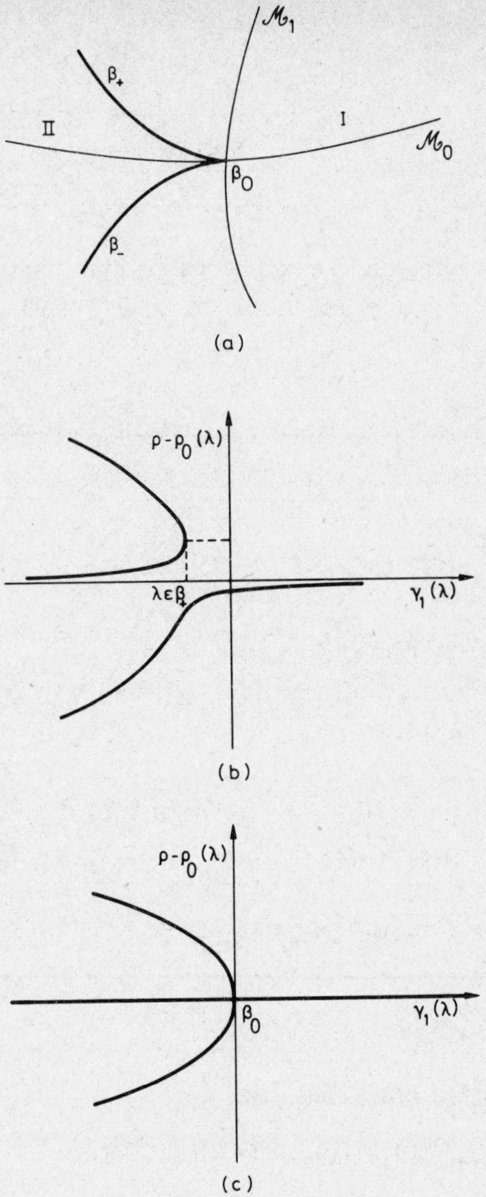

Fig. 5. (a) The bifurcation set
(b) The solution set along $\gamma_0(\lambda)$ = constant > 0
(c) The solution set along $\gamma_0(\lambda) = 0$

5.4.3. Remark. The condition of theorem 2 can be reformulated as follows : there exist $\tilde{\lambda}_1 \in \Lambda$ and $\tilde{\lambda}_2 \in \Lambda$ such that :

$$\det \begin{vmatrix} D_\lambda\gamma_0(0).\tilde{\lambda}_1 & D_\lambda\gamma_1(0).\tilde{\lambda}_1 \\ D_\lambda\gamma_0(0).\tilde{\lambda}_2 & D_\lambda\gamma_1(0).\tilde{\lambda}_2 \end{vmatrix} \neq 0 \ . \tag{7}$$

The derivatives appearing in (7) can be calculated as follows. It follows from the definition of $\gamma_0(\lambda)$ and $\gamma_1(\lambda)$ (see (3.12) and (3.14)), from $\rho_0(0) = 0$ and from (H3) that

$$D_\lambda\gamma_0(0).\tilde{\lambda} = D_\lambda F(0,0).\tilde{\lambda} \qquad , \qquad \forall\tilde{\lambda} \in \Lambda \tag{8}$$

and

$$D_\lambda\gamma_1(0).\tilde{\lambda} = D_\lambda D_\rho F(0,0).\tilde{\lambda} \qquad , \qquad \forall\tilde{\lambda} \in \Lambda \ . \tag{9}$$

Using the definition of $F(\rho,\lambda)$ one finds :

$$D_\lambda\gamma_0(0).\tilde{\lambda} = Q_0 D_\lambda M(0,0).\tilde{\lambda} \qquad , \qquad \forall\tilde{\lambda} \in \Lambda \ . \tag{10}$$

To calculate $D_\lambda\gamma_1(0).\tilde{\lambda}$ from (9), one needs $D_\lambda v^*(0,0)$; this can be obtained by differentiating the relation defining $v^*(u,\lambda)$; one finds :

$$D_\lambda v^*(0,0).\tilde{\lambda} = -K(I-Q)D_\lambda M(0,0).\tilde{\lambda} \ .$$

Finally one obtains

$$\begin{aligned} D_\lambda\gamma_1(0).\tilde{\lambda} = {} & Q_0 D_\lambda D_x M(0,0).(u_0,\tilde{\lambda}) \\ & - Q_0 D_x^2 M(0,0).(u_0,K(I-Q)D_\lambda M(0,0).\tilde{\lambda}) \quad , \quad \forall\lambda \in \Lambda \ . \end{aligned} \tag{11}$$

5.4.4. Corollary. Consider the generic problem described in subsection 2.2. Suppose that $m_0 \in C^3(\Omega;Z)$ satisfies (H1) and (H3) at x_0. Then the bifurcation set in a sufficiently small neighbourhood ω_1 of m_0 in $C^3(\Omega;Z)$ is given by :

$$\beta = \{m \in \omega_1 \mid \gamma_1(m) \leqslant 0, \ \gamma_0(m) = \sigma_\pm(m,\eta(m))(\eta(m))^3\} \ . \tag{12}$$

P r o o f. We only have to show that the condition (7) is satisfied (with $\Lambda = C^3(\Omega;Z)$). From (10) and (11) we find for this case :

$$D_m\gamma_0(m_0).\tilde{m} = Q_0\tilde{m}(x_0) \quad , \quad \forall\tilde{m} \in C^3(\Omega;Z)$$

and

$$D_m\gamma_1(m_0).\tilde{m} = Q_0D_x\tilde{m}(x_0).u_0 - Q_0D_x^2m_0(x_0).(u_0,K(I-Q)\tilde{m}(x_0)) \ ,$$
$$\forall\tilde{m} \in C^3(\Omega;Z) \ .$$

Now make the following choice for $\tilde{m}_1$ and $\tilde{m}_2$:

$$\tilde{m}_1(x) = z_0 \quad , \quad \forall x \in \Omega$$

and

$$\tilde{m}_2(x) = \tilde{m}_2(x_0 + \rho u_0 + v) = \rho z_0 \ , \quad \forall x \in \Omega \ .$$

The matrix in (7) becomes the identity matrix. This proves the corollary. □

5.4.5. We can use corollary 4 to reformulate the condition of theorem 2. We suppose that not only $M(x,\lambda)$, but also $D_\lambda M(x,\lambda)$ is of class C^3. We define the mapping $\tau : \omega \subset \Lambda \to C^3(\Omega;Z)$ by

$$\tau(\lambda) = M(.,\lambda) \quad , \quad \forall\lambda \in \omega \ . \tag{13}$$

Let $m_0 = \tau(\lambda_0)$, and let $\tilde{\omega}$ be a neighbourhood of m_0 in $C^3(\Omega;Z)$ such that for each $m \in \tilde{\omega}$ the map $\psi(m) = (\gamma_0(m),\gamma_1(m))$ is well-defined. We put $\beta_0 = \{m\in\tilde{\omega} \mid \psi(m) = 0\}$; β_0 is a C^1-submanifold of $C^3(\Omega;Z)$ with codimension equal to two.

Using this framework, the condition of theorem 2 can be restated as saying that the mapping $\psi\circ\tau : \omega_1 = \tau^{-1}(\tilde{\omega}) \to \mathbb{R}^2$ is submersive at the point λ_0. The proof of theorem 2 shows that this means that

$$C^3(\Omega;Z) = T_{m_0}\beta_0 + R(D_\lambda\tau(\lambda_0)) \ ,$$

where $T_{m_0}\beta_0$ is the tangent space to β_0 at m_0. So the condition says nothing else than that the map τ is transversal to β_0 at $\lambda = \lambda_0$. It is clear that this can only be satisfied if dim $\Lambda \geqslant 2$, since β_0 has codimension 2. However, if dim $\Lambda \geqslant 2$, then the transversality condition will be satisfied "generically", i.e. for almost all $\tau : \Lambda \to C^3(\Omega;Z)$. For this reason the transversality condition will also be called the *generic condition*, and the corresponding bifurcation set a *generic bifurcation set*.

5.4.6. Similarly as it was the case with the results of section 2, also the results of sections 3 and 4 show a clear relationship with some results from singularity theory; this relationship was already commented by Chow, Hale and Mallet-Paret in [39] and [40].

Suppose that $M(x,\lambda)$ is of class C^∞, that dim $\Lambda = p < \infty$ and that M satisfies (H1) and (H3), say at (0,0). Then also the bifurcation function $F(\rho,\lambda)$ is of class C^∞, and can be considered as an unfolding of the function $F_0(\rho) = F(\rho,0)$, which has the form $F_0(\rho) = a_0\rho^3$ + higher order terms. The transversality condition can be written as :

$$\text{rank}(a_{ij}) = 2 \qquad (i=1,\ldots,p \; ; \; j=1,2) \; , \tag{14}$$

where, for $i = 1,\ldots,p$:

$$a_{i1} = \frac{\partial F}{\partial \lambda_i}(0,0) = \frac{\partial \gamma_0}{\partial \lambda_i}(0,0)$$

and (15)

$$a_{i2} = \frac{\partial^2 F}{\partial \rho \partial \lambda_i}(0,0) = \frac{\partial \gamma_1}{\partial \lambda_i}(0,0) \; .$$

Assume that (14) is satisfied, then we may suppose, without loss of generality, that

$$\frac{\partial(\gamma_0,\gamma_1)}{\partial(\lambda_1,\lambda_2)} \neq 0 \; .$$

This implies that the map $(\rho,\lambda_1,\ldots,\lambda_p) \mapsto (\tilde{\rho},\mu_1,\ldots,\mu_p)$ given by

$$\begin{aligned} \tilde{\rho} &= \rho - \rho_0(\lambda) \\ \mu_1 &= \gamma_0(\lambda) \\ \mu_2 &= \gamma_1(\lambda) \\ \mu_i &= \lambda_i \; , \qquad i = 3,\ldots,p \end{aligned} \tag{16}$$

is a C^∞-diffeomorphism in a neighbourhood of the origin in $\mathbb{R}\times\mathbb{R}^p$. Then (3.26) shows that in the new variables $(\tilde{\rho},\mu)$ the bifurcation function takes the form :

$$F_1(\tilde{\rho},\mu) = \mu_1 + \mu_2\tilde{\rho} + \tilde{\rho}^3 R(\rho,\mu) \ , \tag{17}$$

with $R(0,0) = a_0$. Comparing with the basic unfolding result of singularity theory this shows that the transversality condition (14) is the necessary and sufficient condition for $F(\rho,\lambda)$ to be a universal unfolding of $F_0(\rho)$. When this is the case, then there is a smooth change of coordinates near the origin of the form :

$$\nu_i = \phi_i(\lambda) \qquad , \qquad i = 1,\dots,p$$
$$\bar{\rho} = \zeta(\rho,\lambda)$$

such that in the variables $(\bar{\rho},\nu)$, $F(\rho,\lambda)$ takes the normal form :

$$F_2(\bar{\rho},\nu) = \bar{\rho}^3 + \nu_2\bar{\rho} + \nu_1 \ . \tag{18}$$

The genericity of the condition (14) means the following. Consider the space of all C^∞-mapping $G : \mathbb{R}\times\mathbb{R}^p \to \mathbb{R}$. This space can be given the structure of a Baire topological space by introducing the so-called Whitney topology. Consider the subset U of all $G(\rho,\lambda)$ such that $G(\rho,0) = F_0(\rho)$; U is a linear submanifold. Then it follows from Thom's transversality theorem that the subset of all $G \in U$ which satisfy (14) is generic in U, i.e. forms an open and dense subset of U, equiped with the Whitney topology. (For the results on singularity theory, quoted in this subsection, we refer to Golubitsky and Guillemin [72], Bröcker [26] and Martinet [261]).

5.4.7. It is interesting to compare the bifurcation sets for the equations $F_1(\tilde{\rho},\mu) = 0$ and $F_2(\bar{\rho},\nu) = 0$, with F_1 and F_2 given respectively by (17) and (18). It follows from our foregoing results that the bifurcation set for $F_1(\tilde{\rho},\mu) = 0$ is given by

$$B_1 = \{\mu\in\mathbb{R}^p \mid \mu_2\leq 0,\ \mu_1 = \sigma_\pm(\mu,(-\mu_2)^{1/2})(-\mu_2)^{3/2}\} \ , \tag{19}$$

with $\sigma_{\pm}(0,0) = \pm 2(27a_0)^{-1/2}$. The bifurcation set for $F_2(\bar{\rho},\nu) = 0$ is given by

$$\beta_2 = \{\nu \in \mathbb{R}^p \mid \nu_2 \leqslant 0,\ 27\nu_1^2 + 4\nu_2^3 = 0\} , \tag{20}$$

as can easily be verified.

By a rescaling of $\tilde{\rho}$ one can achieve that $a_0 = 1$. Then one obtains (20) from (19) by replacing $\sigma_{\pm}(\mu)$ by its value at $\mu = 0$; similarly, one obtains (18) from (17) by replacing the remainder term $R(\rho,\mu)$ by its value at (0,0).

5.4.8. We have obtained our bifurcation results for the equation (1) in a general setting, under relatively weak smoothness conditions, and using only classical tools such as rescaling techniques and the implicit function theorem. To the contrary, the approach via singularity theory requires that the equation is C^∞ and that the number of parameters is finite, while the proof uses the less elementary preparation theorem. (In a recent contribution, Michor [164] has extended the preparation theorem to the case where the parameter belongs to a Banach space). However, in more complicated situations the unified treatment via singularity theory seems to have some advantage; indeed, once the general theory is established, the application of the results of singularity theory to particular examples reduces to merely algebraic manipulations. For example, one could try to use our method to study bifurcation in the general case that the bifurcation function satisfies :

$$F(0,0) = D_\rho F(0,0) = \ldots = D_\rho^{k-1} F(0,0) = 0 \quad , \quad D_\rho^k F(0,0) \neq 0 \ .$$

However, the treatment becomes very complicated for $k > 3$. Using singularity theory, the problem reduces to the study of the solutions of a polynomial equation of the form :

$$\rho^k + \sum_{i=0}^{k-2} \alpha_i(\lambda)\rho^i = 0 ,$$

with $\alpha_i(\lambda_0) = 0$.

For a somewhat different approach to bifurcation theory (in particular imperfect bifurcation) via singularity theory, see the basic papers by Golubitsky and Schaeffer ([74] and [75]).

5.4.9. Non-generic bifurcation. We conclude this section with the description of a particular situation where the transversality condition (7) is not satisfied. This case is of some interest when the equation has additional symmetry, as we will see in the next section.

The hypothesis is that $\gamma_0(\lambda) = 0$ for all $\lambda \in \omega_1$. Then the bifurcation equation has the solution $\rho = \rho_0(\lambda)$ for each $\lambda \in \omega_1$. All other solutions near $(0,0)$ can be written in the form $(\rho,\lambda) = (\rho_0(\lambda)+\tilde{\rho},\lambda)$, where $(\tilde{\rho},\lambda)$ has to be a solution of the equation

$$F_1(\tilde{\rho},\lambda) \equiv \int_0^1 D_\rho F(\rho_0(\lambda)+s\tilde{\rho},\lambda)ds = 0 . \tag{21}$$

Indeed, $F(\rho_0(\lambda)+\tilde{\rho},\lambda) = \tilde{\rho}F_1(\tilde{\rho},\lambda)$, and (21) is equivalent with $F(\rho_0(\lambda)+\tilde{\rho},\lambda) = 0$ if $\tilde{\rho} \neq 0$. From (21) we see that :

$$\begin{aligned} &F_1(0,\lambda) = \gamma_1(\lambda) \quad , \quad D_{\tilde{\rho}}F_1(0,\lambda) = 0 \ , \\ &\frac{1}{2!}D^2_{\tilde{\rho}}F_1(0,\lambda) = a(\lambda) \ , \quad \forall\lambda \in \omega_1 \ . \end{aligned} \tag{22}$$

We can apply theorem 2.5 to equation (21); remembering that the bifurcation equation has always the solution $\rho = \rho_0(\lambda)$, while (21) gives us the solutions $\rho \neq \rho_0(\lambda)$, we obtain the following result.

5.4.10. Theorem. Suppose that the equation (1) satisfies the hypotheses (H1) and (H3) at $(0,0)$, while also $\gamma_0(\lambda) = 0$ for all $\lambda \in \omega_1$. Assume that we can find some $\tilde{\lambda} \in \Lambda$ such that

$$D_\lambda\gamma_1(0).\tilde{\lambda} \neq 0 . \tag{23}$$

Then the bifurcation set for (1) is given by $\beta_0 = \{\lambda \in \omega_1 \mid \gamma_0(\lambda) = \gamma_1(\lambda) = 0\} = \{\lambda \in \omega_1 \mid \gamma_1(\lambda) = 0\}$, which is a C^1-submanifold of Λ, with codimension 1.
If $\gamma_1(\lambda) > 0$ then (1) has one simple solution in a sufficiently small neighbourhood Ω_1 of the origin in X; this unique solution corresponds to the solution $\rho = \rho_0(\lambda)$ of the bifurcation equation. If $\gamma_1(\lambda) = 0$ (i.e. $\lambda \in \beta_0$) then the solution of (1) corresponding to $\rho = \rho_0(\lambda)$ remains the unique solution in Ω_1, but is now a triple solution. If $\gamma_1(\lambda) < 0$, then (1) has three simple solutions in Ω_1, one of which corresponds to $\rho_0(\lambda)$. □

The following figures give a sketch of the bifurcation set and the solutions.

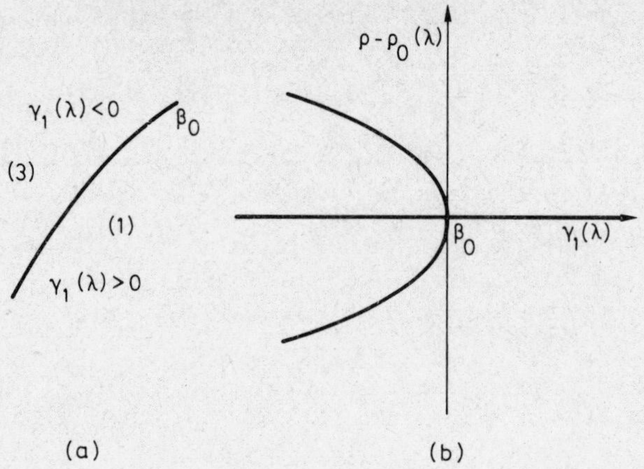

Fig. 6. (a) Bifurcation set
(b) Solution set.

5.5. GENERIC CONDITIONS AND SYMMETRY

5.5.1. In this section we study how the symmetry of the equation

$$M(x,\lambda) = 0 \tag{1}$$

can affect the bifurcation set. Our hypotheses are the same as in the preceding sections : we assume that M satisfies (H1) and (H3) at $(x_0,\lambda_0) = (0,0)$. We will also assume that M is equivariant with respect to some $(G,\Gamma,\tilde{\Gamma})$, where G is a compact group. The projections P and Q used in the Liapunov-Schmidt reduction of (1) are chosen such that they commute with the symmetry operators (see chapter 3). The bifurcation function

$$H(u,\lambda) \equiv QM(u+v^*(u,\lambda),\lambda) \quad , \quad u \in \ker L \ , \ \lambda \in \omega \tag{2}$$

satisfies :

$$H(\Gamma(s)u,\lambda) = \tilde{\Gamma}(s)H(u,\lambda) \quad , \quad \forall s \in G \ , \ \forall (u,\lambda) \ . \tag{3}$$

The relation between $H(u,\lambda)$ and $F(\rho,\lambda)$ is given by :

$$H(\rho u_0,\lambda) = F(\rho,\lambda)z_0 \quad , \quad \forall(\rho,\lambda) \ . \tag{4}$$

5.5.2. The one-dimensional subspaces $\ker L = R(P)$ and $R(Q)$ remain invariant under the operators $\Gamma(s)$, respectively $\tilde{\Gamma}(s)$. It follows that there exist scalar functions $\alpha : G \to \mathbb{R}$ and $\tilde{\alpha} : G \to \mathbb{R}$ such that :

$$\Gamma(s)u_0 = \alpha(s)u_0 \quad , \quad \tilde{\Gamma}(s)z_0 = \tilde{\alpha}(s)z_0 \quad , \quad \forall s \in G \ . \tag{5}$$

It is clear that both α and $\tilde{\alpha}$ are one-dimensional representations of G. Consequently, $\{\alpha(s) \mid s \in G\}$ is a compact subset of $\mathbb{R}\setminus\{0\}$, which is closed under multiplication and inversion; the same holds for $\{\tilde{\alpha}(s) \mid s \in G\}$. We conclude that :

$$|\alpha(s)| = |\tilde{\alpha}(s)| = 1 \quad , \quad \forall s \in G \ . \tag{6}$$

It follows from (3), (4) and (5) that

$$\begin{aligned} F(\alpha(s)\rho,\lambda)z_0 &= H(\alpha(s)\rho u_0,\lambda) = H(\Gamma(s)\rho u_0,\lambda) \\ &= \tilde{\Gamma}(s)H(\rho u_0,\lambda) = \tilde{\alpha}(s)F(\rho,\lambda)z_0 \ , \end{aligned}$$

i.e.

$$F(\alpha(s)\rho,\lambda) = \tilde{\alpha}(s)F(\rho,\lambda) \quad , \quad \forall s \in G \ , \ \forall(\rho,\lambda) \ . \tag{7}$$

5.5.3. <u>Lemma</u>. Let M be equivariant with respect to $(G,\Gamma,\tilde{\Gamma})$, and satisfy (H1) and (H3) at $(0,0)$. Then :

$$\alpha(s) = \tilde{\alpha}(s) \quad , \quad \forall s \in G \ . \tag{8}$$

P r o o f. It follows from (7) that

$$(\alpha(s))^3 D_\rho^3 F(0,0) = \tilde{\alpha}(s) D_\rho^3 F(0,0) \quad , \quad \forall s \in G \ .$$

Since $D^3_\rho F(0,0) \neq 0$, we conclude that $(\alpha(s))^3 = \tilde{\alpha}(s)$, which together with (6) implies (8). □

5.5.4. Define

$$G_0 = \{s \in G \mid \alpha(s) = 1\} \ ; \tag{9}$$

and assume that $G_0 = G$, i.e. :

$$\Gamma(s)u = u \quad , \quad \forall s \in G \ , \ \forall u \in \ker L \ . \tag{10}$$

Then it follows from theorem 3.6.1 that all solutions (x,λ) of (1) in a neighbourhood of the origin in $X\times\Lambda$ will satisfy :

$$\Gamma(s)x = x \quad , \quad \forall s \in G \ . \tag{11}$$

The bifurcation problem can then be reformulated and discussed using the spaces X_0 and Z_0 instead of X and Z (see chapter 3). In this reduced formulation symmetry plays no further role, and the general discussion of the preceding section applies.

5.5.5. Now suppose that G_0 is a proper subgroup of G, i.e. $\alpha(s) = -1$ for some $s \in G$. Then (6) implies that G_0 will be a *normal* subgroup of G, i.e. : $s^{-1} \circ G_0 \circ s = G_0$ for all $s \in G$. Application of theorem 3.6.1 shows that all sufficiently small solutions (x,λ) of (1) will satisfy

$$\Gamma(s)x = x \quad , \quad \forall s \in G_0 \ . \tag{12}$$

This allows us to restrict our attention to solutions (x,λ) of (1), with $x \in X_0$, i.e. we can consider M as a mapping of the type $M : X_0\times\Lambda \to Z_0$, where

$$X_0 = \{x \in X \mid \Gamma(s)x = x, \ \forall s \in G_0\} \ , \ Z_0 = \{z \in Z \mid \tilde{\Gamma}(s)z = z, \ \forall s \in G_0\} \ . \tag{13}$$

The action of the group G on X_0 and Z_0 reduces to the action of a two-element group $\tilde{G} = \{e,i\}$, with $i^2 = e$; the group $\tilde{G}$ is isomorphic to the quotient group G/G_0. So the action of $\tilde{G}$ on X_0 and Z_0 is given by :

$$\Gamma(e)x = x \quad , \quad \Gamma(i)x = \Gamma(s)x \quad , \quad \forall x \in X_0 \ , \tag{14}$$

and

$$\tilde{\Gamma}(e)z = z \quad , \quad \tilde{\Gamma}(i)z = \tilde{\Gamma}(s)z \quad , \quad \forall z \in Z_0 \ , \tag{15}$$

where s is an arbitrary element of $G\backslash G_0$. We have

$$\Gamma(i)u_0 = -u_0 \quad , \quad \tilde{\Gamma}(i)z_0 = -z_0 \ , \tag{16}$$

and (7) shows that $F(\rho,\lambda)$ is odd in ρ :

$$F(-\rho,\lambda) = -F(\rho,\lambda) \quad , \quad \forall(\rho,\lambda) \ . \tag{17}$$

It follows that

$$\rho_0(\lambda) = 0 \quad , \quad \gamma_0(\lambda) = 0 \quad , \quad \gamma_1(\rho) = D_\rho F(0,\lambda) \quad , \quad \forall\lambda \in \omega_1 \ . \tag{18}$$

So we have a case of non-generic bifurcation, as described at the end of the preceding section.

5.5.6. Theorem. Assume M satisfies (H1) and (H3), and is equivariant with respect to some $(G,\Gamma,\tilde{\Gamma})$, where G is a compact group. Define the subgroup G_0 by (9), and suppose the following :

(i) $G_0 \neq G$;

(ii) there is some $\tilde{\lambda} \in \Lambda$ such that

$$D_\lambda D_x M(0,0).(u_0,\tilde{\lambda}) \notin R(D_x M(0,0)) \ .$$

Then the bifurcation set for (1) is given by

$$\beta = \{\lambda \in \omega_1 \mid \gamma_1(\lambda) = 0\} = \{\lambda \in \omega_1 \mid D_\rho F(0,\lambda) = 0\} \ . \tag{19}$$

For all $\lambda \in \omega_1$ the equation (1) has a solution $v^*(0,\lambda) \in X_0$, corresponding to the zero solution of the bifurcation equation. If $\gamma_1(\lambda) < 0$ there are also two other solutions, which are invariant under the action of the subgroup G_0, and which can be obtained one from the other by application of $\Gamma(s)$, $s \in G \backslash G_0$; these two solutions correspond to nontrivial solutions $\pm\rho$ of the bifurcation equation. □

5.5.7. The proof of theorem 6 follows immediately from the foregoing remarks and theorem 4.10. The condition (ii) corresponds to the transversality condition (4.23) of theorem 4.10. The following more direct proof, which uses the oddness of the bifurcation function $F(\rho,\lambda)$, was suggested to us by J.K. Hale.

We have, from (17) :

$$F(\rho,\lambda) = \rho F_1(\rho,\lambda) \quad , \quad F_1(\rho,\lambda) = \int_0^1 D_\rho F(\sigma\rho,\lambda)d\sigma \ . \tag{20}$$

$F_1(\rho,\lambda)$ is of class C^2, and even in ρ. For $\rho \neq 0$ the bifurcation equation $F(\rho,\lambda) = 0$ is equivalent to :

$$F_1(\rho,\lambda) = 0 \ . \tag{21}$$

Define $G(\tilde{\rho},\lambda)$ by :

$$G(\tilde{\rho},\lambda) = F_1(|\tilde{\rho}|^{1/2},\lambda) \ . \tag{22}$$

Since $F_1(\rho,\lambda) = \gamma_1(\lambda) + \rho^2 R(\rho,\lambda)$, with $R(0,0) = a_0$, it is easily seen that G is of class C^1. Moreover, $G(0,0) = 0$ and $D_{\tilde{\rho}}G(0,0) = a_0 > 0$.

Now $F(\rho,\lambda) = G(\rho^2,\lambda)$, and (ρ,λ) is a solution of (21) if and only if (ρ^2,λ) is a solution of $G(\tilde{\rho},\lambda) = 0$. So we look for solutions of :

$$G(\tilde{\rho},\lambda) = 0 \quad , \quad \tilde{\rho} \geqslant 0 \ . \tag{23}$$

We can use the implicit function theorem to solve the equation in (23) : for each λ near zero the equation $G(\tilde{\rho},\lambda) = 0$ has a unique solution $\tilde{\rho} = \tilde{\rho}(\lambda)$. It is easily seen that this solution has the form :

$$\tilde{\rho}(\lambda) = -\tilde{\sigma}(\lambda)\gamma_1(\lambda) \quad , \quad \forall\lambda \in \omega_1 \ , \tag{24}$$

where $\tilde{\sigma}(\lambda)$ is a C^0-function, with $\tilde{\sigma}(0) = a_0^{-1}$ ($\tilde{\sigma}$ is of class C^1 in the region $\gamma_1(\lambda) \neq 0$). It follows that $\tilde{\sigma}(\lambda) > 0$ for all λ near zero.

Now bifurcation occurs because of the condition $\tilde{\rho} \geqslant 0$ in (23). We have $\tilde{\rho}(\lambda) > 0$ if $\gamma_1(\lambda) < 0$, and $\tilde{\rho}(\lambda) < 0$ if $\gamma_1(\lambda) > 0$. It follows that (21) has two nonzero solutions if $\gamma_1(\lambda) < 0$, and no such solutions if $\gamma_1(\lambda) \geqslant 0$. The bifurcation points are given by $\gamma_1(\lambda) = 0$, and the bifurcation takes place at the zero solution.

5.5.8. Remark. There are more general situations than the one described in this section where the transversality condition of section 4 is not satisfied because of some symmetry properties of the operator M. To give an example, let $m_0 = M(.,0)$, and suppose that m_0 is equivariant with respect to some $(G,\Gamma,\tilde{\Gamma})$. Under the hypotheses (H1) and (H3) we can still determine the subgroup G_0; suppose that $G_0 \neq G$.

Suppose that $\Lambda = \mathbb{R}^p$ (as in section 4), and define $m_i = \frac{\partial M}{\partial \lambda_i}(.,0)$. Finally, suppose that for each $i \in \{1,2,\ldots,p\}$ there is some $s_i \in G \backslash G_0^i$ such that :

$$m_i(\Gamma(s_i)x) = \tilde{\Gamma}(s_i)m_i(x) \quad , \qquad \forall x \in \Omega . \tag{25}$$

Then, using (4.15) and (4.10) we find :

$$a_{i1}z_0 = Qm_i(0) = Qm_i(\Gamma(s_i)(0)) = \tilde{\Gamma}(s_i)Qm_i(0) = \tilde{\Gamma}(s_i)(a_{i1}z_0) = -a_{i1}z_0 ,$$

$$i = 1,\ldots,p .$$

We conclude that $a_{i1} = 0$ for all $i = 1,\ldots,p$, and the generic condition (4.14) will not be satisfied, although we do not necessarily have that $\gamma_0(\lambda) = 0$ for all λ.

5.6. APPLICATION : THE VON KARMAN EQUATIONS FOR A RECTANGULAR PLATE

5.6.1. The problem. In this section we consider the buckling problem for a simply supported rectangular plate, subjected to a compressive thrust at its short edges and to a normal load, proportional to a small parameter ν. In the appropriate Hilbert space H (see section 2.4 for the details) the equation for this problem takes the form :

$$(I-\lambda A)w + C(w) = \nu p , \tag{1}$$

where p is some fixed element of H; the parameter λ measures the magnitude of the compressive thrust.

We want to study the bifurcation of solutions of (1) for (w,λ,ν) in a neighbourhood of $(0,\lambda_0,0)$, where λ_0 is a characteristic value of the operator A. These characteristic values were determined in section 2.4, where we found

$$\lambda_{mn} = \frac{\pi^2(m^2+n^2\ell^2)^2}{4\ell^2 m^2} \quad , \qquad m,n = 1,2,\dots , \tag{2}$$

with corresponding eigenfunctions

$$\phi_{mn} = \frac{4\ell^{3/2}}{\pi^2(m^2+n^2\ell^2)} \sin\frac{m\pi}{2\ell}(x+\ell) \sin\frac{n\pi}{2}(y+1) \quad , \quad m,n = 1,2,\dots . \tag{3}$$

Fix some (m,n) such that $\lambda_{m,n} \neq \lambda_{m',n'}$ for all $(m',n') \neq (m,n)$. For notational convenience we denote λ_{mn} and ϕ_{mn} by λ_0 and ϕ_0. Let

$$L = I - \lambda_0 A . \tag{4}$$

Then $\ker L = \mathrm{span}\{\phi_0\}$ and $R(L) = \{w \in H \mid (w,\phi_0) = 0\}$. For the projections P and Q used in the Liapunov-Schmidt reduction we take the orthogonal projection onto ker L :

$$Pw = Qw = (w,\phi_0)\phi_0 \quad , \qquad \forall w \in H . \tag{5}$$

5.6.2. Equivariance group when $p = 0$. In subsection 3.6.16 we discussed already the symmetry of (1) when there is no normal load, i.e. when $p = 0$. Each of the operators A, C and P commute with the operators of the finite group :

$$G = \{I,\tau_x,\tau_y,\tau_{xy},-I,-\tau_x,-\tau_y,-\tau_{xy}\} , \tag{6}$$

where, as in (3.6.44) :

$$(\tau_x w)(x,y) = w(-x,y) \quad , \quad (\tau_y w)(x,y) = w(x,-y) \quad , \quad \tau_{xy} = \tau_x\tau_y . \tag{7}$$

Let

$$G_1 = \{\tau \in G \mid \tau\phi_0 = \phi_0\} \ ; \tag{8}$$

this is precisely the subgroup G_0 defined by (5.9), here denoted by G_1 for notational convenience further on. This subgroup depends on our choice of (m,n). Since :

$$\tau_x\phi_{mn} = (-1)^{m+1}\phi_{mn} \quad , \quad \tau_y\phi_{mn} = (-1)^{n+1}\phi_{mn} \tag{9}$$

it follows that :

$$G_1 = \{\tau_1, \tau_2, \tau_3, \tau_4\} \ , \tag{10}$$

where

$$\tau_1 = I \ , \ \tau_2 = (-1)^{m+1}\tau_x \ , \ \tau_3 = (-1)^{n+1}\tau_y \ , \ \tau_4 = (-1)^{m+n}\tau_{xy} \ . \tag{11}$$

Using this notation we have :

$$G = G_1 \cup (-G_1) \ , \tag{12}$$

$$\tau_j^2 = \tau_1 \quad , \quad \tau_1\tau_j = \tau_j \quad , \qquad j = 1,2,3,4 \tag{13}$$

and

$$\tau_2\tau_3 = \tau_4 \quad , \quad \tau_3\tau_4 = \tau_2 \quad , \quad \tau_2\tau_4 = \tau_3 \ . \tag{14}$$

For general $p \in H$ equation (1) will only be equivariant with respect to the isotropy subgroup of p, that is the subgroup

$$G_p = \{\tau \in G \mid \tau p = p\} \ .$$

We want to study how the bifurcation properties of (1) depend on the symmetry of p, i.e. how they depend on G_p. First we consider the irreducible representations of G_1.

5.6.3. Irreducible representations of G_1. If $\tau \in G$, then also $-\tau \in G$; consequently, the action of G on an element $w \in H$ is uniquely determined by the action of the subgroup G_1 on w. For this reason we restrict our attention to the subgroup G_1.

Let $\Gamma : G_1 \to L(\mathbb{R}^n)$ be an irreducible representation of G_1. It follows from (13) that for each $\tau \in G_1$, $\Gamma(\tau_1)$ has an eigenvalue equal to ± 1. Moreover, the group G_1 is commutative. Then corollary 2.6.10 implies that $\Gamma(\tau) = \pm I$ for each $\tau \in G_1$. Since the representation is irreducible, it follows that n = 1. Such an irreducible representation $\Gamma : G_1 \to \mathbb{R} = L(\mathbb{R})$ is uniquely determined by the real numbers :

$$\alpha_j = \Gamma(\tau_j) \quad , \quad j = 1,2,3,4 \ ;$$

moreover, $\alpha_j = \pm 1$ for each j, and $\alpha_2\alpha_3 = \alpha_4$, $\alpha_3\alpha_4 = \alpha_2$ and $\alpha_2\alpha_4 = \alpha_3$, by (14). These relations are sufficient to show explicitly that there are only four different irreducible representations of G, denoted by Γ_i $(i = 1,2,3,4)$, and determined by the numbers

$$\Gamma_i(\tau_j) = \alpha_{ij} \in \mathbb{R} \quad , \quad i,j = 1,2,3,4 \tag{15}$$

given in the following table :

	τ_1	τ_2	τ_3	τ_4
Γ_1	1	1	1	1
Γ_2	1	1	-1	-1
Γ_3	1	-1	1	-1
Γ_4	1	-1	-1	1

5.6.4. <u>Remark</u> : It follows from the table above that

$$\frac{1}{4} \sum_{k=1}^{4} \alpha_{ik}\alpha_{jk} = \delta_{ij} \ , \tag{16}$$

$$\frac{1}{4} \sum_{i=1}^{4} \alpha_{ij}\alpha_{ik} = \delta_{jk} \ , \tag{17}$$

and

$$\frac{1}{4}\sum_{i=1}^{4} \alpha_{ij} = \delta_{1j} . \tag{18}$$

The following approach shows that these relations are not just a coincidence, but follow from more general relations that have to be satisfied by the matrix elements appearing in irreducible representations of finite groups. One can see Miller [166] for the general theory.

Let $\Gamma : G_1 \to \mathbb{R}$ and $\tilde{\Gamma} : G_1 \to \mathbb{R}$ be two irreducible representations of G_1; let $\alpha_j = \Gamma(\tau_j)$ and $\tilde{\alpha}_j = \tilde{\Gamma}(\tau_j)$. Define

$$a = \frac{1}{4}\sum_{k=1}^{4} \Gamma(\tau_k)\tilde{\Gamma}(\tau_k^{-1}) = \frac{1}{4}\sum_{k=1}^{4} \alpha_k\tilde{\alpha}_k .$$

Then :

$$\alpha_j a = \frac{1}{4}\sum_{k=1}^{4} \Gamma(\tau_j\tau_k)\tilde{\Gamma}(\tau_k^{-1}) = \frac{1}{4}\sum_{k=1}^{4} \Gamma(\tau_j\tau_k)\tilde{\Gamma}((\tau_j\tau_k)^{-1})\tilde{\Gamma}(\tau_j)$$

$$= a\tilde{\alpha}_j \qquad , \qquad j = 1,2,3,4 .$$

It follows that $a = 0$ if $\Gamma \neq \tilde{\Gamma}$, i.e. if $\alpha_j \neq \tilde{\alpha}_j$ for some j. In case $\Gamma = \tilde{\Gamma}$, then :

$$a = \frac{1}{4}\sum_{k=1}^{4} \Gamma(\tau_k\tau_k^{-1}) = \frac{1}{4}\sum_{k=1}^{4} \Gamma(\tau_1) = 1 ,$$

since $\Gamma(\tau_1) = 1$ for each representation. We conclude that

$$\frac{1}{4}\sum_{k=1}^{4} \alpha_k\tilde{\alpha}_k = \delta_{\Gamma,\tilde{\Gamma}} , \tag{19}$$

where $\delta_{\Gamma,\tilde{\Gamma}} = 1$ if $\Gamma = \tilde{\Gamma}$, and $= 0$ if $\Gamma \neq \tilde{\Gamma}$.

Now let X be the space of all functions $x : G_1 \to \mathbb{R}$. Under pointwise addition and multiplication with a scalar, X is a 4-dimensional vectorspace, which becomes a Hilbert space if we use the inner product :

$$\langle x,y\rangle = \frac{1}{4}\sum_{j=1}^{4} x(\tau_j)y(\tau_j) \qquad , \qquad x,y \in X . \tag{20}$$

Now we define a representation $\Gamma_0 : G_1 \to L(X)$ as follows :

$$(\Gamma_0(\tau_j)x)(\tau_k) = x(\tau_k\tau_j) \quad , \quad \forall x \in X \ , \ \forall j,k = 1,2,3,4 \ . \tag{21}$$

It is clear that Γ_0 is orthogonal with respect to the inner product (20).

If $\Gamma : G_1 \to \mathbb{R}$ is an irreducible representation, then we can consider Γ as an element of X, and :

$$(\Gamma_0(\tau_j)\Gamma)(\tau_k) = \Gamma(\tau_k\tau_j) = \Gamma(\tau_j)\Gamma(\tau_k) \quad , \quad \forall j,k = 1,2,3,4 \ .$$

This means that

$$\Gamma_0(\tau_j)\Gamma = \Gamma(\tau_j)\Gamma \ , \tag{22}$$

i.e. span$\{\Gamma\}$ is an invariant subspace of X, which transforms under Γ_0 according to the irreducible representation Γ. Since by (19) different irreducible representations are orthogonal when considered as elements of X, it follows that G_1 has at most four different irreducible representations.

Let V be the subspace of X spanned by all the irreducible representations of G_1; the foregoing shows that V is invariant under the representation Γ_0. If $V \neq X$, then also $V^\perp$ is invariant under Γ_0, and $V^\perp$ contains a one-dimensional subspace span$\{x\}$, which is invariant under Γ_0 and transforms according to some irreducible representation Γ. It follows that

$$(\Gamma_0(\tau_j)x)(\tau_k) = x(\tau_k\tau_j) = \Gamma(\tau_j)x(\tau_k) \quad , \quad j,k = 1,2,3,4 \ .$$

Taking $\tau_k = \tau_1$ we see that

$$x(\tau_j) = \Gamma(\tau_j)x(\tau_1) \quad , \quad j = 1,2,3,4 \ ,$$

i.e. $x = x(\tau_1)\Gamma$, and consequently $x \in V$. Since also $x \in V^\perp$, it follows that $x = 0$, which contradicts dim span$\{x\} = 1$. We conclude that $V = X$; consequently G_1 has precisely four different irreducible representations, which we denote by Γ_i $(i = 1,2,3,4)$. Then (19) takes the form (16). Another way of writing (16) is in the form $AA^T = I$, where A is the 4×4-matrix $1/2(\alpha_{ij})$. This implies $A^TA = I$, i.e. we also have (17).

Denote by $\{x_j \mid j = 1,2,3,4\}$ the canonical basis of X : $x_j(\tau_k) = \delta_{jk}$. Using this basis it is easily seen that

$$\text{trace } \Gamma_0(\tau_j) = 4\delta_{1j} \qquad , \qquad j = 1,2,3,4 \ . \tag{23}$$

Expressing $\Gamma_0(\tau_j)$ with respect to the basis $\{\Gamma_i \mid i = 1,2,3,4\}$ of X, and using (22), we find

$$\text{trace } \Gamma_0(\tau_j) = \sum_{i=1}^{4} \alpha_{ij} \ . \tag{24}$$

A combination of (23) and (24) gives (18).

5.6.5. For each i = 1,2,3,4 we define a symmetric operator $P_i \in L(H)$ by

$$P_i = \frac{1}{4} \sum_{k=1}^{4} \alpha_{ik}\tau_k \ . \tag{25}$$

It follows from (18) that

$$\sum_{i=1}^{4} P_i = \frac{1}{4} \sum_{i=1}^{4} \sum_{k=1}^{4} \alpha_{ik}\tau_k = \sum_{k=1}^{4} \delta_{1k}\tau_k = \tau_1 = I \ . \tag{26}$$

We also have

$$\tau_j P_i = \frac{1}{4} \sum_{k=1}^{4} \alpha_{ik}\tau_j\tau_k = \frac{1}{4} \sum_{k=1}^{4} \alpha_{ij}(\alpha_{ij}\alpha_{ik})(\tau_j\tau_k)$$

$$= \alpha_{ij} \frac{1}{4} \sum_{\ell=1}^{4} \alpha_{i\ell}\tau_\ell = \alpha_{ij}P_i \ . \tag{27}$$

This implies

$$P_k P_i = \frac{1}{4} \sum_{j=1}^{4} \alpha_{kj}\tau_j P_i = \frac{1}{4} \sum_{j=1}^{4} \alpha_{kj}\alpha_{ij}P_i = \delta_{ki}P_i \ . \tag{28}$$

We conclude from (26) and (28) that the P_i are orthogonal projections on mutually orthogonal subspaces $H_i = P_i(H)$, which together span H : $H = H_1 \oplus H_2 \oplus H_3 \oplus H_4$. If $w \in H$ is such that $\tau_j w = \alpha_{ij} w$ for each j, then (25) and (16) show that $P_i w = w$; this, together with (27) proves that

$$H_i = \{w \in H \mid \tau_j w = \alpha_{ij} w, \ j = 1,2,3,4\}$$

$$= \{w \in H \mid \tau w = w, \ \forall w \in G_i\} \ , \tag{29}$$

where the subgroup G_i of G is defined by

$$G_i = \{\alpha_{ij}\tau_j \mid j = 1,2,3,4\} . \tag{30}$$

So the elements of H_i are precisely those $w \in H$ which transform according to the representation Γ_i; H_i also consists of those $w \in H$ which remain invariant under the subgroup G_i of G.

5.6.6. Remark. We will see in the remainder of this section that the introduction of the irreducible representations Γ_i facilitates very much the calculation of certain terms in the bifurcation equation. This is mainly due to the simple form of the representations, the splitting of H into orthogonal subspaces associated with each Γ_i, and also to the fact that the nonlinear term in the equation (1) is homogeneous.

From a general point of view one may ask the question how useful group representation theory really is when dealing with nonlinear bifurcation problems with symmetry. At first sight there seem to be a few severe handicaps : group representation theory is a linear theory, which can only be fully developed when considering complex representations. From the other side bifurcation theory is clearly nonlinear, and usually deals with problems in real spaces. The foregoing chapters seem to indicate that only concepts such as "equivariance with respect to some subgroup" or "invariant subspace" will be relevant. (See also Ruelle [185], Poenaru [174], Golubitsky and Schaeffer [75]).

However, once the problem has been pushed down to that of solving a finite-dimensional bifurcation equation, group theory can actually be of great help in two respects. First, it can sometimes be used to obtain a further reduction of the equations; in the next chapter we will give an example where the specific form of the representation is used to reduce a 5-dimensional problem to a 2-dimensional one. Second, the fact that the bifurcation equations remain equivariant may result in a serious restriction on the terms which can appear in the Taylor development of the bifurcation function; here group representation theory is the appropriate tool when one has to find out which terms will vanish and which terms remain. This has been beautifully illustrated by the work of D.H. Sattinger ([195-197],[265]). Also the remainder of this section illustrates this point.

5.6.7. From (26) it follows that each $p \in H$ can be written in the form :

$$p = \sum_{i=1}^{4} p_i \quad , \quad p_i = P_i p \in H_i \ . \tag{31}$$

We will use this form of p when dealing with the equation (1). Since P is an orthogonal projection on span $\phi_0 \subset H_1$, it follows that

$$Pp = PP_1 p = Pp_1 \quad , \quad \forall p \in H \ . \tag{32}$$

We put

$$w = \rho\phi_0 + v \quad , \quad \rho \in \mathbb{R} \ , \ v \in \ker P \tag{33}$$

and

$$\lambda = \lambda_0(1+\mu)$$

in (1), and project on R(L), using I-P. We find

$$Lv - \lambda_0\mu Av + (I-P)C(\rho\phi_0+v) = \nu((I-P)p_1 + \sum_{i=2}^{4} p_i) \ . \tag{34}$$

For (ρ,μ,ν) in a neighbourhood of $(0,0,0)$ the equation (34) has a unique solution $v = v^*(\rho,\mu,\nu)$ near $v = 0$.

5.6.8. Lemma. There is a constant $c > 0$ such that for (ρ,μ,ν) in a sufficiently small neighbourhood of the origin we have :

$$\|v^*(\rho,\mu,\nu)\| \leq c(|\rho|^3+|\nu|) \ . \tag{35}$$

P r o o f. If no such constant c exists, then we can find a sequence $\{(v_n,\rho_n,\mu_n,\nu_n) \mid n\in\mathbb{N}\}$, converging to $(0,0,0,0)$ as $n \to \infty$, and such that for each $n \in \mathbb{N}$, (v_n,ρ_n,μ_n,ν_n) solves (24), $\|v_n\| \neq 0$ and

$$\frac{|\rho_n|^3}{\|v_n\|} \to 0 \quad , \quad \frac{\|\nu_n\|}{\|v_n\|} \to 0 \quad , \quad \text{as } n \to \infty \ .$$

Expressing that (v_n,ρ_n,μ_n,ν_n) is a solution of (34), dividing by $\|v_n\|$ and taking the limit for $n \to \infty$ gives :

$$\lim_{n\to\infty} L \frac{v_n}{\|v_n\|} = 0 .$$

Since L has a bounded pseudo-inverse $K : R(L) \to \ker P$, it follows that

$$\lim_{n\to\infty} \frac{v_n}{\|v_n\|} = 0 ,$$

which is impossible. □

5.6.9. <u>The bifurcation equation</u>. Using (32) and the solution $v^*(\rho,\mu,\nu)$ of (34) gives us the bifurcation equation for (1) in the form :

$$F(\rho,\mu,\nu) \equiv -\mu\rho + (C(\rho\phi_0+v^*(\rho,\mu,\nu)),\phi_0) - (p_1,\phi_0) = 0 . \tag{36}$$

From lemma 6 and the fact that C(w) is homogeneous of degree three it follows easily that

$$F(0,0,0) = D_\rho F(0,0,0) = D_\rho^2 F(0,0,0) = 0 \tag{37}$$

and

$$a_0 = \frac{1}{3!} D_\rho^3 F(0,0,0) = (C(\phi_0),\phi_0) > 0 . \tag{38}$$

(See lemma 2.4.25). We conclude that equation (1) satisfies the hypotheses (H1) and (H3) of this chapter.

From (36) and (4.8)-(4.9) we see that

$$\begin{aligned} D_\mu\gamma_0(0,0) &= 0 \quad , \quad D_\nu\gamma_0(0,0) = -(p,\phi_0) , \\ D_\mu\gamma_1(0,0) &= -1 \quad , \quad D_\nu\gamma_1(0,0) = 0 . \end{aligned} \tag{39}$$

The transversality condition (4.7) will be satisfied if $(p,\phi_0) = (p_1,\phi_0) \neq 0$. If this is the case, then, in a (μ,ν)-plane, the bifurcation set for (1) will have the generic cusp-form which we found in subsection 4.2.

A necessary condition to have the generic situation is that $p_1 = P_1 p \neq 0$. This implies that when we consider normal loads having a particular symmetry (e.g. $p \in H_2$) then it may very well be that this symmetry prevents the generic condition to be satisfied. So, for certain symmetries of the normal load we expect to have non-generic bifurcation sets. In the remainder of this section we will study how the bifurcation set depends on the components p_i ($i=1,2,3,4$) of p; more in particular, we are interested in the non-generic case that $p \in H_2 \oplus H_3 \oplus H_4$ (i.e. $p_1 = 0$), and in the transition between the generic and the non-generic case (i.e. p_1 small). In order to do so we will suppose that in equation (1) $p \in H$ has the form

$$p = \varepsilon p_1 + \sum_{i=2}^{4} p_i \quad , \qquad p_i \in H_i \ (i=1,2,3,4) \ , \ \varepsilon \in \mathbb{R} \ . \tag{40}$$

We consider ε as an additional parameter; the elements $p_i \in H_i$ are fixed, and we suppose that p_1 is such that

$$k = (p_1, \phi_0) \neq 0 \ . \tag{41}$$

5.6.10. Let us return to the auxiliary equation (34). If we use the form (40) for p (i.e. replace p_1 in (34) by εp_1), put $\mu = \varepsilon = 0$ and neglect the cubic term, then the remaining equation has the solution

$$v_0^* = \nu \sum_{i=2}^{4} Kp_i \ , \tag{42}$$

where K is the pseudo-inverse of L. Therefore we put

$$v = \nu \sum_{i=2}^{4} Kp_i + \tilde{v} \qquad , \qquad \tilde{v} \in \ker P \tag{43}$$

in (34); we obtain the following equation for $\tilde{v}$:

$$L\tilde{v} - \lambda_0 \mu A \tilde{v} + (I-P)C(\rho\phi_0 + \nu \sum_{i=2}^{4} Kp_i + v)$$
$$= \varepsilon\nu(I-P)p_1 + \lambda_0 \mu\nu \sum_{i=2}^{4} AKp_i \ . \tag{44}$$

We denote by $\tilde{v}^*(\rho,\mu,\nu,\varepsilon)$ the unique solution of (44); such solution exists for (ρ,μ,ν) small and ε bounded. A proof similar to the one of lemma 8 shows that we have the following estimate for $\tilde{v}^*$.

5.6.11. <u>Lemma</u>. There is a constant $c > 0$ such that for $(\rho,\mu,\nu,\varepsilon)$ in a sufficiently small neighbourhood of the origin we have

$$\|\tilde{v}^*(\rho,\mu,\nu,\varepsilon)\| \leq c(|\rho|^3+|\nu|^3+|\mu||\nu|+|\varepsilon||\nu|) \ . \qquad \square \tag{45}$$

The bifurcation function $F(\rho,\mu,\nu,\varepsilon)$ takes the form

$$F(\rho,\mu,\nu,\varepsilon) = -\mu\rho + (C(\rho\phi_0 + \nu \sum_{i=2}^{4} Kp_i + \tilde{v}^*(\rho,\mu,\nu,\varepsilon)),\phi_0) - \varepsilon\nu k \ . \tag{46}$$

5.6.12. Lemma. There is a constant $c > 0$ such that all sufficiently small solutions $(\rho,\mu,\nu,\varepsilon)$ of the bifurcation equation $F(\rho,\mu,\nu,\varepsilon) = 0$ satisfy

$$|\rho| \leq c(|\mu|^{1/2} + (|\varepsilon||\nu|)^{1/3} + |\nu|) \ . \tag{47}$$

P r o o f. If no such constant c can be found, then there is a sequence $\{(\rho_n,\mu_n,\nu_n,\varepsilon_n) \mid n\in\mathbb{N}\}$, converging to $(0,0,0,0)$ as $n \to \infty$ such that, for each $n \in \mathbb{N}$, we have $F(\rho_n,\mu_n,\nu_n,\varepsilon_n) = 0$, $\rho_n \neq 0$ and

$$\frac{|\mu_n|}{|\rho_n|^2} \to 0 \ , \quad \frac{|\varepsilon_n||\nu_n|}{|\rho_n|^3} \to 0 \ , \quad \frac{|\nu_n|}{|\rho_n|} \to 0 \quad \text{as } n \to \infty \ .$$

Dividing $F(\rho_n,\mu_n,\nu_n,\varepsilon_n) = 0$ by $|\rho_n|^3$, and taking the limit for $n \to \infty$ gives us $(C(\phi_0),\phi_0) = 0$, which contradicts (38). $\square$

5.6.13. Let

$$F_0(\rho,\mu,\nu,\varepsilon) = -\mu\rho + (C(\rho\phi_0 + \nu \sum_{i=2}^{4} Kp_i),\phi_0) - \varepsilon\nu k \ . \tag{48}$$

Since $C(w) = \frac{1}{2}B(w,B(w,w))$, where $B(.,.)$ is a bounded bilinear operator on H, it is easily seen that there is a constant $c > 0$ such that

$$\|C(w_1)-C(w_2)\| \leq c\|w_1-w_2\|(\|w_1\|^2+\|w_2\|^2) \quad , \quad \forall w_1,w_2 \in H \ . \tag{49}$$

Using (49) and the estimate (45) for $\tilde{v}^*$, it is straightforward to show that

$$|F(\rho,\mu,\nu,\varepsilon)-F_0(\rho,\mu,\nu,\varepsilon)| \leq c[(|\rho|+|\nu|)^5 + (|\mu|+|\varepsilon|)(|\rho|+|\nu|)^3] \ , \tag{50}$$

for some $c > 0$ and for all $(\rho,\mu,\nu,\varepsilon)$ in a neighbourhood of the origin.

We now determine a more explicit form of the function $F_0(\rho,\mu,\nu,\varepsilon)$. Using (2.4.18) and (2.4.19) we find

$$F_0(\rho,\mu,\nu,\varepsilon) = a_0\rho^3 - \mu\rho - \varepsilon\nu k + \frac{3}{2}\rho^2\nu \sum_{i=2}^{4} (B(Kp_i,B(\phi_0,\phi_0)),\phi_0)$$
$$+ \rho\nu^2 \sum_{i=2}^{4}\sum_{j=2}^{4} [(B(Kp_i,B(Kp_j,\phi_0)),\phi_0)$$
$$+ \frac{1}{2}(B(\phi_0,B(Kp_i,Kp_j)),\phi_0)]$$
$$+ \frac{1}{2}\nu^3 \sum_{i=2}^{4}\sum_{j=2}^{4}\sum_{k=2}^{4} (B(Kp_i,B(Kp_j,Kp_k)),\phi_0) \ . \tag{51}$$

As a consequence of symmetry considerations several terms under the summation signs in (51) will vanish. In order to study this in detail we will use the irreducible representations of the group G_1.

5.6.14. Let $w_i \in H_i \cap R(L)$; then $\tau K w_i = K\tau w_i = K w_i$ for each $\tau \in G_i$. This shows that K maps $H_i \cap R(L)$ into H_i. Next, we have for each $\tau \in G$:

$$\tau B(u,B(v,w)) = B(\tau u,B(\tau v,\tau w)) \quad , \quad \forall u,v,w \in H \ . \tag{52}$$

This implies that for $w_i \in H_i$, $w_j \in H_j$ and $w_k \in H_k$ we have

$$\tau_\ell B(w_i,B(w_j,w_k)) = B(\tau_\ell w_i,B(\tau_\ell w_j,\tau_\ell w_k))$$
$$= \alpha_{i\ell}\alpha_{j\ell}\alpha_{k\ell} B(w_i,w_j,w_k) \ . \tag{53}$$

From this relation we see that $B(w_i,B(w_j,w_k))$ transforms under the operators $\tau_\ell \in G_1$ according to the representation $\Gamma : G_1 \to \mathbb{R}$ given by :

$$\Gamma(\tau_\ell) = \alpha_{i\ell}\alpha_{j\ell}\alpha_{k\ell} = \Gamma_i(\tau_\ell)\Gamma_j(\tau_\ell)\Gamma_k(\tau_\ell) \quad , \quad \ell = 1,2,3,4 \ ; \tag{54}$$

this represenation is denoted by $\Gamma_i \otimes \Gamma_j \otimes \Gamma_k$. The next table gives the product representations $\Gamma_i \otimes \Gamma_j$ as a function of Γ_i and Γ_j :

$\otimes$	Γ_1	Γ_2	Γ_3	Γ_4
Γ_1	Γ_1	Γ_2	Γ_3	Γ_4
Γ_2	Γ_2	Γ_1	Γ_4	Γ_3
Γ_3	Γ_3	Γ_4	Γ_1	Γ_2
Γ_4	Γ_4	Γ_3	Γ_2	Γ_1

5.6.15. Let us now consider the different terms appearing in (51). First, since $\phi_0 \in H_1$, the table above shows that $B(Kp_i, B(\phi_0, \phi_0)) \in H_i$ for $i = 1,2,3$. Since the subspaces H_i are mutually orthogonal, we have $(B(Kp_i, B(\phi_0, \phi_0)), \phi_0) = 0$.

Next, $B(Kp_i, B(Kp_j, \phi_0))$ transforms according to $\Gamma_i \otimes \Gamma_j$, which equals Γ_1 if and only if $i = j$. It follows that

$$(B(Kp_i, B(Kp_j, \phi_0)), \phi_0) = (B(Kp_i, B(Kp_i, \phi_0)), \phi_0)\delta_{ij} \ ;$$

the same conclusion holds for $(B(\phi_0, B(Kp_i, Kp_j)), \phi_0)$.

Finally, $B(Kp_i, B(Kp_j, Kp_k))$ transforms according to $\Gamma_i \otimes \Gamma_j \otimes \Gamma_k$; for $i,j,k \in \{2,3,4\}$ this equals Γ_1 if and only if (i,j,k) forms a permutation of $(2,3,4)$.

Taking all this together, we find the following expression for $F_0(\rho, \mu, \nu, \varepsilon)$

$$F_0(\rho, \mu, \nu, \varepsilon) = a_0\rho^3 - \mu\rho - \varepsilon\nu k + b\rho\nu^2 + c\nu^3 \ , \tag{55}$$

where

$$b = \sum_{i=2}^{4} [\, \| B(Kp_i, \phi_0) \|^2 + \tfrac{1}{2}(B(Kp_i, Kp_i), B(\phi_0, \phi_0))] \ , \tag{56}$$

and

$$c = (B(Kp_2,Kp_3),B(Kp_4,\phi_0)) + (B(Kp_3,Kp_4),B(Kp_2,\phi_0)) + (B(Kp_4,Kp_2),B(Kp_3,\phi_0)) \ . \tag{57}$$

5.6.16. It follows from (55) and (50) that for $(\mu,\nu,\varepsilon) \to (0,0,0)$ we have

$$\rho_0(\mu,\nu,\varepsilon) = O(|\nu|^3 + (|\varepsilon|+|\mu|)|\nu|) \ , \tag{58}$$

$$\gamma_0(\mu,\nu,\varepsilon) = c\nu^3 - \varepsilon\nu k + O(|\nu|^5 + (|\varepsilon|+|\mu|)(|\nu|^3 + |\mu||\nu|)) \ , \tag{59}$$

and

$$\gamma_1(\mu,\nu,\varepsilon) = -\mu + b\nu^2 + O((|\nu|^3 + (|\mu|+|\varepsilon|)|\nu|)^2) \ . \tag{60}$$

Moreover, it is easily seen from the equation (44) determining $\tilde{v}(\rho,\mu,\nu,\varepsilon)$ that

$$\tilde{v}(-\rho,\mu,-\nu,\varepsilon) = -\tilde{v}(\rho,\mu,\nu,\varepsilon) \ , \tag{61}$$

which implies that

$$F(-\rho,\mu,-\nu,\varepsilon) = -F(\rho,\mu,\nu,\varepsilon) \ . \tag{62}$$

Using the definitions of the functionals ρ_0,γ_0,γ_1, etc... as given in section 3, we also find the following identities

$$\begin{aligned} \rho_0(\mu,-\nu,\varepsilon) &= -\rho_0(\mu,\nu,\varepsilon) \ , \\ \gamma_0(\mu,-\nu,\varepsilon) &= -\gamma_0(\mu,\nu,\varepsilon) \ , \\ \gamma_1(\mu,-\nu,\varepsilon) &= \gamma_1(\mu,\nu,\varepsilon) \end{aligned} \tag{63}$$

and

$$\sigma_+(\mu,-\nu,\varepsilon) = -\sigma_-(\mu,\nu,\varepsilon) \ .$$

If $(\mu,\nu) \in \mathbb{R}^2$ is a bifurcation point for the equation (1) (with p as in (40) and ε fixed), then also $(\mu,-\nu)$ is a bifurcation point; i.e. the bifurcation set is symmetric with respect to the μ-axis. This follows directly from the definition of a bifurcation point and the observation that if (w,μ,ν) is a solution of (1), then so is $(-w,\mu,-\nu)$.

We now discuss the bifurcation set under different hypotheses for the load p given by (40).

5.6.17. The case $p_2 = p_3 = p_4 = 0$. In this case one should consider $\tilde{\nu} = \varepsilon\nu$ as one single parameter. Since $b = c = 0$ one finds

$$\gamma_0(\mu,\tilde{\nu}) = -k\tilde{\nu} + O(|\tilde{\nu}|^3 + |\mu||\tilde{\nu}|) \tag{64}$$

and

$$\gamma_1(\mu,\tilde{\nu}) = -\mu + O(|\tilde{\nu}|^2) \ . \tag{65}$$

Since $k \neq 0$ we have a case of generic bifurcation. The bifurcation set has a cusp-form and is approximately given by

$$\mu > 0 \quad , \quad \tilde{\nu} = \pm\frac{2}{k}(27a_0)^{-1/2}\mu^{3/2} \ . \tag{66}$$

Also, the equation (1) is equivariant with respect to the subgroup G_1 (i.e. $G_p = G_1$). All small solutions will belong to H_1; therefore the whole problem can be discussed in the subspace H_1.

5.6.18. The case $\varepsilon = 0$. For $\varepsilon = 0$ we have

$$\gamma_0(\mu,\nu,0) = c\nu^3 + O(|\nu|(|\mu|+|\nu|^2)^2) \tag{67}$$

and

$$\gamma_1(\mu,\nu,0) = -\mu + b\nu^2 + O(|\nu|^2(|\mu|+|\nu|^2)^2) \ . \tag{68}$$

The generic condition is not satisfied, although we do *not* have that $\gamma_0(\mu,\nu,0) = 0$ for all (μ,ν). Using the approach of the previous sections it is easily shown that for this case the bifurcation set is given by

$$\gamma_1(\mu,\nu,0) \leq 0 \ , \ \gamma_0(\mu,\nu,0) = \sigma_\pm(\mu,\nu,(-\gamma_1(\mu,\nu,0))^{1/2})(-\gamma_1(\mu,\nu,0))^{1/2} \ . \tag{69}$$

This is approximately given by

$$\mu \geq b\nu^2 \quad , \quad \nu^3 c = \pm 2(27a_0)^{-1/2}(\mu - b\nu^2)^{3/2} \quad , \tag{69}$$

or by

$$\mu = (b + 3a_0^{1/2}(\tfrac{1}{2}c)^{2/3})\nu^2 \quad . \tag{70}$$

The bifurcation set is approximately a parabola, tangent to the ν-axis.

When p_2, p_3 and p_4 are all three different from zero, then the equation is only equivariant with respect to the trivial group (i.e. $G_p = \{\tau_1\}$). In that case the only symmetry is the one expressed in the formulas (61)-(63). When at least one of the p_i $(i = 2,3,4)$ is zero, then $c = 0$, and the problem has some additional symmetry properties, which we discuss now for the case $p_4 = 0$; the cases $p_2 = 0$ and $p_3 = 0$ are similar.

5.6.19. The case $\varepsilon = 0$ and $p_4 = 0$. In this case the equation is equivariant with respect to the subgroup $G_p = G_2 \cap G_3 = \{\tau_1, -\tau_4\}$, while ϕ_0 changes sign under the action of $-\tau_4$. We are in a situation as studied in section 5. According to theorem 5.6 the bifurcation set is given by the equation $\gamma_1(\mu,\nu,0) = 0$; it is approximated by

$$\mu = b\nu^2 \quad . \tag{71}$$

The bifurcation takes place at the solution corresponding to $\rho = 0$; this solution is invariant under the action of the group $\{\tau_1, -\tau_4\}$.

The solutions have further symmetry properties if for example also $p_3 = 0$ (i.e. $p \in H_2$). Then the equation is equivariant with respect to the subgroup G_2. The solution corresponding to $\rho = 0$ is invariant under G_2, while the other bifurcating solutions remain invariant under $G_2 \cap G_1 = \{\tau_1, \tau_2\}$.

5.6.20. The general case. The case that ε is different from zero but small describes the transition between the generic situation, where the bifurcation set has a cusp-form, and the non-generic case where the bifurcation diagram shows a parabolic curve. The generic condition is satisfied for fixed $\varepsilon \neq 0$. The bifurcation set is approximately given by

$$\mu \geq b\nu^2 \quad , \quad c\nu^3 - \varepsilon\nu k = \pm 2(27a_0)^{-1/2}(\mu - b\nu^2)^{3/2} \quad , \tag{72}$$

or

$$\mu = b\nu^2 + 3\left(\frac{a_0}{4}\right)^{1/3} (c\nu^3 - \varepsilon\nu k)^{2/3} . \tag{73}$$

For small ε this is a curve as in Figure 7 : the overall-shape is parabola-like, but there is a small cusp-like singularity at the origin. As $\varepsilon \to 0$ the singularity gradually disappears, and we obtain a parabola in the limit $\varepsilon = 0$.

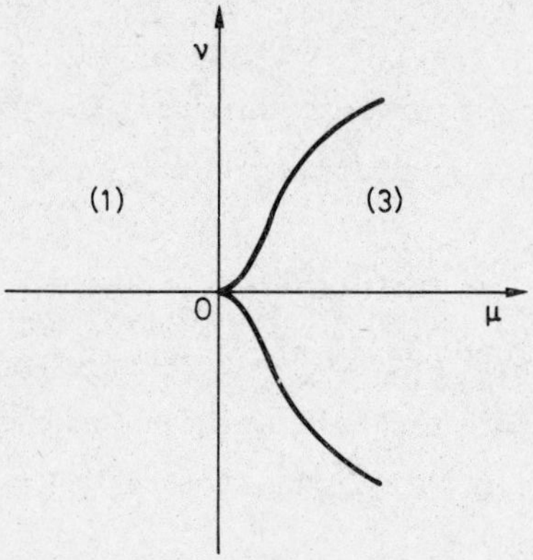

Fig. 7. The bifurcation set for ε small.

6 Symmetry and bifurcation at multiple eigenvalues

6.1. INTRODUCTION

In chapter 5 we studied the bifurcation equation and the bifurcation set in the case that $\dim \ker L = \operatorname{codim} R(L) = 1$; the main goal of this chapter is to prove certain results for the case that $\dim \ker L = \operatorname{codim} R(L) > 1$, and to illustrate these results on a few examples. We will, in a certain sense, generalize a classical theorem of Crandall and Rabinowitz [50] on bifurcation from simple eigenvalues to situations where, due to symmetry, the linearized problem has a higher-dimensional solution space. To start with, let us describe the result of Crandall and Rabinowitz.

Let X and Z be real Banach spaces, and $M : X \times \mathbb{R} \to Z$ a C^2-map satisfying $M(0,\sigma) = 0$ for all $\sigma \in \mathbb{R}$. Consider the equation

$$M(x,\sigma) = 0 . \tag{1}$$

Assume that $L = D_x M(0,0)$ is a Fredholm operator with zero index and $\dim \ker L = 1$; let $\ker L = \operatorname{span}\{u_0\}$ and suppose that $D_\sigma D_x M(0,0).u_0 \notin R(L)$. Then the theorem states that all nontrivial solutions of (1) in a neighbourhood of the origin are given by a continuous branch of the form $\{(x^*(\rho),\sigma^*(\rho)) \mid |\rho| \leq \rho_0\}$; the additional parameter ρ can be regarded as the amplitude of the solution; indeed, we have $x^*(\rho) = \rho u_0 + O(|\rho|)$.

For general problems of the form (1) the condition $\dim \ker L = 1$ is generic : it will be satisfied for "almost all" Fredholm operators $L \in \mathcal{L}(X,Z)$ with zero index and nontrivial kernel. Therefore one could argue that the theorem of Crandall and Rabinowitz handles almost all problems of the form (1). However, if we impose on (1) the additional condition that M is equivariant with respect to a given group action, then this genericity argument fails. Generically one will have that the action of the group on ker L is irreducible, but this does not necessarily imply that $\dim \ker L = 1$. We conclude that in general equivariant problems will result in higher dimensional bifurcation equations. However, the disadvantage of the higher dimension is at least partly compensated by the fact that the bifurcation equations are also equivariant.

In this chapter we will derive a number of results under the generic hypotheses described in the preceding paragraph, and illustrate these results with some examples. In section 2 we formulate the hypotheses and obtain the main abstract results; as in the other chapters, we will allow M to depend on further parameters. In section 3 we obtain the theorem of Crandall and Rabinowitz as a special case. In section 4 we analyse the bifurcation equations in the case that the basic group is O(2), and we give a few examples. In section 5 we consider the problem of bifurcation of subharmonic solutions for periodic ordinary differential equations. Finally, in section 6 we give a particular application where the symmetry group is the rotation group O(3). Another application, to the case of the symmetry group SO(2) and the problem of the Hopf bifurcation will be treated in the next chapter.

6.2. THE ABSTRACT RESULTS

6.2.1. The hypotheses. Let X, Z and Λ be real Banach spaces, and M : $\Omega \subset X \times \Lambda \times \mathbb{R} \to Z$ a mapping, defined and of class C^2 in a neighbourhood Ω of the origin. Consider the equation

$$M(x,\lambda,\sigma) = 0 . \tag{1}$$

We assume that

$$M(0,0,0) = 0 \tag{2}$$

and want to study the bifurcation of solutions of (1) in a neighbourhood of the origin.

Note that in the parameter space $\Lambda \times \mathbb{R}$ we have singled out one particular scalar parameter σ, which has to play a special role in the analysis which follows. In a number of applications (e.g. the Hopf bifurcation) this parameter has a specific physical meaning, distinguishing it from the other parameters. In a remark further on we will show how to reformulate the hypotheses in a way which does not put special emphasis on the parameter σ.

Now we make the following hypotheses on M.

(H1) There exist representations $\Gamma : G \to L(X)$ and $\tilde{\Gamma} : G \to L(Z)$ of a compact topological group G over X and Z, respectively, such that M is equivariant with respect to $(G,\Gamma,\tilde{\Gamma})$; i.e., we have for each $s \in G$ and for all $(x,\lambda,\sigma) \in \Omega$:

$$\begin{aligned} &\text{(i)} \quad (\Gamma(s)x,\lambda,\sigma) \in \Omega \ ; \\ &\text{(ii)} \quad M(\Gamma(s)x,\lambda,\sigma) = \tilde{\Gamma}(s)M(x,\lambda,\sigma) \ . \end{aligned} \tag{3}$$

(H2) $L = D_x M(0,0,0)$ is a Fredholm operator with zero index, and the representation Γ_0 of G, induced by Γ on ker L, is irreducible.

(H3) $D_\sigma M(0,0,0) = 0$ and $D_\sigma D_x M(0,0,0).u_0 \notin R(L)$ for some $u_0 \in \ker L \setminus \{0\}$.

6.2.2. Remarks.

(i) It follows from (H1) that

$$L\Gamma(s) = \tilde{\Gamma}(s)L \quad , \qquad \forall s \in G \ . \tag{4}$$

Consequently, $\Gamma(s)(\ker L) = \ker L$, for each $s \in G$. This implies that Γ induces a representation Γ_0 on ker L, defined by :

$$\Gamma_0(s) = \Gamma(s)\big|_{\ker L} \quad , \qquad \forall s \in G \ . \tag{5}$$

This representation is finite-dimensional, and by (H2) we assume that it is irreducible.

(ii) The condition (H3) is similar to the hypothesis used by Crandall and Rabinowitz in [50].

(iii) In case one does not want to single out one particular parameter σ, one can start with a C^2-map $\tilde{M} : X \times \tilde{\Lambda} \to Z$, such that $\tilde{M}(0,0) = 0$, and satisfying (H1) and (H2). Then (H3) should be replaced by :

(H3)' $D_{\tilde{\lambda}} D_x \tilde{M}(0,0).(\tilde{\lambda}_0,u_0) \notin R(L)$ for some $u_0 \in \ker L \setminus \{0\}$ and some $\tilde{\lambda}_0 \in \ker D_{\tilde{\lambda}} \tilde{M}(0,0) \setminus \{0\}$.

In particular, (H3)' implies that $\ker D_{\tilde\lambda}\tilde M(0,0) \neq \{0\}$. Then let Λ be a subspace of $\tilde\Lambda$, complementary to $\text{span}\{\tilde\lambda_0\}$. We can write each $\tilde\lambda \in \tilde\Lambda$ in the form $\tilde\lambda = \lambda+\sigma\tilde\lambda_0$, with $\lambda \in \Lambda$ and $\sigma \in \mathbb{R}$. Then $\tilde M(x,\lambda) = 0$ is equivalent to an equation of the form (1), with $M(x,\lambda,\sigma) = \tilde M(x,\lambda+\sigma\tilde\lambda_0)$. It is easy to verify that M satisfies (H1)-(H3).

6.2.3. The Liapunov-Schmidt reduction. Because of (H2) it is possible to apply the Liapunov-Schmidt reduction to equation (1). Of course we will do this in a way which is compatible with the symmetry given by (H1) (see chapter 3).

Let $P \in L(X)$ and $Q \in L(Z)$ be projections such that :

$$R(P) = \ker L \quad , \quad \ker Q = R(L) \tag{6}$$

and

$$P\Gamma(s) = \Gamma(s)P \quad , \quad Q\tilde\Gamma(s) = \tilde\Gamma(s)Q \quad , \quad \forall s \in G . \tag{7}$$

Let $v^*(u,\lambda,\sigma)$ be the unique solution of

$$(I-Q)M(u+v,\lambda,\sigma) = 0 , \tag{8}$$

where $u \in \ker L$ and $v \in \ker P$; $v^*(u,\lambda,\sigma)$ is defined and of class C^2 for (u,λ,σ) in a neighbourhood ω of the origin in $\ker L \times \Lambda \times \mathbb{R}$, and satisfies :

$$v^*(0,0,0) = 0 \quad , \quad D_u v^*(0,0,0) = 0 \quad , \quad D_\sigma v^*(0,0,0) = 0 \tag{9}$$

and

$$v^*(\Gamma(s)u,\lambda,\sigma) = \Gamma(s)v^*(u,\lambda,\sigma) \quad , \quad \forall s \in G , \forall (u,\lambda,\sigma) \in \omega . \tag{10}$$

The equation (1) reduces to the bifurcation equation

$$\tilde F(u,\lambda,\sigma) \equiv QM(u+v^*(u,\lambda,\sigma),\lambda,\sigma) = 0 . \tag{11}$$

The map $\tilde F : \omega \to R(Q)$ is of class C^2 and satisfies :

$$\tilde F(0,0,0) = 0 \quad , \quad D_u\tilde F(0,0,0) = 0 \quad , \quad D_\sigma\tilde F(0,0,0) = 0 , \tag{12}$$

and

$$\tilde{F}(\Gamma(s)u,\lambda,\sigma) = \tilde{\Gamma}(s)\tilde{F}(u,\lambda,\sigma) \quad , \quad \forall s \in G \text{ , } \forall (u,\lambda,\sigma) \in \omega \text{ .} \tag{13}$$

In order to use (13) we need some further information on how $\tilde{\Gamma}(s)$ acts on the elements of R(Q). We study this point now.

6.2.4. Lemma. The representation $\tilde{\Gamma}_0 : G \to L(R(Q))$ defined by

$$\tilde{\Gamma}_0(s) = \tilde{\Gamma}(s) \mid_{R(Q)} \quad , \quad \forall s \in G \tag{14}$$

is irreducible and equivalent to the representation Γ_0.

P r o o f. Consider $B \in L(\ker L, R(Q))$ defined by :

$$Bu = QD_\sigma D_x M(0,0,0).u \quad , \quad \forall u \in \ker L \text{ .} \tag{15}$$

We have

$$B\Gamma_0(s) = \tilde{\Gamma}_0(s)B \quad , \quad \forall s \in G \text{ .} \tag{16}$$

By (H3) we have $B \neq 0$; by (H2) Γ_0 is irreducible and dim ker L = codim R(L) = dim R(Q). Then the result follows from a direct application of corollary 2.6.9. □

The same corollary also shows the following.

6.2.5. Lemma. The operator B defined by (15) is an isomorphism between ker L and R(Q). □

Consequently we have that $Bu \neq 0$ for all $u \in \ker L \setminus \{0\}$; that is :

$$D_\sigma D_x M(0,0,0).u \notin R(L) \quad , \quad \forall u \in \ker L \setminus \{0\} \text{ .}$$

Remark that the hypothesis (H3) asks that this is true for at least one $u \in \ker L \setminus \{0\}$.

Using lemma 5 we see that the bifurcation equation (11) is equivalent with the equation

$$F(u,\lambda,\sigma) \equiv B^{-1}QM(u+v^*(u,\lambda,\sigma),\lambda,\sigma) = 0 . \tag{17}$$

The relation between the equations (1) and (17), and the properties of the mapping F are summarized in the following theorem.

6.2.6. Theorem. Assume (H1)-(H3). Then there exist a neighbourhood $\Omega_1 \subset \Omega$ of the origin in $X \times \Lambda \times \mathbb{R}$, a neighbourhood ω of the origin in $\ker L \times \Lambda \times \mathbb{R}$, and mappings

$$F : \omega \to \ker L \quad , \quad (u,\lambda,\sigma) \mapsto F(u,\lambda,\sigma)$$

and

$$\Phi : \omega \to X \times \Lambda \times \mathbb{R} \ , \quad (u,\lambda,\sigma) \mapsto \Phi(u,\lambda,\sigma) = (x^*(u,\lambda,\sigma),\lambda,\sigma) \ ,$$

both of class C^2, such that the following holds :

(i) Φ is one-to-one on its image; if $(x,\lambda,\sigma) \in R(\Phi)$, then $(x,\lambda,\sigma) = \Phi(Px,\lambda,\sigma)$;

(ii) for each $(u,\lambda,\sigma) \in \omega$:

$$F(u,\lambda,\sigma) = 0 \ \Rightarrow \ M(\Phi(u,\lambda,\sigma)) = 0 \ ;$$

(iii) for each $(x,\lambda,\sigma) \in \Omega_1$:

$$M(x,\lambda,\sigma) = 0 \ \Rightarrow \ (x,\lambda,\sigma) \in R(\Phi) \text{ and } F(\Phi^{-1}(x,\lambda,\sigma)) = 0 \ ;$$

(iv) $$F(0,0,0) = 0 \ , \ D_uF(0,0,0) = 0 \ , \ D_\sigma F(0,0,0) = 0 \tag{18}$$

and

$$D_\sigma D_u F(0,0,0) = I_{\ker L} \ ; \tag{19}$$

(v) $$F(\Gamma_0(s)u,\lambda,\sigma) = \Gamma_0(s)F(u,\lambda,\sigma) \ , \quad \forall s \in G \ , \ \forall (u,\lambda,\sigma) \in \omega \ ; \tag{20}$$

(vi) $$x^*(0,0,0) = 0 \ , \ D_u x^*(0,0,0) = I_{\ker L} \tag{21}$$

and

$$x^*(\Gamma(s)u,\lambda,\sigma) = \Gamma(s)x^*(u,\lambda,\sigma) . \tag{22}$$

6.2.7. Theorem 6 shows that there is a one-to-one relationship between the solutions in a neighbourhood of the origin of the equations (1) and (17), respectively. The bifurcation function F has inherited all smoothness and symmetry properties of M. Every statement about solutions of (17) immediately translates to a similar statement about solutions of (1). For example, if (u,λ,σ) is a solution of (17), and $\Gamma(s)u = u$ for some $s \in G$, then also $\Gamma(s)x = x$ for the corresponding solution of (1) : this follows from (22).

From now on we will concentrate on the equation (17). By theorem 2.6.3 we can find a basis in ker L such that, with respect to this basis, Γ_0 is represented by orthogonal matrices. We give ker L the Euclidean structure induced by this basis, and denote the corresponding norm and inner product by $\|.\|$ and $\langle .,. \rangle$. We will also use polar coordinates, defined by :

$$u = \rho\theta \quad , \quad \rho = \|u\| \quad , \theta \in S = \{u \in \ker L \mid \|u\| = 1\} . \tag{23}$$

6.2.8. <u>Theorem</u>. Suppose that Γ_0 is not the trivial representation. Then

$$F(0,\lambda,\sigma) = 0 \qquad , \qquad \forall(\lambda,\sigma) . \tag{24}$$

The corresponding solutions of (1) satisfy

$$\Gamma(s)x = x \qquad , \qquad \forall s \in G . \tag{25}$$

They are the only such solutions in Ω_1.

P r o o f. It is clear from (20) that $\Gamma_0(s)F(0,\lambda,\sigma) = F(0,\lambda,\sigma)$, for all $s \in G$. Then (24) follows from lemma 2.6.11. Putting $u = 0$ in (22) gives (25). Finally, if $(x,\lambda,\sigma) \in \Omega_1$ is a solution of (1) satisfying (25), then also $\Gamma_0(s)Px = Px$, $\forall s \in G$. Again by lemma 2.6.11 it follows that $Px = 0$, i.e. $x = x^*(0,\lambda,\sigma)$. □

6.2.9. <u>Remark</u>. If the operator M in (1) is such that

$$M(0,\lambda,\sigma) = 0 \qquad , \qquad \forall(\lambda,\sigma) \tag{26}$$

then it follows immediately from the definitions that

$$v^*(0,\lambda,\sigma) = 0 \quad , \quad x^*(0,\lambda,\sigma) = 0 \quad , \quad F(0,\lambda,\sigma) = 0 \, . \tag{27}$$

In case M does not satisfy (26), then one can define $M_1(x,\lambda,\sigma)$ by

$$M_1(x,\lambda,\sigma) = M(x^*(0,\lambda,\sigma) + x,\lambda,\sigma) \, . \tag{28}$$

It is easily verified that M_1 also satisfies the hypotheses (H1)-(H3). In case Γ_0 is not the trivial representation, then it follows from lemma 8 and theorem 6 that $M_1(0,\lambda,\sigma) = 0$, $\forall(\lambda,\sigma)$; in that case one can assume (26) without loss of generality.

6.2.10. Assume (26). Using polar coordinates for u in ker L ($u = \rho\theta$) we can write

$$F(\rho\theta,\lambda,\sigma) = \rho H(\rho,\theta,\lambda,\sigma) \, , \tag{29}$$

where

$$H(\rho,\theta,\lambda,\sigma) = \int_0^1 D_u F(\tau\rho\theta,\lambda,\sigma).\theta d\tau \, . \tag{30}$$

The mapping $H(\rho,\theta,\lambda,\sigma)$ is defined and of class C^1 for all $(\rho,\theta,\lambda,\sigma) \in \mathbb{R} \times \ker L \times \Lambda \times \mathbb{R}$ such that $(\tau\rho\theta,\lambda,\sigma) \in \omega$ for all $\tau \in [0,1]$. Further on we will restrict to $\rho \geq 0$ and $\theta \in S$. The mapping H satisfies

$$H(0,\theta,0,0) = 0 \quad , \quad D_\sigma H(0,\theta,0,0) = \theta \quad , \quad \forall\theta \in \ker L \, , \tag{31}$$

$$H(-\rho,-\theta,\lambda,\sigma) = -H(\rho,\theta,\lambda,\sigma) \, , \tag{32}$$

and

$$H(\rho,\Gamma_0(s)\theta,\lambda,\sigma) = \Gamma_0(s)H(\rho,\theta,\lambda,\sigma) \, . \tag{33}$$

6.2.11. <u>Lemma</u>. Assume (26). Then the scalar equation

$$h(\rho,\theta,\lambda,\sigma) \equiv \langle\theta,H(\rho,\theta,\lambda,\sigma)\rangle = 0 \tag{34}$$

has, for each $\theta \in S$ and each (ρ,λ) in a neighbourhood of the origin a unique solution $\sigma = \sigma^*(\rho,\theta,\lambda)$ near $\sigma = 0$. The mapping σ^* is of class C^1, and satisfies :

$$\sigma^*(0,\theta,0) = 0 \qquad , \qquad \forall\theta \in S \tag{35}$$

$$\sigma^*(-\rho,-\theta,\lambda) = \sigma^*(\rho,\theta,\lambda) \tag{36}$$

and

$$\sigma^*(\rho,\Gamma_0(s)\theta,\lambda) = \sigma^*(\rho,\theta,\lambda) \qquad , \qquad \forall s \in G . \tag{37}$$

P r o o f. It follows from (31) that $h(0,\theta,0,0) = 0$ and $D_\sigma h(0,\theta,0,0) = \langle\theta,\theta\rangle = 1$, for all $\theta \in S$. Using the implicit function theorem and the compactness of S we obtain the first part of the lemma and (35); (36) follows from (32) and the uniqueness of the solution $\sigma^*(\rho,\theta,\lambda)$. In a similar way (37) follows from

$$h(\rho,\Gamma_0(s)\theta,\lambda,\sigma) = h(\rho,\theta,\lambda,\sigma) \qquad , \qquad \forall s \in G , \tag{38}$$

which is a consequence of (33) and the orthogonality of Γ_0. □

6.2.12. <u>Theorem</u>. Assume (H1)-(H3) and (26). Let $\theta_0 \in S$ be such that

$$F(\rho\theta_0,\lambda,\sigma) \in \mathrm{span}\{\theta_0\} \qquad , \qquad \forall(\rho,\lambda,\sigma) . \tag{39}$$

Then the equation (1) has a nontrivial branch of solutions given by

$$\{((\Gamma(s)x^*(\rho\theta_0,\lambda,\sigma^*(\rho,\theta_0,\lambda)),\lambda,\sigma^*(\rho,\theta_0,\lambda)) \mid s\in G,\ (\rho,\lambda) \text{ near } (0,0) \text{ in } \mathbb{R}\times\Lambda\} .$$

P r o o f. It follows from (39) that

$$H(\rho,\theta_0,\lambda,\sigma) = h(\rho,\theta_0,\lambda,\sigma).\theta_0 \qquad , \qquad \forall(\rho,\lambda,\sigma) . \tag{40}$$

Then the equation $F(\rho\theta_0,\lambda,\sigma) = 0$ is, for $\rho \neq 0$, equivalent to the equation (34). The result follows from lemma 11 and the Liapunov-Schmidt method. Remark that if the hypothesis (39) is satisfied for some $\theta_0 \in S$, then it is also satisfied for each $\Gamma(s)\theta_0$, $s \in G$. □

6.2.13. Theorem. Assume (H1)-(H3), (26) and dim ker L = 1. Then all nontrivial solutions of (1) in a neighbourhood of the origin are given by

$$\{(\tilde{x}(\rho,\lambda),\lambda,\tilde{\sigma}(\rho,\lambda)) \mid 0<|\rho|\leq\rho_0,\ \lambda \text{ near } 0\}\ , \tag{41}$$

where : $\tilde{x}(\rho,\lambda) = x^*(\rho\theta_1,\lambda,\sigma^*(\rho,\theta_1,\lambda))$, $\tilde{\sigma}(\rho,\lambda) = \sigma^*(\rho,\theta_1,\lambda)$ and ker L = span$\{\theta_1\}$.

P r o o f. It is clear that the condition dim ker L = 1 implies (39) with $\theta_0 = \theta_1$. Then the result follows from theorem 12 and the Liapunov-Schmidt method. □

Theorem 13 contains as a particular case the classical theorem of Crandall and Rabinowitz [50] mentioned in the introduction to this chapter; we will give more details in the next section.

Another consequence of theorem 12 can be obtained by considering those $\theta \in S$ for which the condition (39) is a consequence of the equivariance of the bifurcation function. In order to give a precise formulation, let us define the isotropy subgroup of $\theta \in S$ as the subgroup :

$$G(\theta) = \{s\in G \mid \Gamma_0(s)\theta = \theta\}\ . \tag{42}$$

6.2.14. Theorem. Assume (H1)-(H3), and suppose that Γ_0 is not the trivial representation. Let $\theta_0 \in S$ be such that

$$\{u\in \ker L \mid \Gamma_0(s)u = u,\ \forall s\in G(\theta_0)\} = \text{span}\{\theta_0\}\ . \tag{43}$$

Then the conclusion of theorem 12 holds. Moreover, the solution $\tilde{x}(\rho,\lambda) = x^*(\rho\theta_0,\lambda,\sigma^*(\rho,\theta_0,\lambda))$ of (1) satisfies

$$\Gamma(s)\tilde{x}(\rho,\lambda) = \tilde{x}(\rho,\lambda)\quad ,\quad \forall s \in G(\theta_0)\ ,\ \forall(\rho,\lambda)\ . \tag{44}$$

P r o o f. It follows from (20) that $\Gamma_0(s)F(\rho\theta_0,\lambda,\sigma) = F(\rho\theta_0,\lambda,\sigma)$, $\forall s \in G(\theta_0)$. Then (43) implies (39). □

6.2.15. Let us now return to the general situation. The case that Γ_0 is trivial is contained in theorem 13, so we can assume that Γ_0 is not trivial.

Then $F(0,\lambda,\sigma) = 0$ for all (λ,σ), and the bifurcation equation is, for $\rho \neq 0$, equivalent to the equation

$$H(\rho,\theta,\lambda,\sigma) = 0 \ . \tag{45}$$

By lemma 11 we can solve the "radial" part of (45). Bringing the solution $\sigma = \sigma^*(\rho,\theta,\lambda)$ into (45) we obtain a reduced bifurcation equation

$$G(\rho,\theta,\lambda) \equiv H(\rho,\theta,\lambda,\sigma^*(\rho,\theta,\lambda)) = 0 \ . \tag{46}$$

The mapping G is defined and of class C^1 for $\theta \in S$ and (ρ,λ) in a neighbourhood of the origin, and has the following properties :

$$G(0,\theta,0) = 0 \ , \qquad \forall\theta \in S \ , \tag{47}$$

$$\langle\theta,G(\rho,\theta,\lambda)\rangle = 0 \ , \qquad \forall(\rho,\theta,\lambda) \ , \tag{48}$$

$$G(-\rho,-\theta,\lambda) = -G(\rho,\theta,\lambda) \ , \qquad \forall(\rho,\theta,\lambda) \ , \tag{49}$$

and

$$G(\rho,\Gamma_0(s)\theta,\lambda) = \Gamma_0(s)G(\rho,\theta,\lambda) \ , \qquad \forall s \in \mathcal{G} \ . \tag{50}$$

6.2.16. Lemma. Assume that there is some $s \in \mathcal{G}$ such that $\Gamma_0(s) = -I_{\ker L}$. Then

$$\sigma^*(-\rho,\theta,\lambda) = \sigma^*(\rho,\theta,\lambda) \tag{51}$$

and

$$G(-\rho,\theta,\lambda) = G(\rho,\theta,\lambda) \ . \tag{52}$$

P r o o f. By the hypothesis Γ_0 is not trivial; so we can define σ^* and G. The relations (51) and (52) follow from (36) and (49), combined with (37) and (50). □

6.2.17. Theorem. Assume (H1)-(H3), and suppose Γ_0 is not the trivial representation. Then, for $\theta \in S$ and (ρ,λ,σ) near the origin, with $\rho \neq 0$, we have $F(\rho\theta,\lambda,\sigma) = 0$ if and only if $\sigma = \sigma^*(\rho,\theta,\lambda)$ and $G(\rho,\theta,\lambda) = 0$. □

This result reduces our original problem (1) to that of solving the equation (46), in which the mapping G has the properties (47)-(50). Because of (48) one can, for fixed (ρ,λ), consider the mapping $\theta \mapsto G(\rho,\theta,\lambda)$ as an equivariant vectorfield over the compact manifold S; the solutions of our original problem correspond to stationary points of this vectorfield. In certain cases the equivariance of the vectorfield may imply the existence of some stationary points : this is the situation described in theorem 14. In the remainder of this chapter and in the next chapter we will discuss several examples of such problems.

Another consequence of the equivariance of the bifurcation function may be that some of the derivatives $D_u^j F(0,\lambda,\sigma)$ will vanish. The following lemma considers the consequences of such a situation for the mappings σ^* and G.

6.2.18. <u>Lemma</u>. Let M be of class C^{k+1}, satisfy (H1)-(H3), and such that either $M(0,\lambda,\sigma) = 0$ for all (λ,σ), or that Γ_0 is not the trivial representation. Let $F(u,\lambda,\sigma)$ be the bifurcation function given by theorem 6. Suppose that

$$F(0,0,0) = 0 \quad , \quad D_u F(0,0,0) = 0 \; , \quad \dots \quad , \; D_u^k F(0,0,0) = 0 \; . \tag{53}$$

Then $\sigma^*(\rho,\theta,\lambda)$ and $G(\rho,\theta,\lambda)$ are of class C^k, and we have for each $\theta \in S$:

$$\sigma^*(0,\theta,0) = D_\rho \sigma^*(0,\theta,0) = \dots = D_\rho^{k-1} \sigma^*(0,\theta,0) = 0 \; , \tag{54}$$

$$D_\rho^k \sigma^*(0,\theta,0) = -\frac{1}{k+1} \langle \theta, D_u^{k+1} F(0,0,0).(\theta,\theta,\dots,\theta)\rangle \; , \tag{55}$$

$$G(0,\theta,0) = D_\rho G(0,\theta,0) = \dots = D_\rho^{k-1} G(0,\theta,0) = 0 \; , \tag{56}$$

and

$$D_\rho^k G(0,\theta,0) = \frac{1}{k+1} D_u^{k+1} F(0,0,0).(\theta,\theta,\dots,\theta) - \frac{\theta}{k+1} \langle \theta, D_u^{k+1} F(0,0,0).(\theta,\theta,\dots,\theta)\rangle \; . \tag{57}$$

P r o o f. This follows from some straightforward calculations, using the equation defining $\sigma^*(\rho,\theta,\lambda)$ and the definition of $G(\rho,\theta,\lambda)$. □

It follows that under the conditions of lemma 18 we have

$$\sigma^*(\rho,\theta,0) = \rho^k \tilde{\sigma}(\rho,\theta) \quad , \quad G(\rho,\theta,0) = \rho^k \tilde{G}(\rho,\theta) \; ,$$

where $\tilde{\sigma}(0,\theta)$ and $\tilde{G}(0,\theta)$ are, up to a factor $(k!)^{-1}$, given by the right-hand sides of (55) and (57). So, when one restricts to $\lambda = 0$, equation (46) becomes equivalent to the equation

$$\tilde{G}(\rho,\theta) = 0 . \tag{58}$$

Since one has an explicit expression for $\tilde{G}(0,\theta)$ one may try to solve (58) under appropriate hypotheses for $\theta = \theta^*(\rho)$, using the implicit function theorem. However, since $\tilde{G}(\rho,\theta)$ describes in fact a ρ-dependant vectorfield on the manifold S, the formulation of such a result becomes rather involved, and a more direct approach, starting from the bifurcation equation (17) seems to be appropriate.

6.2.19. Let us suppose that $M(x,\sigma)$ is of class C^{k+2}, satisfies (H1)-(H3) and is such that dim ker L > 1. We will consider the equation

$$M(x,\sigma) = 0 . \tag{59}$$

We will use the same notations as before, but suppress the λ-dependance.

Problem (1) is equivalent to that of solving the bifurcation equation

$$F(\rho\theta,\sigma) = 0 , \tag{60}$$

where $\rho \in \mathbb{R}$ and $\theta \in S$. We have $F(0,\sigma) = 0$ for all σ, $D_uF(0,0) = 0$, $D_\sigma D_uF(0,0) = I_{\ker L}$, and F is equivariant. Suppose that also :

$$D_u^2F(0,0) = 0 \quad , \quad \ldots \quad , \quad D_u^kF(0,0) = 0 . \tag{61}$$

Later on we will see some examples which show that some of these equalities may be a consequence of the equivariance of F. It follows that for each solution (ρ,θ,σ) of (60) with (ρ,σ) sufficiently small, $\rho \neq 0$ and $\theta \in S$, we have

$$|\sigma| \leqslant C|\rho|^k , \tag{62}$$

for some constant $C > 0$. Indeed, for each solution of (60) with $\rho \neq 0$ we have $\sigma = \sigma^*(\rho,\theta)$; then (62) follows from lemma 18. Therefore we write

$$\sigma = \rho^k \zeta \tag{63}$$

and try to solve the equation

$$F(\rho\theta, \rho^k \zeta) = 0 . \tag{64}$$

It follows from a straightforward calculation using (61), that

$$F(\rho\theta, \rho^k \zeta) = \rho^{k+1} \tilde{F}(\rho, \theta, \zeta) , \tag{65}$$

where $\tilde{F}$ is of class C^1, and

$$\tilde{F}(0, \theta, \zeta) = \frac{1}{(k+1)!} D_u^{k+1} F(0,0).(\theta, \theta, \ldots, \theta) + \zeta\theta . \tag{66}$$

For $\rho \neq 0$ the problem (64) is equivalent to that of solving the equation

$$\tilde{F}(\rho, \theta, \zeta) = 0 \tag{67}$$

with the side-condition

$$\langle \theta, \theta \rangle = 0 . \tag{68}$$

6.2.20. <u>Lemma</u>. Let $\theta_0 \in S$ and $\zeta_0 \in \mathbb{R}$ be such that

$$\tilde{F}(0, \theta_0, \zeta_0) = \frac{1}{(k+1)!} D_u^{k+1} F(0,0).(\theta_0, \theta_0, \ldots, \theta_0) + \zeta_0 \theta_0 = 0 . \tag{69}$$

Let $P_0 \in L(\ker L)$ be the orthogonal projection on the orthogonal complement of θ_0 :

$$P_0 u = u - \langle \theta_0, u \rangle \theta_0 \quad , \quad \forall u \in \ker L . \tag{70}$$

Suppose that the problem

$$\begin{aligned} &\zeta_0 \tilde{u} + \frac{1}{k!} P_0 D_u^{k+1} F(0,0).(\theta_0, \ldots, \theta_0, \tilde{u}) = 0 , \\ &\langle \theta_0, \tilde{u} \rangle = 0 . \end{aligned} \tag{71}$$

has only the trivial solution $\tilde{u} = 0$.

Then there exist mappings $\theta^* : \mathbb{R} \to \ker L$ and $\zeta^* : \mathbb{R} \to \mathbb{R}$, defined and of class C^1 for $|\rho|$ sufficiently small, such that

$$\tilde{F}(\rho,\theta^*(\rho),\zeta^*(\rho)) = 0 \quad , \quad \langle\theta^*(\rho),\theta^*(\rho)\rangle = 1 \quad , \quad \forall\rho . \tag{72}$$

Moreover, $\theta^*(0) = \theta_0$, $\zeta^*(0) = \zeta_0$ and for (ρ,θ,ζ) in a neighbourhood of $(0,\theta_0,\zeta_0)$, (ρ,θ,ζ) is a solution of (67)-(68) if and only if $\theta = \theta^*(\rho)$ and $\zeta = \zeta^*(\rho)$.

P r o o f. We apply the implicit function theorem on the system (67)-(68), where a priori θ is an element of ker L; the condition (68) makes sure that for a solution we have $\theta \in S$. The condition for the application of the implicit function theorem is that the problem

$$\theta_0\tilde{\zeta} + \zeta_0\tilde{u} + \frac{1}{k!} D_u^{k+1}F(0,0).(\theta_0,\ldots,\theta_0,\tilde{u}) = 0$$

$$\langle\theta_0,\tilde{u}\rangle = 0$$

has only the trivial solution $(\tilde{u},\tilde{\zeta}) = (0,0)$. Using the projection P_0 it is easily seen that this is equivalent to the condition that the problem (71) has only the trivial solution $\tilde{u} = 0$. □

6.2.21. The equation (69) is sometimes called the *reduced bifurcation equation*. Its solutions are related to the solutions of the bifurcation equation (60) in the following way. Suppose that at each solution $(\theta_0,\zeta_0) \in S\times\mathbb{R}$ of the reduced bifurcation equation the condition of lemma 20 is satisfied. Since S is compact, and since for fixed $\theta_0 \in S$ (69) has at most one solution ζ_0, this will in the first place imply that there are only a finite number of solutions of (69). Moreover, using again the compactness of S and (62) it is easily seen that all nontrivial solutions (ρ,θ,σ) near $\{0\}\times S\times\{0\}$ of the bifurcation equation (60) will be obtained from the different solution branches of (67)-(68) given by lemma 20.

Since F is equivariant, it follows that

$$D_u^{k+1}F(0,0).(\Gamma_0(s)\theta,\ldots,\Gamma_0(s)\theta) = \Gamma_0(s)D_u^{k+1}F(0,0).(\theta,\ldots,\theta) \quad , \quad \forall s \in G. \tag{73}$$

This implies that if (θ_0,ζ_0) solves (69), then so does $(\Gamma_0(s)\theta_0,\zeta_0)$, for each $s \in G$. Also, the condition (73) imposes certain restrictions on the actual form of the multilinear operator $D_u^{k+1}F(0,0)$; this may be of great help when studying the solution set of (69).

We conclude this section with a remark on the equivariance conditions in the hypotheses (H1)-(H3).

6.2.22. Remark. Most of the results in this section remain valid if we replace the equivariance hypothesis (H1) by the following weakened version :

(H1)' $M(x,0,\sigma)$ is equivariant with respect to $(G,\Gamma,\tilde{\Gamma})$, while for $\lambda \neq 0$, $M(x,\lambda,\sigma)$ is only equivariant with respect to some closed subgroup G_1 of G.

Then also the bifurcation function $F(u,\lambda,\sigma)$ will for $\lambda = 0$ be equivariant with respect to G, and for $\lambda \neq 0$ with respect to G_1. One will still have that $F(0,\lambda,\sigma) = 0$ for all (λ,σ) if the condition of theorem 8, namely that Γ_0 is not the trivial representation, is replaced by the following

$$\Gamma_0(s)u = u \quad , \quad \forall s \in G_1 \Rightarrow u = 0 .$$

6.3. SYMMETRY AND BIFURCATION FROM A SIMPLE EIGENVALUE

In this section we briefly review the Crandall-Rabinowitz theorem in its classical form, and show what further information on the nontrivial solution branch can be gained from the symmetry of the equation. We start with a definition.

6.3.1. Definition. Let X and Z be real Banach spaces, and let $A,B \in L(X,Z)$. Then we say that $\lambda \in \mathbb{R}$ is a B-simple eigenvalue of A if

$$\dim \ker(A-\lambda B) = \operatorname{codim} R(A-\lambda B) = 1$$

and

$$Bu_0 \notin R(A-\lambda B) \quad , \quad \forall u_0 \in \ker(A-\lambda B) \setminus \{0\} .$$

If $X = Z$, A compact and $B = I$, we have the usual concept of a simple eigenvalue of A (i.e. λ is an eigenvalue of A, and both the algebraic and geometric multiplicities of λ are equal to 1).

6.3.2. Let X and Z be real Banach spaces, Ω a neighbourhood of the origin in X, and $M : \Omega \times]-1,1[\to Z$ a C^2-map such that :

(i) $M(0,\sigma) = 0 \quad , \quad \forall \sigma \in]-1,1[$;

(ii) 0 is a $D_\sigma D_X M(0,0)$-simple eigenvalue of $L = D_X M(0,0)$;

(iii) M is equivariant with respect to some $(G,\Gamma,\tilde{\Gamma})$, where G is a compact topological group while $\Gamma : G \to L(X)$ and $\tilde{\Gamma} : G \to L(Z)$ are representations of G over X, respectively Z.

We consider the equation

$$M(x,\sigma) = 0 . \tag{1}$$

Let $\ker L = \text{span}\{u_0\}$. Since $\Gamma(s)(\ker L) = \ker L$ for all $s \in G$, there is a representation $\alpha : G \to \mathbb{R}$ such that

$$\Gamma(s)u_0 = \alpha(s)u_0 \quad , \quad \forall s \in G . \tag{2}$$

Since α is a real one-dimensional representation of the compact group G, we have $|\alpha(s)| = 1$ for all $s \in G$. Define

$$G_0 = \{s \in G \mid \alpha(s) = 1\} . \tag{3}$$

6.3.3. Theorem. Suppose that M satisfies (i)-(iii). Let $P \in L(X)$ be a projection such that $R(P) = \ker L$ and $P\Gamma(s) = \Gamma(s)P$, $\forall s \in G$. Then there exist a neighbourhood W of $(0,0)$ in $X \times \mathbb{R}$, an interval $]-\rho_0,\rho_0[$, and mappings $\tilde{\sigma} :]-\rho_0,\rho_0[\to \mathbb{R}$ and $\psi :]-\rho_0,\rho_0[\to \ker P$, both of class C^1 and satisfying $\tilde{\sigma}(0) = 0$ and $\psi(0) = 0$, such that for each $(x,\sigma) \in W$ the following statements are equivalent :

(i) $M(x,\sigma) = 0$;

(ii) $(x,\sigma) \in \{(\rho u_0 + \rho\psi(\rho), \tilde{\sigma}(\rho)) \mid |\rho| < \rho_0\} \cup \{(0,\lambda) \in W\}$.

Furthermore, each solution $(x,\lambda) \in W$ of (1) satisfies

$$\Gamma(s)x = x \quad , \quad \forall s \in G_0 . \tag{4}$$

Finally, if $G_0 \neq G$, then

$$\tilde{\sigma}(-\rho) = \tilde{\sigma}(\rho) \quad , \quad \forall |\rho| < \rho_0 \tag{5}$$

and

$$\Gamma(s)\psi(\rho) = -\psi(-\rho) \ , \quad \forall s \in G \backslash G_0 \ , \ \forall |\rho| < \rho_0 . \tag{6}$$

P r o o f. It is clear that (i)-(iii) imply (H1)-(H3), with dim ker L = 1 and $F(0,\sigma) = 0$ for all σ. The first part of the theorem follows from theorem 2.13; since $\tilde{x}(\rho) = \rho u_0 + v^*(\rho u_0, \tilde{\sigma}(\rho))$,, with $\tilde{\sigma}(0) = 0$, $v^*(0,\sigma) = 0$ and $D_u v^*(0,0) = 0$, we can define $\psi(\rho)$ by

$$\psi(\rho) = \int_0^1 D_u v^*(\tau\rho u_0, \tilde{\sigma}(\rho)).u_0 d\tau . \tag{7}$$

(4) follows from (2.22). In case $G_0 \neq G$, we can apply lemma 16 to obtain (5); finally, we get (6) from (5), (2.10) and $v^*(\rho u_0, \tilde{\sigma}(\rho)) = \rho\psi(\rho)$, $\forall |\rho| < \rho_0$. □

6.3.4. <u>Application</u>. Consider the buckling problem for a clamped plate with shape Ω. Let λ_0 be a simple characteristic value of the corresponding operator A (so $\lambda_0 \neq 0$; see section 2.4), and let $\lambda = \lambda_0 + \alpha$. The equation describing the equilibrium state of the plate takes the form

$$(I - \lambda_0 A)w - \sigma A w + C(w) = 0 . \tag{8}$$

We will consider solutions (w,σ) of (8) in a neighbourhood of the origin in $H \times \mathbb{R}$. A is compact and self-adjoint, and by assumption λ_0 is a simple characteristic value of A; this implies that the hypotheses (i) and (ii) are satisfied for (8). As for hypothesis (iii), (8) is equivariant with respect to the group $G = \{e,\tau\}$, using the representation

$$\Gamma(e)w = w \quad , \quad \Gamma(\tau)w = -w \quad , \quad \forall w \in H . \tag{9}$$

From (5) and (6) we obtain for this example :

$$\tilde{\sigma}(-\rho) = \tilde{\sigma}(\rho) \quad , \quad \psi(-\rho) = \psi(\rho) \quad , \quad \forall|\rho| < \rho_0 . \tag{10}$$

This gives us the following result about the number of solutions of (8).

6.3.5. Theorem. Let $\lambda_0 > 0$ be a simple characteristic value of A. Then there exist numbers $\sigma_0 > 0$ and $\rho_0 > 0$ such that (8) has no nontrivial solutions satisfying $\|w\| \leq \rho_0$ for $\sigma \in [-\sigma_0, 0]$, while for $\sigma \in]0, \sigma_0]$ the same equation has exactly two such solutions, one being the negative of the other.

In case $\lambda_0 < 0$, there are no such solutions for $\sigma \in [0, \sigma_0]$, and two such solutions for $\sigma \in [-\sigma_0, 0[$.

P r o o f. Nontrivial solutions of (8) near the origin have the form

$$w = \rho u_0 + \rho\psi(\rho) \quad , \quad \sigma = \tilde{\sigma}(\rho) , \tag{11}$$

for some $0 < |\rho| \leq \rho_0$; here u_0 spans $\ker(I-\lambda_0 A)$, while ψ and $\tilde{\sigma}$ satisfy (10). Since all operators in (8) are C^∞, the same holds for ψ and $\tilde{\sigma}$.

Now express that (11) solves (8), and take the inner product with u_0; we obtain

$$\tilde{\sigma}(\rho) = \lambda_0 \rho^2 (C(u_0+\psi(\rho)), u_0)_H . \tag{12}$$

Since $(C(u_0), u_0)_H > 0$ by (2.4.22) and lemma 2.4.25, it follows that $\tilde{\sigma}(0) =$ and $D^2_\rho \tilde{\sigma}(0) > 0$. This proves the theorem. □

The same result was obtained by Berger [17], using similar arguments.

6.3.6. The cylindrical plate. A similar result can be proved for the cylindrical plate. The equation is then no longer invariant under the transformation $w \leftrightarrow -w$, so that we will have to use other symmetry operators.

Let λ_{m0} be a characteristic value of the linearized problem (see section 2.4), corresponding to an axial symmetric eigenfunction. If m is even, then this eigenfunction changes sign under the operator Γ_x (see subsection 3.6.17). The nontrivial solutions of the equation

$$(I-\lambda_{m0}A+\alpha^2A^2)w - \sigma Aw + \alpha Q(w) + C(w) = 0 \qquad (13)$$

have, for $\|w\|$ and $|\sigma|$ sufficiently small, the form

$$w = \rho u_m + \rho\psi(\rho) \quad , \qquad \sigma = \tilde{\sigma}(\rho) \ , \qquad (14)$$

for some $0 < |\rho| \leq \rho_0$, where $u_m = u_{m0}$ is the eigenfunction corresponding to λ_{m0}, and $\psi(\rho) \in (\text{span}\{u_m\})^\perp$. The maps σ and ψ satisfy

$$\tilde{\sigma}(-\rho) = \tilde{\sigma}(\rho) \quad , \quad \psi(-\rho) = -\Gamma_x\psi(\rho) \quad , \quad \forall|\rho| \leq \rho_0 \ ; \qquad (15)$$

we also have the relation

$$\tilde{\sigma}(\rho) = \mu_{m0}\{\alpha\rho(Q(u_m+\psi(\rho)),u_m)_H + \rho^2(C(u_m+\psi(\rho)),u_m)_H\}, \quad \forall|\rho| \leq \rho_0 \ . \qquad (16)$$

Here μ_{m0} is the characteristic value of A corresponding to λ_{m0} (see (2.4.70)). Using (15) and the self-adjointness of Γ_x, it follows that $(Q(u_m+\psi(\rho)),u_m)_H$ is odd in ρ, while $(C(u_m+\psi(\rho)),u_m)_H$ is even. Here, the sign of $D^2_\rho\tilde{\sigma}(0)$ will in general depend on λ_{m0} and on α.

For example, if $\alpha > 0$ is sufficiently small, then $D^2_\rho\tilde{\sigma}(0) > 0$. On condition that $D^2_\rho\tilde{\sigma}(0) \neq 0$, one can formulate a result similar to that of theorem 5.

6.4. APPLICATION : EQUATIONS WITH O(2)-SYMMETRY

In this section we will apply some of the results of section 2, and in particular theorem 2.12, to equations which are equivariant with respect to a representation of the group O(2) or some of its subgroups.

6.4.1. The hypotheses. We consider the equation

$$M(x,\lambda,\sigma) = 0 \qquad (1)$$

where we assume that M satisfies the hypotheses (H1)-(H3) of section 2, with $G = O(2)$, the group of rotations and reflections in the plane.

In order to study the bifurcation equation (2.17) for this case, it is important to know the transformation properties of the elements of ker L under the operators $\Gamma(\alpha)$ and Γ_τ, representing $\phi(\alpha) \in O(2)$ and $\tau \in O(2)$. (For the notation, see section 2.6; here we use τ and Γ_τ instead of σ and Γ_σ). The irreducible representations of O(2) are given by theorem 2.6.19. In case dim ker L = 1 and ker L = span$\{u_0\}$, then there are two possibilities, depending on whether $\Gamma_\tau u_0 = u_0$ or $\Gamma_\tau u_0 = -u_0$. In both cases we can apply theorem 6.2.13 to obtain the solution set of (1) and the symmetry of the solutions.

Therefore we will suppose that dim ker L = 2; from the irreducibility hypothesis (H2) and theorem 2.6.19 it follows that ker L has a basis $\{u_1,u_2\}$, and there exists an integer $k \in \mathbb{N} \setminus \{0\}$ such that

$$\begin{aligned} \Gamma(\alpha)u_1 &= \cos k\alpha . u_1 - \sin k\alpha . u_2 , \\ \Gamma(\alpha)u_2 &= \sin k\alpha . u_1 + \cos k\alpha . u_2 , \\ \Gamma_\tau u_1 &= u_1 \quad , \quad \Gamma_\tau u_2 = -u_2 . \end{aligned} \tag{2}$$

Since Γ is orthogonal with respect to the basis $\{u_1,u_2\}$, we will give ker L the Euclidean structure associated with this basis. We have $S = \{\beta_1u_1+\beta_2u_2 \mid \beta_1^2+\beta_2^2 = 1\}$.

6.4.2. It is immediate from (2) that for each $u \in \ker L$ we can find some $\rho \geq 0$ and some $\alpha \in \mathbb{R}$ such that

$$u = \rho\Gamma(\alpha)u_1 \quad , \quad \rho = \|u\| . \tag{3}$$

If $F(u,\lambda,\sigma)$ is the bifurcation function associated with problem (1) (see (2.17)), then

$$F(\rho\Gamma(\alpha)u_1,\lambda,\sigma) = \Gamma(\alpha)F(\rho u_1,\lambda,\sigma) ,$$

and $F(\rho\Gamma(\alpha)u_1,\lambda,\sigma) = 0$ if and only if $F(\rho u_1,\lambda,\sigma) = 0$. Therefore it is sufficient to solve the equation

$$F(\rho u_1,\lambda,\sigma) = 0 , \tag{4}$$

for (ρ,λ,σ) near $(0,0,0)$, and $\rho \geq 0$.

We also have

$$G(u_1) = \{\phi(\tfrac{2\pi}{k}p),\ \tau\circ\phi(\tfrac{2\pi}{k}p) \mid p = 0,1,\dots,k-1\} \tag{5}$$

and

$$\{u\in \ker L \mid \Gamma(s)u = u,\ \forall s\in G(u_1)\} = \operatorname{span}\{u_1\}\ . \tag{6}$$

Since the representation of O(2) given by (2) is not trivial, we have $F(0,\lambda,\sigma) = 0$ for all (λ,σ). It follows from the theory of section 2 that

$$F(\rho u_1,\lambda,\sigma) = \rho h(\rho,\lambda,\sigma)u_1 \tag{7}$$

where

$$h(\rho,\lambda,\sigma)u_1 = \int_0^1 D_uF(\tau\rho u_1,\lambda,\sigma)u_1 d\tau\ . \tag{8}$$

Finally, using $\Gamma_0(\frac{\pi}{k}) = -I_{\ker L}$, we see that

$$h(-\rho,\lambda,\sigma) = h(\rho,\lambda,\sigma)\ . \tag{9}$$

For $\rho \neq 0$ equation (4) is equivalent to the equation

$$h(\rho,\lambda,\sigma) = 0\ . \tag{10}$$

By lemma 6.2.11 this equation can be solved for $\sigma = \sigma^*(\rho,u_1,\lambda) = \tilde{\sigma}(\rho,\lambda)$. This gives the following result.

6.4.3. <u>Theorem</u>. Let M satisfy (H1)-(H3), with $G = O(2)$. Assume that $\dim \ker L = 2$ and that with respect to an appropriate basis $\{u_1,u_2\}$ of ker L we have (2). Let $\sigma = \tilde{\sigma}(\rho,\lambda)$ be the solution of (10). Finally, let

$$x_0(\lambda,\sigma) = x^*(0,\lambda,\sigma) \quad\text{and}\quad \tilde{x}(\rho,\lambda) = x^*(\rho u_1,\lambda,\tilde{\sigma}(\rho,\lambda))\ .$$

Then there exist a neighbourhood W of the origin in $X\times\Lambda\times\mathbb{R}$, a neighbourhood U_1 of the origin in $\Lambda\times\mathbb{R}$ and a neighbourhood U_2 of the origin in $\mathbb{R}\times\Lambda$ such that

$$\{(x,\lambda,\sigma)\in W \mid M(x,\lambda,\sigma)=0\}$$

$$= \{(x_0(\lambda,\sigma),\lambda,\sigma) \mid (\lambda,\sigma)\in U_1\}$$

$$\cup \{(\Gamma(\alpha)\tilde{x}(\rho,\lambda),\lambda,\tilde{\sigma}(\rho,\lambda)) \mid (\rho,\lambda)\in U_2,\ \rho>0,\ \alpha\in\mathbb{R}\} . \tag{11}$$

The mapping x_0 is of class C^2, while $\tilde{x}$ and $\tilde{\sigma}$ are of class C^1. Moreover :

$$\tilde{x}(0,\lambda) = x_0(\lambda,\tilde{\sigma}(0,\lambda)) , \tag{12}$$

$$\Gamma(\alpha)x_0(\lambda,\sigma) = \Gamma_\tau x_0(\lambda,\sigma) = x_0(\lambda,\sigma) \quad , \quad \forall\alpha\in\mathbb{R} , \tag{13}$$

$$\Gamma(\tfrac{2\pi}{k}p)\tilde{x}(\rho,\lambda) = \Gamma_\tau\tilde{x}(\rho,\lambda) = \tilde{x}(\rho,\lambda) \quad , \quad p = 0,1,\ldots,k-1 , \tag{14}$$

and

$$\tilde{\sigma}(-\rho,\lambda) = \tilde{\sigma}(\rho,\lambda) . \qquad \square \tag{15}$$

The theorem shows that for fixed λ near 0 the solutions of (1) near the origin are completely determined by the "bifurcation diagram" B_λ, that is the subset of the (ρ,σ)-plane given by

$$B_\lambda = \{(\rho,\sigma)\in U \mid \text{either } \rho=0 \text{ or } \sigma=\tilde{\sigma}(\rho,\lambda)\} . \tag{16}$$

This diagram is symmetric with respect to the σ-axis.

6.4.4. Application : buckling of a circular plate. As an application of theorem 3 we consider the buckling problem for a clamped circular plate, subjected to a uniform radial pressure along the boundary, and to a small radially symmetric normal load. The equations for the equilibrium state of such a plate were discussed in section 2.4; we obtained the equation

$$(I-\lambda A)w + C(w) = p \tag{17}$$

in the Hilbert space $H = W_0^{2,2}(\Omega)$, where $\Omega = \{(x,y)\in\mathbb{R}^2 \mid x^2+y^2<1\}$. Using polar coordinates (r,θ) in Ω, we can introduce the following representation of $O(2)$ over H

$$\begin{aligned}(\Gamma(\alpha)w)(r,\theta) &= w(r,\theta+\alpha) ,\\ (\Gamma_\tau w)(r,\theta) &= w(r,-\theta) ,\end{aligned} \qquad \forall w\in H , \ \forall\alpha\in\mathbb{R} . \tag{18}$$

Define the subspace

$$H_0 = \{w \in H \mid \Gamma(\alpha)w = w, \ \forall \alpha \in \mathbb{R}\} \ ; \tag{19}$$

then

$$w \in H_0 \ \Rightarrow \ \Gamma_\tau w = w \ . \tag{20}$$

This shows that H has no one-dimensional subspace transforming under Γ according to the case (ii) of theorem 2.6.19.

Since we assume that the normal load is radially symmetric, we take $p \in H_0$ in (17). Then (17) is equivariant with respect to $(O(2),\Gamma)$. Let, for each $\lambda \in \mathbb{R}$

$$L_\lambda = I - \lambda A \ , \tag{21}$$

and let λ_0 be a characteristic value of A, i.e. $\ker L_{\lambda_0} \neq 0$. We want to solve (17) for (w,λ,p) in an neighbourhood of $(0,\lambda_0,0)$ in $H \times \mathbb{R} \times H_0$.

As a first step we consider the transformation properties of the elements of $\ker L_{\lambda_0}$. In subsection 2.4.14 we discussed already the existence of characteristic values corresponding to radially symmetric eigenfunctions. Here we reconsider the general problem of the determination of the characteristic values of A and of the corresponding eigenfunctions.

6.4.5. Characteristic values of A. From lemma 2.4.10 and subsection 2.4.13 we know that A is a compact, self-adjoint and positive operator in H. Therefore A has a countable set of positive characteristic values, each corresponding to a finite-dimensional eigenspace. The set of characteristic values has no finite accumulation point.

Let $\lambda \in \mathbb{R}$ be a characteristic value. Since A commutes with the symmetry operators $\Gamma(\alpha)$ and Γ_τ, these induce on $\ker L_\lambda$ a finite-dimensional representation of O(2). Let E be an invariant subspace of $\ker L_\lambda$ on which the representation is irreducible. By theorem 2.6.19 this implies that $\dim E = 1$ or 2.

First assume $\dim E = 1$. It follows from theorem 2.6.19 that $E \subset H_0$, i.e. $\Gamma(\alpha)u = u$, $\forall \alpha \in \mathbb{R}$, $\forall u \in E$. Our earlier remark shows that also $\Gamma_\tau u = u$. This implies that λ is a characteristic value corresponding to a radially symmetric eigenfunction. We have seen in subsection 2.4.14 that there is an infinite sequence of such characteristic values, denoted by λ_{0j}, $j = 1,2,\ldots$, with $\lambda_{0j} \to \infty$ as $j \to \infty$. For each j the corresponding radially symmetric eigenfunction $u_{0j}(r)$ is uniquely determined up to a constant factor.

Next, suppose that dim E = 2. Again by theorem 2.6.19 E has a basis $\{u_1,u_2\}$ such that (2) holds for some $k \in \mathbb{N}\setminus\{0\}$. Using the definition (18) it follows easily that

$$u_1(r,\theta) = u(r)\,cos\,k\theta \quad , \quad u_2(r,\theta) = u(r)\,sin\,k\theta \tag{22}$$

where $u(r) = u_1(r,0)$. Remark that $u_2(r,0) = 0$. We can determine u(r) from the fact that u_1 and u_2 belong to ker L_λ. Using the theory of section 2.4 this means that u_1 and u_2 solve the following boundary value problem

$$\begin{aligned} &\Delta^2 w + \lambda w = 0 \quad , \quad &&\text{in } \Omega , \\ &w = \frac{\partial w}{\partial n} = 0 \quad , \quad &&\text{along } \partial\Omega . \end{aligned} \tag{23}$$

Bringing (22) into (23) gives a fourth order differential equation for u(r); this equation depends on k and λ, and is singular at r = 0. Fix some k; then the condition that the equation has a nontrivial solution u(r) satisfying the appropriate boundary conditions at r = 1 determines a sequence of characteristic values λ_{kj}; again $\lambda_{kj} \to \infty$ as $j \to \infty$. The corresponding radial functions $u_{kj}(r)$ are uniquely determined up to a constant factor.

Let us now assume that $\lambda_{kj} \neq \lambda_{k'j'}$ if $k \neq k'$. An actual verification of this fact requires a detailed study of the equations. Under our hypotheses on λ and E we see that λ must belong to the set of characteristic values $\{\lambda_{kj} \mid j = 1,2,\ldots\}$, and that E must coincide with the corresponding eigenspace.

We conclude that the set of characteristic values of A can be written as $\{\lambda_{kj} \mid k\in\mathbb{N},\ j\in\mathbb{N}\setminus\{0\}\}$. If we write $L_{kj} = I - \lambda_{kj}A$, then

$$\ker L_{0j} = \mathrm{span}\{u_{0j}(r)\} \tag{24}$$

and

$$\ker L_{kj} = \mathrm{span}\{u_{kj}(r)cosk\theta,\ u_{kj}(r)sink\theta\} \quad , \quad k \geq 1 . \tag{25}$$

6.4.6. The classification of the characteristic values of A according to the corresponding representation of O(2) can also be obtained by a somewhat different approach, as follows.

For each $k \in \mathbb{N}$ we define an operator $P_k : H \to H$ by

$$
\begin{aligned}
P_0 w &= \frac{1}{2\pi}\int_0^{2\pi} \Gamma(\alpha) w \, d\alpha \ , \\
P_k w &= \frac{1}{\pi}\int_0^{2\pi} \cos k\alpha \ \Gamma(\alpha) w \, d\alpha \quad , \qquad k \geq 1 ;
\end{aligned}
\tag{26}
$$

It is straightforward to verify the following properties of these operators:

(i) P_k is a continuous projection in H, for each $k \in \mathbb{N}$; (use an argument as in the proof of theorem 2.5.13);

(ii) $P_k\Gamma(\alpha) = \Gamma(\alpha)P_k \ , \ P_k\Gamma_\tau = \Gamma_\tau P_k \ , \quad \forall k \in \mathbb{N} \ , \ \forall\alpha \in \mathbb{R}$; (27)

(iii) P_k is self-adjoint ;

(iv) $P_k P_\ell = 0$ if $k \neq \ell$;

(v) if $H_k = R(P_k)$, then P_k is the orthogonal projection on H_k, and the subspaces H_k are mutually orthogonal :

(vi) $w = \sum_{k\in\mathbb{N}} P_k w \quad , \quad \forall w \in H$;

(vii) if E is a subspace of H on which Γ induces an irreducible representation of O(2), then $E \subset H_k$ for some $k \in \mathbb{N}$; if $k = 0$, then the representation is trivial; if $k \geq 1$, then E has a basis $\{u_1,u_2\}$ such that with respect to this basis Γ is given by (2).

Since A commutes with the symmetry operators, it follows that

$$
AP_k = P_k A \quad , \quad \forall k \in \mathbb{N} \ , \tag{28}
$$

that is, each of the subspaces H_k is invariant under the operator A. Let A_k be the restriction of A to H_k. Then A_k is a compact, self-adjoint and positive operator on H_k. Its characteristic values are precisely the λ_{kj}, j = 1,2,... introduced in the previous subsection.

6.4.7. Now fix some characteristic value λ_{kj}, with $k \neq 0$. We want to solve equation (17) for (w,λ,p) in a neighbourhood of $(0,\lambda_{kj},0)$ in $H \times \mathbb{R} \times H_0$. Let $\lambda = (1+\sigma)\lambda_{kj}$; then (17) takes the form (1), with

$$M(w,p,\sigma) = L_{kj}w - \sigma\lambda_{kj}Aw + C(w) - p . \tag{29}$$

Let $u_1(r,\theta) = u_{kj}(r)\mathit{cos}k\theta$ and $u_2(r,\theta) = u_{kj}\mathit{sin}k\theta$. Then

$$\ker L_{kj} = \mathrm{span}\{u_1,u_2\} \quad , \quad R(L_{kj}) = \{w \in H \mid (u_1,w) = (u_2,w) = 0\} . \tag{30}$$

The group action on $\ker L_{kj}$ is given by (2). It is easy to verify the hypothesis of theorem 3; for example, if $u \in \ker L_{kj} \setminus \{0\}$, then $-\lambda_{kj}Au = -u \notin R(L_{kj})$, which proves (H3). Theorem 3 gives us the existence of a radially symmetric solution $w_0(\lambda,p)$ for each (λ,p) near $(\lambda_{kj},0)$ in $\mathbb{R} \times H_0$; all other solutions appear in families of the form

$$\{\Gamma(\alpha)\tilde{w}(\rho,p),(1+\tilde{\sigma}(\rho,p))\lambda_{kj},p) \mid \alpha \in \mathbb{R}\} ,$$

for some (ρ,p) near the origin in $\mathbb{R} \times H_0$. The mappings $\tilde{w}(\rho,p)$ and $\tilde{\sigma}(\rho,p)$ are smooth and have the properties described in theorem 3.

6.4.8. <u>The bifurcation equation</u>. In order to decide for what values of (λ,p) equation (17) has nonradially symmetric solutions, we have to study the behaviour of the function $\tilde{\sigma}(\rho,p)$. Since $\tilde{\sigma}(\rho,p)$ is the solution of the equation (10) we first determine the function $h(\rho,p,\sigma)$.

We assume that the radial function $u_{kj}(r)$ is normalized such that $(u_1,u_1) = (u_2,u_2) = 1$. It follows from the transformation properties of u_1 and u_2 that $(u_1,u_2) = 0$. For the projections P and Q we take

$$Pw = Qw = (u_1,w)u_1 + (u_2,w)u_2 \quad , \quad \forall w \in H . \tag{31}$$

The operator B equals -I in our example. Using (2.17) and (8) we obtain :

$$h(\rho,p,\sigma) = \sigma - \int_0^1 (u_1,D_wC(\tau\rho u_1+v^*(\tau\rho u_1,p,\sigma)).(u_1+D_uv^*(\tau\rho u_1,p,\sigma)u_1))_H d\tau . \tag{32}$$

Here $v^*(u,p,\sigma)$ is the unique solution $v \in (\ker L_{k_j})^{\perp}$ of the equation

$$L_{k_j} v - \sigma\lambda_{k_j} Av + (I-Q)C(u+v) = p \ . \tag{33}$$

It follows that $v^*(0,0,\sigma) = 0$, and, since C is a cubic operator :

$$v^*(-u,-p,\sigma) = -v^*(u,p,\sigma) \ . \tag{34}$$

Consequently $h(\rho,p,\sigma)$ will be even in $p \in H_0$. The solution $\tilde{\sigma}(\rho,p)$ of $h(\rho,p,\sigma) = 0$ satisfies :

$$\tilde{\sigma}(-\rho,p) = \tilde{\sigma}(\rho,-p) = \tilde{\sigma}(\rho,p) \qquad , \qquad \forall(\rho,p) \ . \tag{35}$$

The expression for h takes a somewhat simpler form for $p = 0$. Since $v^*(0,0,\sigma) = 0$ we have

$$v^*(\rho u_1,0,\sigma) = \rho \int_0^1 D_u v^*(\tau\rho u_1,0,\sigma).u_1 d\tau \ ;$$

using the fact that C is cubic this gives

$$h(\rho,0,\sigma) = \sigma - \rho^2 (u_1, C(u_1 + \int_0^1 D_u v^*(\tau\rho u_1,0,\sigma).u_1 d\tau))_H \ .$$

From this expression we obtain

$$D_\rho^2 \tilde{\sigma}(0,0) = 2(u_1,C(u_1))_H > 0 \ . \tag{36}$$

This implies

$$D_\rho^2 \tilde{\sigma}(\rho,p) \geqslant \eta > 0 \tag{37}$$

for all (ρ,p) in a sufficiently small neighbourhood of the origin. Let $\bar{\sigma}(p) = \tilde{\sigma}(0,p)$; then we obtain from (37) the following result.

6.4.9. <u>Theorem</u>. Let λ_{k_j} be a characteristic value of A, with $k \neq 0$. Then there is a neighbourhood W of the origin in H, and a neighbourhood U of $(\lambda_{k_j},0)$ in $\mathbb{R} \times H_0$, such that for $(\lambda,p) \in U$ the equation (17) has nonradially symmetric solutions $w \in W$ if and only if

$$\lambda > \lambda_{k_j}(1+\bar{\sigma}(p)) . \tag{38}$$

When this condition is satisfied, then (17) has exactly one family of non-radially symmetric solutions in W; this family has the form

$$\{\Gamma(\alpha)\bar{w}(\lambda,p) \mid \alpha \in \mathbb{R}\}$$

for some $\bar{w}(\lambda,p) \in W$. The function $\bar{\sigma}(p)$ is smooth, with $\bar{\sigma}(0) = 0$ and $\bar{\sigma}(-p) = \bar{\sigma}(p)$.

P r o o f. Given (σ,p) near the origin in $\mathbb{R} \times H_0$, it follows from (37) that there is some $\rho > 0$ such that $\sigma = \tilde{\sigma}(\rho,p)$ if and only if $\sigma > \bar{\sigma}(p)$. This proves the theorem, if we define $\bar{w}(\lambda,p) = \tilde{w}(\rho,p)$, where $\rho > 0$ is chosen such that $\lambda = (1+\tilde{\sigma}(\rho,p))\lambda_{k_j}$. $\square$

6.4.10. Remark. It is interesting to compare the foregoing results with the results of chapter 3 and more in particular with theorem 3.6.6. Indeed, this theorem implies that every small solution of (17) will remain invariant under Γ_τ after application of an appropriate $\Gamma(\alpha)$; the relation (3) is nothing else than the hypothesis (R) used in chapter 3. Considering solutions which are invariant under Γ_τ, one obtains the reduced bifurcation equation (4). The theory of this chapter shows that (4) consists of one single scalar equation, which can be handled if the "transversality condition" (H3) is satisfied.

6.4.11. A nonlinear boundary value problem. Let Ω be the unit ball in $\mathbb{R}^2$, $R > 0$ and Λ the subspace of $C^3(\bar{\Omega} \times [-R,R])$ consisting of functions $f(x,v)$ such that $f(x,0) = 0$ for all $x \in \bar{\Omega}$, and $f(x,v) = f(y,v)$ if $\|x\| = \|y\|$ (we use the Euclidean norm in $\mathbb{R}^2$). For each $f \in \Lambda$ we want to determine $\sigma \in \mathbb{R}$ such that the nonlinear boundary value problem

$$\begin{aligned} \Delta v + \sigma f(x,v) &= 0 , && \text{in } \Omega \\ v &= 0 , && \text{along } \partial\Omega \end{aligned} \tag{39}$$

has nontrivial solutions. So this is a nonlinear eigenvalue problem. To bring it in the form (1) we let $X = \{v \in C^{2,\alpha}(\bar{\Omega}) \mid v(x) = 0, \forall x \in \partial\Omega\}$ and $Z = C^{0,\alpha}(\bar{\Omega})$, for some $\alpha \in]0,1[$. The mapping $M : X \times \Lambda \times \mathbb{R} \to Z$ given by

$$M(v,f,\sigma)(x) = \Delta v(x) + \sigma f(x,v(x)) \quad , \quad \forall x \in \bar{\Omega} \tag{40}$$

is defined and of class C^2 for all $(f,\sigma) \in \Lambda \times \mathbb{R}$ and for v in a neighbourhood of the origin in X. Our problem can then be written in the form (1). Moreover, M is equivariant with respect to $(O(2),\Gamma)$ if we use a representation $\Gamma : O(2) \to L(Z)$ defined as in (18).

Fix some $f_0 \in \Lambda$ and consider the linearized problem

$$\begin{aligned} \Delta v + \sigma D_v f_0(x,0)v &= 0 \quad , \quad \text{in } \Omega \\ v &= 0 \quad , \quad \text{along } \partial\Omega \, . \end{aligned} \tag{41}$$

For simplicity, let us also assume that

$$D_v f_0(x,0) > 0 \quad , \quad \forall x \in \bar{\Omega} \, . \tag{42}$$

Since $\partial\Omega$ is smooth, and $D_v f_0(x,0)$ continuously differentiable, classical and generalized solutions of (41) coincide. Using the Hilbert space method of section 2.3, in combination with the projection operators P_k defined as in (26) (but with $H = L^2(\Omega)$), one can show that for each $k \in \mathbb{N}$ there is sequence $\{\sigma_{k_j} \mid j = 1,2,...\}$ of positive characteristic values of (41); we have $\sigma_{kj} \to \infty$ as $j \to \infty$. Avoiding accidental degeneracies, the corresponding eigenspaces will be one-dimensional if $k = 0$, and two-dimensional if $k \geq 1$. For $k = 0$, the eigenspace is spanned by a radially symmetric eigenfunction $\psi_{k0}(r)$; if $k \neq 0$, then the eigenspace is spanned by functions $u_1 \in X$ and $u_2 \in X$, of the form :

$$u_1(r,\theta) = \psi_{kj}(r)\cos k\theta \quad , \quad u_2(r,\theta) = \psi_{kj}(r)\sin k\theta \, . \tag{43}$$

6.4.12. Let now $\sigma_0 = \sigma_{kj}$ for some $k \neq 0$; then

$$\ker D_u M(0,f_0,\sigma_0) = \text{span}\{u_1,u_2\} \; ; \tag{44}$$

also, since (41) is formally self-adjoint :

$$R(D_u M(0,f_0,\sigma_0)) = \{v \in Z \mid (u_1,v) = (u_2,v) = 0\} \ . \tag{45}$$

In (45) we use the bilinear form :

$$(u,v) = \int_\Omega u(x)v(x)dx \quad , \qquad \forall u,v \in Z \ . \tag{46}$$

It is easily seen that the hypotheses (H1) and (H2) are satisfied. For our example (H3) reduces to the condition :

$$\int_\Omega D_u f_0(x,0)u_1^2(x)dx \neq 0 \ , \tag{47}$$

which is satisfied because of (42).

Application of theorem 3 then shows that for each f near f_0 in Λ the problem (39) has the following solutions near $(0,\sigma_0)$:

(i) for each σ near σ_0 there is a unique solution which is radially symmetric; since $f(x,0) = 0$ for each $f \in \Lambda$, this unique radially symmetric solution is the trivial solution $v = 0$;

(ii) for each sufficiently small $\rho > 0$ there is an eigenvalue $\tilde{\sigma}_f(\rho)$ such that, for $\sigma = \tilde{\sigma}_f(\rho)$, (39) has a unique family of nonradially symmetric solutions, of the form :

$$\{\Gamma(\alpha)\tilde{v}_f(\rho) \mid \alpha \in \mathbb{R}\} \ .$$

The function $\tilde{\sigma}_f(\rho)$ is of class C^1 in ρ and f, and $\tilde{\sigma}_{f_0}(0) = \sigma_0$; the solution $\tilde{v}_f(\rho)$ satisfies

$$\Gamma(\tfrac{2\pi}{k}p)\tilde{v}_f(\rho) = \Gamma_\tau \tilde{v}_f(\rho) = \tilde{v}_f(\rho) \quad , \quad p = 0,1,\ldots k-1 \ . \tag{48}$$

6.4.13. Now we modify the problem (39) adding a small inhomogeneous term

$$\begin{aligned} \Delta v + \sigma f(x,v) &= g(x) \quad , \qquad x \in \Omega \ , \\ v(x) &= 0 \qquad\quad , \qquad x \in \partial\Omega \ . \end{aligned} \tag{49}$$

Here again we take $\Omega = \{(x_1,x_2) \in \mathbb{R}^2 \mid x_1^2 + x_2^2 < 1\}$, $f \in \Lambda$, and we suppose that

g is small and $g \in Z_\ell$ for some $\ell \in \mathbb{N} \setminus \{0\}$, where

$$Z_\ell = \{z \in C^{0,\alpha}(\bar{\Omega}) \mid \Gamma(s)z = z, \forall s \in \Delta_\ell\} \tag{50}$$

and

$$\Delta_\ell = \{\phi(\tfrac{2\pi}{\ell}j), \tau\circ\phi(\tfrac{2\pi}{\ell}j) \mid j = 0,1,\dots,\ell-1\} . \tag{51}$$

For $g = 0$, (49) reduces to (39), and the equation is then equivariant with respect to the group $O(2)$; for $g \neq 0$ one has only equivariance with respect to the subgroup Δ_ℓ.

Fix some $f_0 \in \Lambda$ which satisfies (42); let $\sigma_0 = \sigma_{kj}$, where σ_{kj} is a characteristic value determined by (41), with $k \neq 0$. We want to find solutions (v,σ) of (49) near $(0,\sigma_0)$, for each (f,g) near $(f_0,0)$ in $\Lambda \times Z_\ell$. To find such solutions we have to solve the bifurcation equation

$$F(u,f,g,\sigma) = 0 , \tag{52}$$

where $F : U \times \Lambda \times Z_\ell \times \mathbb{R} \to U$ is of class C^2, $U = \text{span}\{u_1,u_2\}$, $F(0,f,0,\sigma) = 0$, $D_uF(0,f_0,0,\sigma_0) = 0$, $D_uD_\sigma F(0,f_0,0,\sigma_0) = I_U$,

$$F(\Gamma(s)u,f,0,\sigma) = \Gamma(s)F(u,f,0,\sigma) \quad , \quad \forall s \in O(2) , \tag{53}$$

and

$$F(\Gamma(s)u,f,g,\sigma) = \Gamma(s)F(u,f,g,\sigma) \quad , \quad \forall s \in \Delta_\ell . \tag{54}$$

We will not give a complete discussion of the equation (52), but apply the ideas of theorem 2.12 and theorem 2.14 to obtain certain solutions which are a consequence of the equivariance (54).

6.4.14. In order to apply the approach of theorem 2.12 we need to have $F(0,f,g,\sigma) = 0$ for all (f,g,σ). According to remark 2.22 this will be the case if we have

$$u \in U \quad , \quad \Gamma(s)u = u \quad , \quad \forall s \in \Delta_\ell \quad \Rightarrow \quad u = 0 .$$

Using (2) we see that this will be satisfied if k/ℓ is not an entire number; if we write $k/\ell = k_1/\ell_1$, such that k_1 and ℓ_1 have no common divisors, this means that $\ell_1 \neq 1$.

Now we look for $\theta_0 = \Gamma(\alpha_0)u_1 \in S = \{\Gamma(\alpha)u_1 \mid \alpha\in\mathbb{R}\}$ satisfying the condition (2.43) of theorem 2.14, where we should remember that the group is Δ_ℓ (and not $O(2)$). The condition will be satisfied if α_0 is such that

$$\Gamma_\tau\Gamma(\tfrac{2\pi}{\ell}j)\Gamma(\alpha_0)u_1 = \Gamma(\alpha_0)u_1 ,$$

i.e. when

$$\frac{2\pi}{\ell}j + 2\alpha_0 = \frac{2\pi}{k}m \tag{55}$$

for some $(j,m) \in \mathbb{Z}\times\mathbb{Z}$. Indeed, when (55) is satisfied then $\{\Gamma(\alpha_0)u_1,\Gamma(\alpha_0)u_2\}$ forms a basis of U, and

$$\Gamma_\tau\Gamma(\tfrac{2\pi}{\ell}j)\Gamma(\alpha_0)u_2 = \Gamma_\tau\Gamma(-\alpha_0)u_2 = \Gamma(\alpha_0)\Gamma_\tau u_2 = -\Gamma(\alpha_0)u_2 .$$

Then (2.43) follows easily.

Since $\Gamma(\frac{2\pi}{k})u = u$ for $u \in U$, and $\mathbb{Z} = \{m\ell_1 + jk_1 \mid m,j\in\mathbb{Z}\}$, ($\ell_1$ and k_1 have no common divisors) it follows that we can rewrite the condition (55) in the form

$$\begin{aligned}\alpha_0 &\in \{\tfrac{\pi}{k}m + \tfrac{\pi}{\ell}j\,(\mathrm{mod}\ \tfrac{2\pi}{k}) \mid m,j\in\mathbb{Z}\}\\ &= \{\tfrac{\pi}{k}(\tfrac{m\ell_1 + jk_1}{\ell_1})\ \mathrm{mod}\ \tfrac{2\pi}{k} \mid m,j\in\mathbb{Z}\}\\ &= \{\tfrac{\pi}{k}\cdot\tfrac{p}{\ell_1} \mid p=0,1,\dots,2\ell_1-1\} .\end{aligned}$$

Moreover, we have

$$\begin{aligned}&\{\tfrac{2\pi}{\ell}j\,(\mathrm{mod}\ \tfrac{2\pi}{k}) \mid j\in\mathbb{Z}\}\\ &= \{\tfrac{2\pi}{k}(\tfrac{m\ell_1 + jk_1}{\ell_1})\mathrm{mod}\ \tfrac{2\pi}{k} \mid m,j\in\mathbb{Z}\} = \{\tfrac{2\pi}{k}\cdot\tfrac{p}{\ell_1} \mid p=0,1,\dots,\ell_1-1\} ;\end{aligned}$$

so the condition on α_0 becomes :

$$\alpha_0 \in \{\tfrac{2\pi}{k}\cdot\tfrac{p}{\ell_1} \mid p=0,1,\dots,\ell_1-1\} \cup \{\tfrac{\pi}{k}\cdot\ \tfrac{1}{\ell_1} + \tfrac{2\pi}{k}\cdot\tfrac{p}{\ell_1} \mid p=0,1,\dots,\ell_1-1\} .$$

We conclude that the set of $\theta_0 \in S$ satisfying the condition (2.43) contains $2\ell_1$ elements, and can be written in the form

$$\{\Gamma(\frac{2\pi}{k}\cdot\frac{p}{\ell_1})u_1 \mid p=0,1,\ldots,\ell_1-1\} \cup \{\Gamma(\frac{2\pi}{k}\cdot\frac{p}{\ell_1})\Gamma(\frac{\pi}{k}\cdot\frac{1}{\ell_1})u_1 \mid p=0,1,\ldots,\ell_1-1\}$$

$$= \{\Gamma(s)u_1 \mid s\in\Delta_\ell\} \cup \{\Gamma(s)\Gamma(\frac{\pi}{k}\cdot\frac{1}{\ell_1})u_1 \mid s\in\Delta_\ell\} .$$

Finally, if ℓ_1 is odd then $\frac{\pi}{k}\cdot\frac{1}{\ell_1} = \frac{\pi}{k} \bmod \frac{2\pi}{k}\cdot\frac{1}{\ell_1}$, and since $\Gamma(\frac{\pi}{k})u = -u$ for $u\in U$, the set above takes for this case the form

$$\{\pm\Gamma(s)u_1 \mid s\in\Delta_\ell\} .$$

Then the results of section 2 give us the following theorem.

6.4.15. <u>Theorem</u>. Under the conditions mentionned in the previous subsections, suppose that $\ell_1 \neq 1$. Then the bifurcation equation (52) has, for each $(f,g) \in \Lambda\times Z_\ell$ near $(f_0,0)$, at least the following nontrivial solution branches bifurcating from the trivial branch $\{(0,\sigma) \mid |\sigma-\sigma_{kj}|<\delta\}$:

(i) if ℓ_1 is odd, then there are ℓ_1 branches bifurcating at the same point, and given by

$$\{(\rho\Gamma(s)u_1,\sigma_{f,g}(\rho)) \mid |\rho|<\rho_0,\ s\in\Delta_\ell\} ;$$

(ii) if ℓ_1 is even, then there are two times $\ell_1/2$ branches, given by

$$\{(\rho\Gamma(s)u_1,\tilde\sigma^{(1)}_{f,g}(\rho)) \mid |\rho|<\rho_0,\ s\in\Delta_\ell\}$$

and

$$\{(\rho\Gamma(s)\Gamma(\frac{\pi}{k}\cdot\frac{1}{\ell_1})u_1,\tilde\sigma^{(2)}_{f,g}(\rho)) \mid |\rho|<\rho_0,\ s\in\Delta_\ell\} ;$$

The functions $\tilde\sigma^{(1)}_{f,g}(\rho)$ and $\tilde\sigma^{(2)}_{f,g}(\rho)$ are even in ρ, and both sets of branches bifurcate at the same point $\tilde\sigma^{(1)}_{f,g} = \tilde\sigma^{(2)}_{f,g}$. □

The proof just copies some of the results of section 2, and will not be given explicitly. Remark that if ℓ_1 is even, then there is some $s \in \Delta_\ell$ such that $\Gamma(s)u_1 = -u_1$; if, for a fixed $\theta \in S$, one considers $\{(\rho\theta,\sigma^*(\rho,\theta)) \mid$

$|\rho| < \rho_0\}$ as one branch, this accounts for the fact that in case (ii) one has two times $\ell_1/2$ branches, and for the evenness of the functions $\tilde{\sigma}^{(1)}_{f,g}(\rho)$ and $\tilde{\sigma}^{(2)}_{f,g}(\rho)$.

6.5. BIFURCATION OF SUBHARMONIC SOLUTIONS

6.5.1. Introduction. In subsection 3.6.8 we introduced the problem of bifurcation of subharmonic solutions, that is, the bifurcation of m T-periodic solutions for T-periodic ordinary differential equations in $\mathbb{R}^n$. In this section we will give a few results on such bifurcations. Our presentation has partly been influenced by the treatment given in Iooss and Joseph [255].

When formulating the problem we may assume that through an appropriate time rescale we are reduced to the case $T = 2\pi$. Let $f : \mathbb{R}\times\mathbb{R}^n\times\mathbb{R} \to \mathbb{R}^n$ be of class C^2, 2π-periodic in the first variable t, and such that $f(t,0,\sigma) = 0$ for all (t,σ). Fix some $m \in \mathbb{N}\setminus\{0\}$. Then we want to determine, for all σ near 0, all small $2\pi m$-periodic solutions of the equation

$$\dot{x} = f(t,x,\sigma) \ . \tag{1}$$

6.5.2. Some function spaces. In order to bring our problem in the form (2.1) we will need certain spaces of periodic functions. It will appear that at certain points in the treatment it has some advantage to work with complex valued functions. The basic space will be Z_c, the space of all continuous $2\pi m$-periodic functions $z : \mathbb{R} \to \mathbb{C}^n$. By X_c we denote the subspace of all $z \in Z_c$ which are continuously differentiable. X and Z will be the subspaces of X_c, respectively Z_c, consisting of real-valued functions. Finally, X^0_c and Z^0_c will be the subspaces of X_c and Z_c consisting of 2π-periodic functions. Using appropriate supremum norms all these spaces become Banach spaces.

On Z_c we can define a bilinear form $\langle .,. \rangle : Z_c \times Z_c \to \mathbb{C}$ by

$$\langle z_1,z_2\rangle = \frac{1}{2\pi m}\int_0^{2\pi m} (z_1(t),z_2(t))dt \quad , \quad \forall z_1,z_2 \in Z_c \ , \tag{2}$$

where

$$(a,b) = \sum_{i=1}^{n} \bar{a}_i b_i \qquad , \quad \forall a,b \in \mathbb{C}^n \ . \tag{3}$$

The restriction of $<.,.>$ to $Z \times Z$ is real-valued, and on Z_c^0 it takes the simplified form

$$<w_1,w_2> = \frac{1}{2\pi}\int_0^{2\pi} (w_1(t),w_2(t))dt \quad , \quad \forall w_1,w_2 \in Z_c^0 . \tag{4}$$

Finally, if $A \in L(\mathbb{R}^n)$ then we will denote by $\ker_c A$ the kernel of the canonical extension of A to a linear operator over $\mathbb{C}^n$.

6.5.3. <u>Abstract formulation of the problem</u>. Define $M : X \times \mathbb{R} \to Z$ by

$$M(x,\sigma)(t) = -\dot{x}(t) + f(t,x(t),\sigma) \quad , \quad \forall t \in \mathbb{R} . \tag{5}$$

Then M is of class C^2, $M(0,\sigma) = 0$ for all σ, and our problem takes the form

$$M(x,\sigma) = 0 . \tag{6}$$

Let $D_m = \{\phi(\frac{2\pi}{m}j) \mid j = 0,1,\ldots,m-1\}$ be the abelian subgroup of SO(2) generated by $\phi(2\pi/m)$. We can define a representation of D_m over Z (and X) by putting $\Gamma(\frac{2\pi}{m}j) = \Gamma^j$ $(j = 0,1,\ldots,m-1)$, where $\Gamma = \Gamma(\frac{2\pi}{m})$ is given by

$$(\Gamma z)(t) = z(t + 2\pi) \quad , \quad \forall t \in \mathbb{R} . \tag{7}$$

Since the equation (1) is 2π-periodic, it follows that (6) is equivariant with respect to (D_m,Γ).

6.5.4. <u>The linearization</u>. Let $L = D_x M(0,0)$; L is an element of $L(X,Z)$ explicitly given by

$$(Lx)(t) = -\dot{x}(t) + A(t,0)x(t) \quad , \quad \forall t \in \mathbb{R} \tag{8}$$

where $A : \mathbb{R} \times \mathbb{R} \to L(\mathbb{R}^n)$ is defined by

$$A(t,\sigma) = D_x f(t,0,\sigma) \quad , \quad \forall (t,\sigma) . \tag{9}$$

The operator L has a formal adjoint $L^* \in L(X,Z)$ given by

$$(L^*x)(t) = \dot{x}(t) + A^T(t,0)x(t) \qquad , \qquad \forall t \in \mathbb{R} \; ; \tag{10}$$

we have

$$<x^*, Lx> = <L^*x^*, x> \qquad , \qquad \forall x, x^* \in X \; . \tag{11}$$

Also, L is associated with the 2π-periodic linear differential equation

$$\dot{x} = A(t,0)x \; . \tag{12}$$

We denote by $\Phi(t)$ the transition matrix of (12), that is the fundamental matrix solution of (12) such that $\Phi(0) = I$. Since (12) is 2π-periodic, it follows that

$$\Phi(t+2\pi) = \Phi(t)C \qquad , \qquad \forall t \in \mathbb{R} \; , \tag{13}$$

where $C = \Phi(2\pi)$ is the monodromy matrix of (12). In particular we have $\Phi(2\pi m) = C^m$.

The theory of section 2.2 shows that L is a Fredholm operator with zero index. We have

$$\ker L = \{u(.) = \Phi(.)u_0 \mid u_0 \in \ker(\Phi(2\pi m)-I)\}, \tag{14}$$

$$\ker L^* = \{u^*(.) = (\Phi^T(.))^{-1}u_0^* \mid u_0^* \in \ker(\Phi^T(2\pi m)-I)\} \; , \tag{15}$$

and

$$R(L) = \{z \in Z \mid <u^*, z> = 0, \; \forall u^* \in \ker L^*\} \; . \tag{16}$$

Denote by L_c and L_c^* the canonical extension of L, respectively L^*, to X_c. Then we have

$$\ker L = \{\mathrm{Re}\; v \mid v \in \ker L_c\} = \{\mathrm{Re}\; \Phi(.)v_0 \mid v_0 \in \ker_c(C^m - I)\}$$

$$= \{\mathrm{Re}\; \Phi(.)v_0 \mid v_0 \in \mathrm{span}(\bigcup_{0 \leqslant k < m} \ker_c(C - e^{\frac{2\pi i k}{m}} I))\} \; , \tag{17}$$

with an analogous expression for $\ker L^*$.

We see from (17) that we have to study the eigenvalues of the monodromy matrix C. These eigenvalues are related to the eigenvalues of a linear operator J over X_c^0 which we introduce now.

6.5.5. <u>The operators J and J*</u>. Let $J \in L(X_c^0, Z_c^0)$ and $J^* \in L(X_c^0, Z_c^0)$ be the restrictions of L_c, respectively L_c^*, to the subspace X_c^0 of X_c. J and J* are formal adjoints, i.e. we have

$$\langle w^*, Jw\rangle = \langle J^* w^*, w\rangle \quad , \qquad \forall w, w^* \in X_c^0 . \tag{18}$$

Let $\mu \in \mathbb{C}$, and let $w \in \ker(J-\mu I)$. Then it is easily seen that $v(t) = e^{\mu t} w(t)$ is a solution of (12), from which it follows that $v(t) = \Phi(t)w(0)$. We leave it as an exercise to use this fact in combination with the arguments used in subsection 2.6.3 to prove the following.

6.5.6. <u>Lemma</u>. One has for each $\mu \in \mathbb{C}$:

$$\ker(J-\mu I) = \{w(t) = e^{-\mu t}\Phi(t)w_0 \mid w_0 \in \ker_c(C-e^{2\pi\mu}I)\} , \tag{19}$$

$$\ker(J^*-\bar{\mu}I) = \{w^*(t) = e^{\bar{\mu}t}(\Phi^T(t))^{-1}w_0^* \mid w_0^* \in \ker_c(C^T-e^{2\pi\bar{\mu}}I)\} , \tag{20}$$

and

$$R(J-\mu I) = \{w \in Z_c^0 \mid \langle w^*, w\rangle = 0, \ \forall w^* \in \ker(J^*-\bar{\mu}I)\} . \quad \square \tag{21}$$

We call $\mu \in \mathbb{C}$ an eigenvalue of J if $\ker(J-\mu I)$ is nontrivial; these eigenvalues are also called the *Floquet exponents* of the 2π-periodic equation (12). The lemma shows that $\mu \in \mathbb{C}$ is an eigenvalue of J if and only if $e^{2\pi\mu}$ is an eigenvalue of the monodromy matrix C; these corresponding eigenvalues $e^{2\pi\mu}$ are called the *Floquet multipliers* of (12). This also shows that if μ is an eigenvalue of J, then so are $\mu+i\ell$ and $\bar{\mu}+i\ell$, for any $\ell \in \mathbb{Z}$.

6.5.7. Now we will make some further hypotheses on the equation (1). In order to do so we define, for each $\sigma \in \mathbb{R}$, the operator $J(\sigma) \in L(X_c^0, Z_c^0)$ by

$$(J(\sigma)w)(t) = -\dot{w}(t) + A(t,\sigma)w(t) \quad , \qquad \forall t \in \mathbb{R} . \tag{22}$$

We have $J = J(0)$, and $J(\sigma)$ has similar properties as J. Let $J_\sigma(0) = D_\sigma J(0)$; $J_\sigma(0)$ is an element of $L(X_c^0, Z_c^0)$ which can in fact be extended to all of Z_c^0, and which is explicitly given by

$$(J_\sigma(0)w)(t) = D_\sigma A(t,0)w(t) \quad , \quad \forall t \in \mathbb{R} . \tag{23}$$

We make the following assumption :

(H) (i) There is some $k \in \mathbb{N}$ such that $i\frac{k}{m}$ is a $J_\sigma(0)$-simple eigenvalue of J (see definition 6.3.1). By the remark after lemma 6 we may suppose $0 \leq k \leq \frac{m}{2}$.

(ii) J has no eigenvalues of the form $i\frac{k'}{m}$, with $k' \in \mathbb{N}$, $0 \leq k' \leq \frac{m}{2}$, $k' \neq k$.

The hypothesis (H)(ii) is called a *nonresonance condition*; using (19) it says that the only eigenvalues λ of C such that $\lambda^m = 1$ are given by $\exp(\pm\frac{2\pi i}{m}k)$.

Let us explain now in some more detail the meaning of (H)(i). Let $\mu_0 = \frac{k}{m}i$. Then (H)(i) implies first of all that $\dim \ker(J-\mu_0 I) = 1$, which, by (19), translates into $\dim \ker_c(C - e^{2\pi\mu_0}I) = 1$. Let $\zeta_0 \in \mathbb{C}^n$ and $\zeta_0^* \in \mathbb{C}^n$ be such that

$$\ker_c(C - e^{2\pi\mu_0}I) = \text{span}\{\zeta_0\} \ , \ \ker_c(C^T - e^{2\pi\bar{\mu}_0}I) = \text{span}\{\zeta_0^*\} . \tag{24}$$

Define $\zeta, \zeta^* \in X_c^0$ by

$$\zeta(t) = e^{-\mu_0 t}\Phi(t)\zeta_0 \quad , \quad \zeta^*(t) = e^{\bar{\mu}_0 t}(\Phi^T(t))^{-1}\zeta_0^* \quad , \quad \forall t \in \mathbb{R} ; \tag{25}$$

then we have

$$\ker(J-\mu_0 I) = \text{span}\{\zeta\} \quad , \quad \ker(J^* - \bar{\mu}_0 I) = \text{span}\{\zeta^*\} \tag{26}$$

and

$$R(J-\mu_0 I) = \{w \in Z_c^0 \mid \langle \zeta^*, w \rangle = 0\} . \tag{27}$$

The condition that μ_0 is a $J_\sigma(0)$-simple eigenvalue of J means that $J_\sigma(0)\zeta \notin R(J-\mu_0 I)$, i.e.

$$\langle \zeta^*, J_\sigma(0)\zeta \rangle \neq 0 , \tag{28}$$

or more explicitly

$$\frac{1}{2\pi}\int_0^{2\pi} (\zeta^*(t),D_\sigma A(t,0)\zeta(t))dt \neq 0 . \tag{29}$$

We will normalize ζ_0 and ζ_0^* by the condition

$$<\zeta^*,J_\sigma(0)\zeta> = 2 . \tag{30}$$

6.5.8. Remark. In other treatments of the problem of bifurcation of subharmonic solutions (see e.g. Iooss and Joseph [255]) one usually assumes that $e^{2\pi\mu_0}$ is a simple eigenvalue of C, which means that we have (24) and $\zeta_0 \notin R_C(C-e^{2\pi\mu_0}I)$, i.e. $(\zeta_0^*,\zeta_0) \neq 0$. Then one normalizes by the condition that $(\zeta_0^*,\zeta_0) = 1$. It is easily seen that this implies $<\zeta^*,\zeta> = 1$. Let us show that this implies (H)(i) under an additional condition.

Let $C(\sigma)$ be the monodromy matrix for the linear 2π-periodic equation

$$\dot{x} = A(t,\sigma)x . \tag{31}$$

Under the condition $(\zeta_0^*,\zeta_0) \neq 0$ one can show (for example by using a complex version of the Crandall-Rabinowitz theorem given in section 3) that $C(\sigma)$ has a continuously differentiable eigenvalue branch $e^{2\pi\mu(\sigma)}$, such that $\mu(0) = \mu_0$. Then $\mu(\sigma)$ is an eigenvalue of $J(\sigma)$, and one can construct $\zeta(\sigma) \in X_C^0$ such that $\zeta(0) = \zeta$ and

$$J(\sigma)\zeta(\sigma) = \mu(\sigma)J(\sigma) \qquad , \qquad \forall\sigma . \tag{32}$$

Differentiating (32) at $\sigma = 0$, and taking the inner product with ζ^*, we obtain

$$<\zeta^*,J_\sigma(0)\zeta> = D_\sigma\mu(0)<\zeta^*,\zeta> , \tag{33}$$

since $<\zeta^*,(J-\mu_0 I)D_\sigma\zeta(0)> = <(J^*-\bar{\mu}_0 I)\zeta^*,D_\sigma\zeta(0)> = 0$, by (26).

The additional assumption which one has to make is that $D_\sigma\mu(0) \neq 0$. Usually one even assumes that Re $D_\sigma\mu(0) \neq 0$. This so-called *transversality condition* plays an important role in the stability analysis of the bifurcating subharmonic solutions, a topic which we will not treat here. One can see from (33) that the transversality condition, together with $\langle\zeta^*,\zeta\rangle = (\zeta_0^*,\zeta_0) \neq 0$ implies (H)(i).

6.5.9. We now return to our description of ker L, using (17). It follows from our hypothesis (H) that

$$\operatorname{span}\Big(\bigcup_{0\leq \ell<m} \ker_C\big(C - e^{\frac{2\pi i}{m}\ell} I\big)\Big)$$

$$= \operatorname{span}\big(\ker_C(C - e^{\frac{2\pi i}{m}k} I) \cup \ker_C(C - e^{-\frac{2\pi i}{m}k} I)\big)$$

$$= \operatorname{span}\{\zeta_0,\bar{\zeta}_0\} .$$

Define $\chi \in X_C$ and $\chi^* \in X_C$ by

$$\chi(t) = \Phi(t)\zeta_0 = e^{i\frac{k}{m}t}\zeta(t) \quad , \quad \chi^*(t) = (\Phi^T(t))^{-1}\zeta_0^* = e^{i\frac{k}{m}t}\zeta^*(t) . \tag{34}$$

Then we see from (17) that

$$\ker L = \{\operatorname{Re}(z\chi) \mid z\in\mathbb{C}\} \ , \quad \ker L^* = \{\operatorname{Re}(z\chi^*) \mid z\in\mathbb{C}\} . \tag{35}$$

Let $k/m = k_1/m_1$, where k_1 and m_1 have no common divisors. (Take $m_1 = 1$ if $k = 0$). Since $\zeta(t)$ is 2π-periodic (i.e. $\Gamma\zeta = \zeta$), it follows from (34) and (35) that

$$\Gamma^{m_1}u = u \qquad , \qquad \forall u \in \ker L , \tag{36}$$

that is, the elements of ker L are all $2\pi m_1$-periodic. Then we know from the theory of chapter 3 that all sufficiently small solutions of (6) will also satisfy $\Gamma^{m_1}x = x$, i.e. all such solutions are $2\pi m_1$-periodic. So we can replace k and m in the foregoing theory by k_1 and m_1. This implies that we may assume that k and m have no common divisors, and that $0 \leq k < \frac{m}{2}$. We will consider now several cases, depending on the value of m.

6.5.10. <u>The cases m = 1 and m = 2</u>. If m = 1, then k = 0 and the hypothesis (H) says that C has an eigenvalue equal to 1, and that the corresponding eigenspace is one-dimensional. We may take ζ_0 and ζ_0^* to be real vectors, and it follows that dim ker L = 1. We have a bifurcation problem from a simple eigenvalue. Because of the condition (28) it is possible to apply the Crandall-Rabinowitz theorem discussed in section 3, to obtain one single branch of 2π-periodic solutions of (1) bifurcating from the zero solution. From now on we will suppose $k \neq 0$.

If $m = 2$, then $k = 1$, and C has an eigenvalue equal to -1, and the corresponding eigenspace is again one-dimensional. The elements of ker L satisfy $u(t+2\pi) = -u(t)$. As for the case $m = 1$ one has here also a problem of bifurcation from a simple eigenvalue, on which one can apply theorem 3.3. One finds a single branch of 4π-periodic solutions bifurcating from the trivial solutions. This branch has the symmetry given by the second part of theorem 3.3. We leave the details to the reader.

6.5.11. The case $m \geqslant 3$. When $m \geqslant 3$ and $0 < k < \frac{m}{2}$, then $\exp(\frac{2\pi k}{m}i)$ is complex, and it follows from (35) that dim ker L = 2. We can define a basis $\{u_1,u_2\}$ for ker L by

$$u_1 = \mathrm{Re}\ \chi = \frac{1}{2}(\chi+\bar{\chi}) \quad , \quad u_2 = \mathrm{Im}\ \chi = \frac{1}{2i}(\chi-\bar{\chi}) \ . \tag{38}$$

It also follows from (34) that

$$\Gamma\chi = e^{\frac{2\pi k}{m}i}\chi \quad , \quad \Gamma\chi^* = e^{\frac{2\pi k}{m}i}\chi^* \ . \tag{39}$$

Using (38) we see that

$$\begin{aligned} \Gamma u_1 &= \cos\frac{2\pi k}{m}\, u_1 - \sin\frac{2\pi k}{m}\, u_2 \ , \\ \Gamma u_2 &= \sin\frac{2\pi k}{m}\, u_1 + \cos\frac{2\pi k}{m}\, u_2 \ . \end{aligned} \tag{40}$$

We conclude that the representation Γ of D_m is irreducible on ker L; moreover, (40) shows that Γ is orthogonal with respect to the basis $\{u_1,u_2\}$ of ker L. More generally, we have

$$\langle \Gamma u, \Gamma v\rangle = \langle u,v\rangle \quad , \quad \forall u,v \in Z_c \ . \tag{41}$$

The range of L is given by

$$R(L) = \{w \in Z \mid \langle\chi^*, w\rangle = 0\} . \tag{42}$$

Using (5),(9),(23),(30) and (34) it is easily seen that

$$\langle\chi^*, D_\sigma D_x M(0,0)\chi\rangle = \langle\zeta^*, J_\sigma(0)\zeta\rangle = 2 , \tag{43}$$

while

$$\langle\chi^*, D_\sigma D_x M(0,0)\bar{\chi}\rangle = \frac{1}{m}\sum_{j=0}^{m-1} \langle\Gamma^j\chi^*, \Gamma^j D_\sigma D_x M(0,0)\bar{\chi}\}$$

$$= \left(\frac{1}{m}\sum_{j=0}^{m-1} e^{-\frac{4\pi k}{m}j}\right)\langle\chi^*, D_\sigma D_x M(0,0)\bar{\chi}\rangle = 0 . \tag{44}$$

It follows in particular that $\langle\chi^*, D_\sigma D_x M(0,0)u_1\rangle = 1$, i.e. $D_\sigma D_x M(0,0)u_1 \notin R(L)$.

6.5.12. <u>Projection operators</u>. We define projections P and Q, respectively on ker L and on a complement of R(L), as follows

$$Pw = \mathrm{Re}(\langle\chi^*, D_\sigma D_x M(0,0)w\rangle\chi) \quad , \quad \forall w \in X \tag{45}$$

and

$$Qw = \mathrm{Re}(\langle\chi^*, w\rangle D_\sigma D_x M(0,0)\chi) \quad , \quad \forall w \in Z . \tag{46}$$

Using (43) and (44) it is easily verified that P and Q are indeed projections on the appropriate subspaces, while also

$$P\Gamma = \Gamma P \quad , \quad Q\Gamma = \Gamma Q . \tag{47}$$

If we let $w_i = D_\sigma D_x M(0,0)u_i$ $(i = 1,2)$, then also

$$Bu_i = QD_\sigma D_x M(0,0)u_i = Qw_i = w_i \quad (i = 1,2) . \tag{48}$$

Using the foregoing definition in the general formula (2.17) for the bifurcation function F, we obtain for our problem

$$F(u,\sigma) = \mathrm{Re}(\langle\chi^*, M(u+v^*(u,\sigma),\sigma)\rangle\chi) \quad , \quad \forall(u,\sigma) \in \ker L \times \mathbb{R} , \tag{49}$$

where $v^*(u,\sigma)$ is the unique solution of the appropriate auxiliary equation. From the form (49) for F, in which also $u = \mathrm{Re}(z\chi)$ for some $z \in \mathbb{C}$ (see (35)) it seems reasonable to use complex coordinates when discussing the bifurcation equation, which is an equation in a two-dimensional space. We introduce now such coordinates.

6.5.13. <u>Complex coordinates</u>. Consider the following mapping from $\ker L = U$ into $\mathbb{C}$, the complex plane

$$\psi : U \to \mathbb{C} \quad , \quad u = xu_1 - yu_2 \quad \leftrightarrow \quad \psi(u) = x+iy = z \; ; \tag{50}$$

ψ is a bijection with inverse

$$\psi^{-1}(z) = \mathrm{Re}(z\chi) = xu_1 - yu_2 \quad , \quad \forall z = x+iy \in \mathbb{C} \; . \tag{51}$$

So $\psi(\mathrm{Re}(z\chi)) = z$, $\forall z \in \mathbb{C}$. The mapping ψ is a linear isomorphism if we consider both U *and* $\mathbb{C}$ as two-dimensional *real* vectorspaces.

To each mapping $F : U \to U$ we can associate a mapping $G : \mathbb{C} \to \mathbb{C}$ defined by $G = \psi \circ F \circ \psi^{-1}$; conversely, to each $G ; \mathbb{C} \to \mathbb{C}$ there corresponds a mapping $F : U \to U$ given by $F = \psi^{-1} \circ G \circ \psi$. When doing calculus on the mapping G (for example, finding the Taylor expansion of G) one has to keep in mind that $\mathbb{C}$ is considered to be a real vectorspace. For example, one has

$$DG(z)h = \psi(DF(\psi^{-1}(z))\psi^{-1}(h)) \qquad , \qquad \forall h \in \mathbb{C} \; .$$

More practically, one first defines $\tilde{F} : U \to \mathbb{C}$ such that $F(u) = \mathrm{Re}(\tilde{F}(u)\chi)$, $\forall u \in U$; this $\tilde{F}$ is given by $\tilde{F} = \psi \circ F$. Then $G(z) = \tilde{F}(\mathrm{Re}(z\chi))$, $\forall z \in \mathbb{C}$, and if F (and consequently $\tilde{F}$ and G) are sufficiently smooth, then we have, for each $h = h_1 + ih_2 \in \mathbb{C}$

$$\begin{aligned} DG(z)h &= D\tilde{F}(\mathrm{Re}(z\chi)).\mathrm{Re}(h\chi) = D\tilde{F}(\mathrm{Re}(z\chi)).(h_1u_1 - h_2u_2) \\ &= \tfrac{1}{2}[\,D\tilde{F}(\mathrm{Re}(z\chi))u_1 + iD\tilde{F}(\mathrm{Re}(z\chi))u_2\,](h_1+ih_2) \\ &\quad + \tfrac{1}{2}[D\tilde{F}(\mathrm{Re}(z\chi))u_1 - iD\tilde{F}(\mathrm{Re}(z\chi))u_2\,](h_1-ih_2) \\ &= D_zG(z)h + D_{\bar{z}}G(z)\bar{h} \; , \end{aligned} \tag{52}$$

where the functions $D_zG : \mathbb{C} \to \mathbb{C}$ and $D_{\bar{z}}G : \mathbb{C} \to \mathbb{C}$ are defined by

$$D_zG(z) = \frac{1}{2}[D\widetilde{F}(\mathrm{Re}(z\chi))u_1 + iD\widetilde{F}(\mathrm{Re}(z\chi))u_2] \tag{53}$$

and

$$D_{\bar{z}}G(z) = \frac{1}{2}[D\widetilde{F}(\mathrm{Re}(z\chi))u_1 - iD\widetilde{F}(\mathrm{Re}(z\chi))u_2] \ . \tag{54}$$

These functions are associated to the mappings $D_zF : U \to U$ and $D_{\bar{z}}F : U \to U$ given by

$$D_zF(u) = \mathrm{Re}(D_z\widetilde{F}(u)\chi) \quad , \quad D_{\bar{z}}F(u) = \mathrm{Re}(D_{\bar{z}}\widetilde{F}(u)\chi) \ , \tag{55}$$

where

$$D_z\widetilde{F}(u) = \frac{1}{2}[D\widetilde{F}(u)u_1 + iD\widetilde{F}(u)u_2] \tag{56}$$

and

$$D_z\widetilde{F}(u) = \frac{1}{2}[D\widetilde{F}(u)u_1 - iD\widetilde{F}(u)u_2] \ . \tag{57}$$

The form (52) of $DG(z)$, in which $D_zG(z)$ and $D_{\bar{z}}G(z)$ are again functions from $\mathbb{C}$ into itself allow us to consider higher derivatives of G by differentiating the functions D_zG and $D_{\bar{z}}G$ in precisely the same way we did with G. This gives us for example the following Taylor formula if F (and G) is of class C^k :

$$G(z+h) = G(z) + \sum_{j=1}^{k} \frac{1}{j!}(hD_z + \bar{h}D_{\bar{z}})^j G(z) + R(z,h) \ , \tag{58}$$

where $R(z,h) = o(|h|^k)$ as $|h| \to 0$.

6.5.14. The bifurcation function. From (49) we see that in the complex coordinates just introduced our bifurcation function will take the form

$$G(z,\sigma) = \langle \chi^*, M(\mathrm{Re}(z\chi) + v^*(\mathrm{Re}(z\chi),\sigma),\sigma)\rangle \quad , \quad \forall (z,\sigma) \in \mathbb{C} \times \mathbb{R} \ . \tag{59}$$

We know from the general theory that $G(0,\sigma) = 0$ for all σ, while $DG(0,0) = 0$ and $D_\sigma DG(0,0) = I_{\mathbb{C}}$. (By D we denote differentiation in the variable z, as described in subsection 13). Moreover, the function G is equivariant, i.e.

we have $G(\Gamma z,\sigma) = \Gamma G(z,\sigma)$, where $\Gamma : \mathbb{C} \to \mathbb{C}$ is the action on $\mathbb{C}$ corresponding to (40). It is easily seen from (39) that this action is given by

$$\Gamma z = e^{\frac{2\pi k}{m}i} z \qquad , \qquad \forall z \in \mathbb{C} . \tag{60}$$

Since k and m have no common divisors, we have $\{\frac{k}{m}j \pmod 1 \mid j \in \mathbb{Z}\} = \{\frac{kj+m\ell}{m} \pmod 1 \mid j,\ell \in \mathbb{Z}\} = \{\frac{j}{m} \mid j = 0,1,\ldots,m-1\}$. Consequently we have

$$G(e^{\frac{2\pi i}{m}} z,\sigma) = e^{\frac{2\pi i}{m}} G(z,\sigma) \qquad , \qquad \forall (z,\sigma) . \tag{61}$$

This equivariance condition strongly restricts the possibilities for the form G can take, as we will show now.

Suppose that M in (1) is of class C^p; then also F and G will be of class C^p. Expanding $G(z,\sigma)$ around the point $z = 0$ we obtain from (58)

$$G(z,\sigma) = \sum_{1 \leq j+\ell \leq p} \alpha_{j\ell}(\sigma) z^j \bar{z}^\ell + R_p(z,\sigma) , \tag{62}$$

where $R_p(z,\sigma) = o(|z|^p)$ as $|z| \to 0$. The functions $\alpha_{j\ell} : \mathbb{R} \to \mathbb{C}$ are, up to a constant, given by $D_z^j D_{\bar{z}}^\ell G(0,\sigma)$. It follows from the equivariance condition (61) that $\alpha_{j\ell}(\sigma)$ can be different from zero only if

$$j = \ell + 1 \pmod m . \tag{63}$$

We now consider separately the cases $m = 3$, $m = 4$ and $m \geq 5$.

6.5.15. The case $m = 3$. When $m = 3$ (and consequently $k = 1$) then the rule (63) gives us the following form for the bifurcation function

$$G(z,\sigma) = \alpha(\sigma) z + \beta(\sigma) \bar{z}^2 + R_3(z,\sigma) , \tag{64}$$

where $R_3 = O(|z|^3)$ if we assume f in (1) to be of class C^3. Moreover $\alpha(\sigma) = \sigma + O(|\sigma|^2)$, and a straightforward calculation shows that

$$\begin{aligned} \beta(0) &= \frac{1}{8} \langle \chi^*, D_x^2 M(0,0)(\bar{\chi},\bar{\chi}) \rangle \\ &= \frac{1}{8}\,\frac{1}{6\pi} \int_0^{6\pi} (\chi^*(t), D_x^2 f(t,0,0)(\bar{\chi}(t),\bar{\chi}(t)))dt . \end{aligned} \tag{65}$$

We will assume that $\beta(0) \neq 0$.

As in subsection 2.19 one proves that for each solution (z,σ) of $G(z,\sigma) = 0$ in a sufficiently small neighbourhood of the origin we have $|\sigma| \leq C|z|$ for some constant C. This allows us to put

$$z = \rho e^{i\phi} \quad , \qquad \sigma = \rho\mu \tag{66}$$

in (64); since (ρ,ϕ,μ) and $(-\rho,\phi+\pi,-\mu)$ correspond to the same point (z,σ), we will assume $\mu \geq 0$. Bringing (66) into (64), we can divide by ρ^2 in the resulting equation, and obtain :

$$H(\rho,\phi,\mu) \equiv \frac{1}{\rho^2} G(\rho e^{i\phi},\rho\mu) = \mu e^{i\phi} + \beta(0)z^{-2i\phi} + \tilde{R}(\rho,\phi,\mu) = 0 \ , \tag{67}$$

where $\tilde{R}(\rho,\phi,\mu)$ is of class C^1 and such that $\tilde{R}(0,\phi,\mu) = 0$. In order to solve (67) for small ρ, we have to find a solution of the reduced bifurcation equation $H(0,\phi,\mu) = 0$, which takes the form

$$\mu e^{i\phi} + \beta(0)e^{-2i\phi} = 0 \ . \tag{68}$$

When $\beta(0) \neq 0$ then this equation has the solutions $(\phi_0 + \frac{2\pi}{3}j,\mu_0)$, with $\mu_0 = |\beta(0)|$, $j \in \mathbb{Z}$ and an appropriate ϕ_0. Moreover : $D_{(\phi,\mu)}H(0,\phi_0 + \frac{2\pi}{3}j,\mu_0)$ is an isomorphism from $\mathbb{R}^2$ onto $\mathbb{C}$. We can apply the implicit function theorem on (67) to obtain the following branches of solutions: $(\phi(\rho) + \frac{2\pi}{3}j,\mu(\rho))$, where $\phi(\rho)$ and $\mu(\rho)$ are of class C^1, $\phi(0) = \phi_0$, $\mu(0) = \mu_0$ and $j \in \mathbb{Z}$. We leave it to the reader to construct from these solutions the corresponding branches of 6π-periodic solutions of (1). Remark that the solutions corresponding to different values of j (only $j = 0,1$ and 2 give different solutions) can be obtained one from the other by a phase shift over 2π or 4π, i.e. by application of the symmetry operator Γ.

We conclude that under the hypothesis (H) with $m = 3$ (and $k = 1$) there will generically (i.e. when $\beta(0) \neq 0$) be bifurcation of 6π-periodic solutions.

6.5.16. <u>The case m = 4</u>. When $m = 4$ (and $k = 1$) we find from (63) that the bifurcation function takes the form

$$G(z,\sigma) = \alpha(\sigma)z + \gamma(\sigma)z^2\bar{z} + \delta(\sigma)\bar{z}^3 + R_4(z,\sigma) \ , \tag{69}$$

where $R_4 = O(|z|^4)$, assuming that f in (1) is of class C^4. We still have $\alpha(\sigma) = \sigma + O(|\sigma|^2)$, and one can express $\gamma(0)$ and $\delta(0)$ in the form of integrals, involving $\chi(t)$, $\chi^*(t)$, derivatives of $f(t,x,\sigma)$, and the first approximation of v^*, which can be obtained by solving an inhomogeneous variant of the linear equation (12). We will not give the details here, but only discuss what kind of solutions we can expect. All small solutions of $G(z,\sigma) = 0$ will satisfy $|\sigma| \leq C|z|^2$. Therefore, we put

$$z = \rho e^{i\phi} \quad , \qquad \sigma = \mu\rho^2 \tag{70}$$

in (69). Dividing by ρ^3 we obtain the equation

$$H(\rho,\phi,\mu) \equiv \mu e^{i\phi} + \gamma(0)e^{i\phi} + \delta(0)e^{-3i\phi} + \tilde{R}(\rho,\phi,\mu) \ , \tag{71}$$

where $\tilde{R}$ is of class C^1 and $\tilde{R}(0,\phi,\mu) = 0$. The reduced bifurcation equation has the form

$$H(0,\phi,\mu) = (\mu+\gamma(0))e^{i\phi} + \delta(0)e^{-3i\phi} = 0 \ . \tag{72}$$

This equation will have solutions $(\mu,\phi) \in \mathbb{R}^2$ if and only if $|\delta(0)| \geq |\mathrm{Im}\ \gamma(0)|$. If $|\delta(0)| > |\mathrm{Im}\ \gamma(0)|$ then there are two sets of solutions : $(\phi_0^{(1)} + \frac{\pi}{2}j, \mu_0^{(1)})$ and $(\phi_0^{(2)} + \frac{\pi}{2}j, \mu_0^{(2)})$; at these solutions the Jacobian for (72) is different from zero, and we can apply the implicit function theorem to obtain two times four branches of 8π-periodic solutions. The four branches correspond to different values of j, and correspond to phase shifts over multiples of 2π. Moreover, the bifurcation function $G(z,\sigma)$ is odd in z (by lemma 2.16), so that $\mu(\rho)$ is even in ρ; this gives some supplementary symmetry in the bifurcation diagram.

We conclude that for $m = 4$ there will only be bifurcation of 8π-periodic solutions if the condition $|\delta(0)| \geq |\mathrm{Im}\ \gamma(0)|$ is satisfied.

6.5.17. <u>The case $m \geq 5$</u>. When $m \geq 5$ then the bifurcation function has the form

$$G(z,\sigma) = \alpha(\sigma)z + \gamma(\sigma)z^2\bar{z} + R_4(z,\sigma) \ , \tag{73}$$

with $R_4 = O(|z|^4)$. Again we have $|\sigma| \leq C|z|^2$ for small solutions of the bifurcation equation, so that we can use the rescaling (70). This leads to the following reduced bifurcation equation

$$H(0,\phi,\mu) = \mu e^{i\phi} + \gamma(0)e^{i\phi} = 0 . \tag{74}$$

Generically $\gamma(0)$ will not be real, in which case (74) has no solution $(\mu,\phi) \in \mathbb{R}^2$. For $m \geq 5$ there can only be bifurcation of subharmonic solutions if $\gamma(0) \in \mathbb{R}$, and even then some further conditions have to be satisfied. One speaks about *weak resonance* when there is indeed bifurcation.

6.5.18. Comments. It is an interesting exercise to find all possible irreducible representations of the group D_m; such representations are 1-dimensional if $m \leq 2$, and 1 or 2-dimensional if $m \geq 3$. For fixed m one finds precisely those representations which we found in the foregoing theory by restricting Γ to ker L, and by giving to k all possible values. The hypothesis (H) corresponds precisely to the irreducibility and the transversality condition in the general theory of section 2. Moreover, when the representation is 2-dimensional (i.e. when $m \geq 3$ and k appropriate), then the equivariance of the bifurcation function gives us the selection rule (63), which is sufficient to discuss the qualitative bifurcation picture. One can find more details on bifurcation of subharmonic solutions in the books of Iooss [254] and Iooss and Joseph [255].

6.6. A BIFURCATION PROBLEM WITH O(3)-SYMMETRY

6.6.1. The problem. Let $\Omega = \{x \in \mathbb{R}^3 \mid x_1^2+x_2^2+x_3^2 < 1\}$ be the unit ball in $\mathbb{R}^3$, and consider the following boundary value problem :

$$\begin{aligned} &\Delta v + \mu f(x,v,\lambda) = 0 , && x \in \Omega \\ &v(x) = 0 , && x \in \partial\Omega . \end{aligned} \tag{1}$$

We assume that $f : \bar{\Omega} \times \mathbb{R} \times \Lambda \to \mathbb{R}$ is of class C^3,

$$f(x,0,\lambda) = 0 , \qquad \forall (x,\lambda) \tag{2}$$

and

$$f(x,v,\lambda) = f(y,v,\lambda) \quad \text{if} \quad \|x\| = \|y\| , \tag{3}$$

i.e. f is rotationally symmetric with respect to the x-variable. Λ is any Banach space, and μ is considered as a nonlinear eigenvalue. We want to determine, for each λ near 0, corresponding values of μ for which (1) has a nontrivial solution.

To bring the problem in the form (2.1) we let $X = C_0^{2,\alpha}(\bar{\Omega})$ and $Z = C^{0,\alpha}(\bar{\Omega})$ for some $\alpha \in]0,1[$, and we define $\tilde{M} : X \times \Lambda \times \mathbb{R} \to Z$ by

$$\tilde{M}(v,\lambda,\mu)(x) = \Delta v(x) + \mu f(x,v(x),\lambda) \quad , \quad \forall x \in \bar{\Omega} , \tag{4}$$

where $v \in X$. Then our problem has the form

$$\tilde{M}(v,\lambda,\mu) = 0 . \tag{5}$$

We can define a representation $\Gamma : O(3) \to L(Z)$ of $O(3)$ over Z by

$$(\Gamma(R)v)(x) = v(R^{-1}x) \quad , \quad \forall R \in O(3) . \tag{6}$$

Then it is easily seen that the equation (5) is equivariant with respect to $(O(3),\Gamma)$.

6.6.2. <u>The linearized problem</u>. Let $\tilde{L}(\mu) = D_v\tilde{M}(0,0,\mu)$. In order to determine $\ker \tilde{L}(\mu)$ we have to solve the linear boundary value problem

$$\begin{aligned} &\Delta u + \mu D_v f(x,0,0)u = 0 \quad , \quad x \in \Omega \\ &u(x) = 0 \quad , \quad x \in \partial\Omega . \end{aligned} \tag{7}$$

One knows from the general theory of such equations (see e.g. [71] and section 2.3) that (7) has nontrivial solutions if and only if μ belongs to a countable set of eigenvalues for the problem (7). If μ_0 is such an eigenvalue, then $\ker \tilde{L}(\mu_0)$ will be finite-dimensional. Moreover, since the problem (7) is equivariant, the restriction of Γ to $\ker \tilde{L}(\mu_0)$ will give a finite-dimensional representation of $O(3)$. Generically, one may assume that this representation is irreducible. Then it follows from the theory of section 2.6 that dim ker $L(\mu_0) = 2\ell+1$ for some $\ell \in \mathbb{N}$, and that the elements of ker $L(\mu_0)$ transform under the symmetry operators $\Gamma(R)$ in the same way as the spherical harmonics of order ℓ. Using the completeness of the spherical harmonics on $L_2(S^2)$ one can show that the elements of ker $\tilde{L}(\mu_0)$, when written in spherical coordinates, must have the form $u(r,\theta) = \phi_0(r)Y_\ell(\theta)$, $(r \geq 0$, $\theta \in S^2)$, with a fixed radial symmetric function $\phi_0(r)$ and $Y_\ell \in U_\ell$, the space of spherical harmonics of order ℓ (see (2.6.42)).

Because of the foregoing one can label the eigenvalues as $\mu_{\ell j}$ $(\ell \in \mathbb{N}, j \in \mathbb{N})$ with dim ker $\tilde{L}(\mu_{\ell j}) = 2\ell+1$, and ker $\tilde{L}(\mu_{\ell j}) = \{\phi_{\ell j}(r)Y_\ell(\theta) \mid Y_\ell \in U_\ell\}$, for some radial function $\phi_{\ell j}(r)$ depending on ℓ and j.

6.6.3. Let us fix some eigenvalue $\mu_{\ell j}$, and study (1) for μ near $\mu_{\ell j}$. To do so we put $\mu = \mu_{\ell j}+\sigma$ and $M(v,\lambda,\sigma) = \tilde{M}(v,\lambda,\mu_{\ell j}+\sigma)$. We have to solve the equation

$$M(v,\lambda,\sigma) = 0 \tag{8}$$

for (v,λ,σ) near $(0,0,0)$. We have $L = D_vM(0,0,0) = \tilde{L}(\mu_{\ell j})$; we know from the general theory for equation (7) that L is a Fredholm operator with zero index, and by the preceding considerations we may assume that the restriction of Γ to ker L is irreducible. Therefore the hypotheses (H1) and (H2) of section 2 are satisfied. As for (H3), this will be satisfied if

$$\int_{\bar{\Omega}} D_vf(x,0,0)u^2(x)dx \neq 0 \tag{9}$$

for some $u \in U = \ker L$. However, it follows from (7), with $\mu = \mu_{\ell j}$, that for $u \in \ker \tilde{L}(\mu_{\ell j})$ we have

$$\int_{\bar{\Omega}} D_vf(x,0,0)u^2(x)dx = (\mu_{\ell j})^{-1} \int_{\bar{\Omega}} (\nabla u(x))^2dx \ ,$$

since $\mu = 0$ is not an eigenvalue. It follows that (9) is satisfied.

The theory of section 2 shows that the problem (8) reduces to the bifurcation equation

$$F(u,\lambda,\sigma) = 0 \tag{10}$$

where $F : U\times\Lambda\times\mathbb{R}\to U$ is defined and of class C^2 in a neighbourhood of the origin, and has the following properties :

$$F(0,\lambda,\sigma) = 0 \qquad , \qquad \forall(\lambda,\sigma)\ , \tag{11}$$

$$D_uF(0,0,0) = 0 \qquad , \qquad D_\sigma D_uF(0,0,0) = I_U\ , \tag{12}$$

and

$$F(\Gamma(R)u,\lambda,\sigma) = \Gamma(R)F(u,\lambda,\sigma) \qquad , \qquad \forall R\in O(3)\ . \tag{13}$$

One could write down a precise expression for F; however, the properties (11)-(13) will be sufficient for the qualitative discussion which follows.

6.6.4. Bifurcation of axisymmetric solutions. Let us first look for axisymmetric solutions of (1); these are solutions (v_1,λ,μ) such that, for a certain vector $a\in\mathbb{R}^3\setminus\{0\}$, we have

$$\Gamma(R)v_1 = v_1 \qquad , \qquad \forall R\in G_a \tag{14}$$

where $G_a = \{R\in O(3)\mid Ra=a\}$ is the isotropy group of the vector a. One can find some $R_1\in O(3)$ such that $R_1a=\nu e_3$; then $v = R_1v_1$ will also be a solution, and satisfy

$$\Gamma(\phi_3(\alpha))v = v \qquad , \qquad \forall\alpha\in\mathbb{R}\ . \tag{15}$$

Conversely, let (v,λ,μ) be a solution of (1) satisfying (15).Then, for any $R_1\in O(3)$, $v_1 = \Gamma(R_1)v$ will also be an axisymmetric solution, with axis defined by $a = R_1e_1$. This shows that it is sufficient to find solutions (v,λ,μ) satisfying (15). By the Liapunov-Schmidt reduction we have to find solutions (u,λ,μ) of the bifurcation equation (10) which satisfy $\Gamma(\phi_3(\alpha))u = u$ for all $\alpha\in\mathbb{R}$.

It follows from corollary 2.6.29 and the irreducibility of Γ over U that $\dim\{u\in U\mid\Gamma(\phi_3(\alpha))u=u,\ \forall\alpha\in\mathbb{R}\} = 1$. If this subspace is spanned by $u_3\in U$, then the condition (2.39) of theorem 2.12 is satisfied for $\theta_0 = u_3$. That means, we have $F(\rho u_3,\lambda,\sigma) = \rho h(\rho,\lambda,\sigma)u_3$, with $h(0,0,0) = 0$ and $D_\sigma h(0,0,0) = 1$. The equation $h(\rho,\lambda,\sigma) = 0$ has a solution $\sigma = \sigma^*(\rho,\lambda)$, by the implicit function theorem. This gives the following result.

6.6.5. Theorem. For each eigenvalue $\mu_{\ell j}$ of (7) there exists a map $\sigma^* : \mathbb{R}\times\Lambda \to \mathbb{R}$, defined and of class C^1 in a neighbourhood of the origin, such that for each sufficiently small (ρ,λ), with $\rho > 0$, and for $\mu = \mu_{\ell j} + \sigma^*(\rho,\lambda)$, problem (1) has a family of nontrivial axisymmetric solutions

$$\{\Gamma(R)\tilde{v}(\rho,\lambda) \mid R\in O(3)\} ;$$

one has $\tilde{v}(\rho,\lambda)(\phi_3(\alpha)x) = \tilde{v}(\rho,\lambda)(x)$, $\forall\alpha\in\mathbb{R}$, and $\tilde{v}(\rho,\lambda) = O(\rho)$.

In case $\ell = 0$ or $\ell = 1$ these are the only nontrivial solutions of (1) near $(0,0,\mu_{\ell j})$. In case $\ell = 0$, all these solutions are rotationally symmetric.

P r o o f. It remains to prove the last part of the theorem. For $\ell = 0$ or $\ell = 1$ all $u \in U$ are axisymmetric, as can easily be seen from the definition of spherical harmonics. In case $\ell = 0$ we have dim $U = 1$ and all $u \in U$ are rotationally symmetric. Then the result follows from the theory of chapter 3. □

If we want to find non-axisymmetric solutions, we have to consider eigenvalues $\mu_{\ell j}$ with $\ell \geqslant 2$. For such cases we have dim U = dim $U_\ell \geqslant 5$. In the remainder of this section we will discuss the case $\ell = 2$.

6.6.6. The case $\ell = 2$. When $\ell = 2$, then dim $U = 5$, and (10) forms a five-dimensional problem. However, we can use the orbit structure of U under the action of $O(3)$ to reduce this to a two-dimensional problem. Our presentation partly follows Golubitsky and Schaeffer [251].

Let us first look at the structure of the second order spherical harmonics. Each quadratic polynomial $H_2 : \mathbb{R}^3 \to \mathbb{R}$ can be written in the form $H_2(x) = (x,Ax)$, where $(x,y) = \sum_{i=1}^{3} x_i y_i$ and $A \in L(\mathbb{R}^3)$ is symmetric : $A^T = A$. The condition $\Delta H_2(x) = 0$, appearing in the definition of a spherical harmonic (see section 2.6), then reduces to trace $A = 0$. When we define

$$V = \{A\in L(\mathbb{R}^3) \mid A^T = A \text{ and } \operatorname{tr} A = 0\} , \tag{16}$$

the foregoing implies that there is an isomorphism between U and V. On V the group $O(3)$ is represented by

$$\Gamma(R)A = RAR^T \quad , \quad \forall R \in O(3) \quad , \quad \forall A \in V . \tag{17}$$

Using the isomorphism between U and V, we can consider F as a mapping from $V \times \Lambda \times \mathbb{R}$ into V, with properties corresponding to (11)-(13).

Let $\{e_1, e_2, e_3\}$ be the canonical basis of $\mathbb{R}^3$, and let D be the subspace of all diagonal $A \in V$, i.e.

$$D = \{A \in V \mid \langle e_i, Ae_j \rangle = \langle e_i, Ae_i \rangle \delta_{ij},\ i,j = 1,2,3\} . \tag{18}$$

Since every symmetric $A \in L(\mathbb{R}^3)$ can be diagonalized by an appropriate orthogonal similarity transformation, it follows that

$$V = \{\Gamma(R)A \mid A \in D,\ R \in O(3)\} . \tag{19}$$

This implies that each orbit of solutions of the bifurcation equation has a nonempty intersection with D; consequently, it is sufficient to find solutions of (10) belonging to the subspace D.

Let Σ_0 be the 8-element subgroup of O(3) containing those $R \in O(3)$ of the form $Re_i = \varepsilon_i e_i$ $(i = 1,2,3)$, with $\varepsilon_i = \pm 1$. Then we have

$$D = \{A \in V \mid \Gamma(R)A = A,\ \forall R \in \Sigma_0\} \text{ and } \Sigma_0 = \{R \in O(3) \mid \Gamma(R)A = A,\ \forall A \in D\} . \tag{20}$$

It follows that for $A \in D$ and $R \in \Sigma_0$ we have

$$\Gamma(R)F(A,\lambda,\sigma) = F(\Gamma(R)A,\lambda,\sigma) = F(A,\lambda,\sigma) ;$$

we conclude that F maps $D \times \Lambda \times \mathbb{R}$ into D.

In the two-dimensional space D we will use coordinates $(x,y) \in \mathbb{R}^2$ defined by the isomorphism $\psi : \mathbb{R}^2 \to D$ given by

$$\psi(x,y) = \psi(z) = \begin{pmatrix} \operatorname{Re} z & 0 & 0 \\ 0 & \operatorname{Re}(e^{i\frac{2\pi}{3}} z) & 0 \\ 0 & 0 & \operatorname{Re}(e^{-i\frac{2\pi}{3}} z) \end{pmatrix} , \tag{21}$$

where, for notational convenience, we identify $z = (x,y) \in \mathbb{R}^2$ with $z = x+iy \in \mathbb{C}$. Using this isomorphism (10) takes the form

$$H(z,\lambda,\sigma) = 0 \ , \tag{22}$$

where $H : \mathbb{R}^2 \times \Lambda \times \mathbb{R} \to \mathbb{R}^2$ is defined by $H(z,\lambda,\sigma) = \psi^{-1}F(\psi(z),\lambda,\sigma)$. The mapping H has the following properties, corresponding to (11)-(12) :

$$H(0,\lambda,\sigma) = 0 \ , \ D_zH(0,0,0) = 0 \ , \ D_\sigma D_zH(0,0,0) = I_{\mathbb{R}^2} \ . \tag{23}$$

As for the symmetry properties of H, let $\Sigma = \{R \in O(3) \mid \Gamma(R)(D) = D\}$ be the subgroup of symmetry operators leaving the subspace D invariant; Σ_0 is a normal subgroup of Σ. The quotient group Σ/Σ_0 has 6 elements, which can be identified with those $R \in O(3)$ which merely interchange the coordinate axes, i.e. which have the form $Re_i = e_{\nu(i)}$ $(i = 1,2,3)$ for some permutation ν of $\{1,2,3\}$. The action of such symmetry operator on some $A \in D$ consists in a corresponding permutation of the diagonal elements of A. Via the isomorphism ψ the corresponding action on $\mathbb{R}^2 \cong \mathbb{C}$ is generated by the mappings $\delta : \mathbb{C} \to \mathbb{C}$ and $\tau : \mathbb{C} \to \mathbb{C}$ given by

$$\delta z = e^{i\frac{2\pi}{3}} z \quad , \quad \tau z = \bar{z} \quad , \quad \forall z \in \mathbb{C} \ . \tag{24}$$

Let $\Delta_3 = \{I,\delta,\delta^2,\tau,\tau\circ\delta,\tau\circ\delta^2\}$; (for the notation Δ_3, compare (24) with (4.51)). Then the equivariance of H can be expressed by

$$H(\gamma z,\lambda,\sigma) = \gamma H(z,\lambda,\sigma) \quad , \quad \forall \gamma \in \Delta_3 \ . \tag{25}$$

6.6.7. <u>Lemma</u>. Let $H : \mathbb{R}^2 \times \Lambda \times \mathbb{R} \to \mathbb{R}^2$ be of class C^{k+4}, and suppose H satisfies (25). Then there exist C^k-functions $p : \mathbb{R}^2 \times \Lambda \times \mathbb{R} \to \mathbb{R}$ and $q : \mathbb{R}^2 \times \Lambda \times \mathbb{R} \to \mathbb{R}$ such that

$$\text{(i)} \quad H(z,\lambda,\sigma) = p(z,\lambda,\sigma)z + q(z,\lambda,\sigma)\bar{z}^2 \ ; \tag{26}$$

$$\text{(ii)} \quad p(\gamma z,\lambda,\sigma) = p(z,\lambda,\sigma) \quad , \quad q(\gamma z,\lambda,\sigma) = q(z,\lambda,\sigma) \quad , \quad \forall \gamma \in \Delta_3 \ . \tag{27}$$

P r o o f. Define $h : \mathbb{R}^2 \times \Lambda \times \mathbb{R} \to \mathbb{R}$ by

$$h(z,\lambda,\sigma) = -yH_1(z,\lambda,\sigma) + xH_2(z,\lambda,\sigma) \ , \tag{28}$$

where $H = (H_1,H_2)$. It follows from (25) that

$$h(\delta z,\lambda,\sigma) = h(z,\lambda,\sigma) \quad , \quad h(\tau z,\lambda,\sigma) = -h(z,\lambda,\sigma) \ . \tag{29}$$

From the second of these relations it follows that $h(z,\lambda,\sigma) = 0$ if $z = (x,0)$, i.e. if $z = \bar{z}$. So we have

$$h(z,\lambda,\sigma) = yh_1(z,\lambda,\sigma) = \frac{1}{2i}(z-\bar{z})h_1(z,\lambda,\sigma) \ .$$

Now

$$h(z,\lambda,\sigma) = h(\delta z,\lambda,\sigma) = \frac{1}{2i}(\delta z-\overline{\delta z})h_1(\delta z,\lambda,\sigma) \ ;$$

since, for $z = \bar{z}$, we have $\delta z-\overline{\delta z} = (\delta-\bar{\delta})z \neq 0$ (except when $z = 0$), it follows that $h_1(\delta z,\lambda,\sigma) = 0$ if $z = \bar{z}$; as before, we conclude that

$$h_1(\delta z,\lambda,\sigma) = \frac{1}{2i}(z-\bar{z})h_2(z,\lambda,\sigma) \ .$$

Another application of the same argument gives us finally

$$h(z,\lambda,\sigma) = \frac{i}{2}(z^3-\bar{z}^3)q(z,\lambda,\sigma) \ , \tag{30}$$

for some $q : \mathbb{R}^3 \times \Lambda \times \mathbb{R} \to \mathbb{R}$ satisfying the condition (27).

Let now $\tilde{H}(z,\lambda,\sigma) = H(z,\lambda,\sigma) - q(z,\lambda,\sigma)\bar{z}^2$; then it follows from (28) and (30) that $y\tilde{H}_1 = x\tilde{H}_2$, which implies that $\tilde{H}_1(z,\lambda,\sigma) = p(z,\lambda,\sigma)x$ and $\tilde{H}_2(z,\lambda,\sigma) = p(z,\lambda,\sigma)y$ for some $p : \mathbb{R}^2 \times \Lambda \times \mathbb{R} \to \mathbb{R}$, i.e. $\tilde{H}(z,\lambda,\sigma) = p(z,\lambda,\sigma)z$. This gives us (26), and also the first part of (27) follows immediately. □

6.6.8. Now we bring the form (26) in the equation (22), put $z = \rho e^{i\theta}$, multiply by $\bar{z}$, split into real and imaginary parts, and devide by ρ^2, respectively ρ^3; this gives us the following equations for nonzero solutions :

$$p(\rho e^{i\theta},\lambda,\sigma) + \rho q(\rho e^{i\theta},\lambda,\sigma) \cos 3\theta = 0 \ , \tag{31}$$

$$q(\rho e^{i\theta},\lambda,\sigma) \sin 3\theta = 0 \ . \tag{32}$$

Now it follows from (23) that $p(0,0,0) = 0$ and $D_\sigma p(0,0,0) = 1$. We can apply the implicit function theorem to solve (31) for σ; we obtain $\sigma = \sigma^*(\rho,\theta,\lambda)$, with $\sigma^*(0,\theta,0) = 0$ and $\sigma^*(\rho,\pm\theta + \frac{2\pi}{3}k,\lambda) = \sigma^*(\rho,\theta,\lambda)$, $k = 0,1,2$.

The equation (32) is satisfied for $\theta = \frac{2\pi}{3}k$, $k = 0,1,2$. This corresponds to the lines $y = 0$ and $y = \pm\sqrt{3}x$ in $\mathbb{R}^2$. It is easly seen that these solutions of (31)-(32) generate precisely the branch of axisymmetric solutions described in theorem 5.

6.6.9. Non-axisymmetric solutions. For non-axisymmetric solutions we have $\sin 3\theta \neq 0$; for such solutions (31)-(32) reduce to

$$p(z;\lambda,\sigma) = 0 \quad , \quad q(z;\lambda,\sigma) = 0 . \tag{33}$$

Again the first of these equations can be solved for $\tilde{\sigma} = \tilde{\sigma}(z,\lambda)$, with $\tilde{\sigma}(0,0) = 0$ and $\tilde{\sigma}(\gamma z,\lambda) = \tilde{\sigma}(z,\lambda)$, $\forall\gamma \in \Delta_3$, by (27). It is also clear that for solutions of (33) we have $\tilde{\sigma}(\rho\cos\theta,\rho\sin\theta,\lambda) = \sigma^*(\rho,\theta,\lambda)$.

Bringing the solution $\tilde{\sigma}(z,\lambda)$ of $p = 0$ into the second equation $q = 0$, we obtain a final (scalar) bifurcation equation

$$r(z,\lambda) = 0 , \tag{34}$$

with $r : \mathbb{R}^2 \times \Lambda \to \mathbb{R}$ given by $r(z,\lambda) = q(z,\lambda,\tilde{\sigma}(z,\lambda))$. The smoothness of r depends on the smoothness of the original problem, and, moreover, r is invariant under the symmetry operators

$$r(\gamma z,\lambda) = r(z,\lambda) \quad , \quad \forall\gamma \in \Delta_3 . \tag{35}$$

Using polar coordinates, it is easily seen that $r(z,\lambda)$ has the form

$$r(z,\lambda) = \alpha(\lambda) + \beta(\lambda)\rho^2 + O(\rho^3) , \tag{36}$$

for appropriate functionals $\alpha,\beta : \Lambda \to \mathbb{R}$.

If $\alpha(0) \neq 0$, then (34) has no solutions (z,λ) near the origin, and in this case there is only bifurcation of axisymmetric solutions, as given by theorem 5.

Now suppose that $\alpha(0) = 0$ and $\beta(0) \neq 0$; we will assume $\beta(0) > 0$; the case $\beta(0) < 0$ is similar. Consider the equation

$$\tilde{r}(z,\lambda,\alpha) \equiv \alpha + \beta(0)\rho^2 + [\, r(z,\lambda) - \alpha(\lambda) - \beta(0)\rho^2 \,] = 0 ; \tag{37}$$

it coincides with (34) for $\alpha = \alpha(\lambda)$. There is a neighbourhood of the origin in $\mathbb{R}^2 \times \Lambda \times \mathbb{R}$ and a constant $C > 0$ such that solutions of (37) belonging to that neighbourhood satisfy $\rho^2 \leqslant C|\alpha|$. Therefore we use in (37) the rescaling

$$\rho = \xi\eta \quad , \quad \alpha = \pm\xi^2 \quad , \quad \eta \geqslant 0 . \tag{38}$$

Dividing by ξ^2 in the resulting equation, we find

$$\pm 1 + \beta(0)\eta^2 + O(\xi) = 0 . \tag{39}$$

If $\alpha > 0$, then one has the plus sign in (39), and there are no solutions η when $\xi = 0$; consequently, there are also no solutions for ξ near 0. If $\alpha < 0$, then one has the minus sign, and one can solve for η; one finds $\eta = \eta^*(\theta,\lambda,\xi)$, with $\eta^*(\theta,0,0) = \beta(0)^{-1/2}$.

Returning to (34) we see that if $\alpha(0) = 0$ and $\beta(0) > 0$ we have the following conclusions :

(i) for all λ near 0 with $\alpha(\lambda) \geqslant 0$, (34) has no nontrivial solutions near $z = 0$; for such λ there are only axisymmetric solutions in a neighbourhood of the origin;

(ii) for all λ near 0 with $\alpha(\lambda) < 0$, (34) has a small "circle" of solutions near $z = 0$, given in polar coordinates by

$$\rho = \rho_\lambda(\theta) = |\alpha(\lambda)|^{1/2}\eta^*(\theta,\lambda,|\alpha(\lambda)|^{1/2}) ;$$

one has

$$\rho_\lambda(\pm\theta + \tfrac{2\pi}{3}k) = \rho_\lambda(\theta) \quad , \quad k = 0,1,2$$

and $\rho_\lambda(\theta) \to 0$ as $\alpha(\lambda) \to 0$; these solutions generate a family of non-axisymmetric solutions of the original problem (1).

We conclude that as we change λ in a neighbourhood of the origin, such that $\alpha(\lambda)$ goes from positive to negative, then there is a family of non-axisymmetric solutions of (1) bifurcating from the trivial solution.

7 Bifurcation problems with SO(2)-symmetry and Hopf bifurcation

7.1. INTRODUCTION

In the recent past much research has been devoted to the study of the Hopf bifurcation phenomenon, that is the bifurcation of periodic solutions from equilibrium solutions for autonomous differential equations. As a general reference we can refer to the book by Marsden and McCracken [156], which also contains an extensive bibliography. The original contribution by E. Hopf [98] considered one parameter families of autonomous ordinary differential equations, and studied the local bifurcation of periodic solutions from an equilibrium at the origin; such bifurcation takes place at parameter values for which there occurs a change in the stability properties of the equilibrium point.

Since this first contribution the scope of the theory has been much widened. Several techniques have been used to approach the problem (see e.g. Friedrichs [67], Chafee [34], Schmidt [201], Ruelle [185], Poore [175], Ize [104]); some global bifurcation results have been obtained (Alexander and Yorke [4], Chow, Mallet-Paret and Yorke [44],[45]); also, Hopf bifurcation has been studied for functional differential equations (Chafee [35], Hale [79],[86], Hale and De Oliveira [91],[170]), partial differential equations (Iooss [100], Joseph and Sattinger [105], Sattinger [192]) and abstract evolution equations (Crandall and Rabinowitz [52], Kielhöfer [118], Ize [104]); finally, Golubitsky and Langford [250] used singularity theory to classify perturbations of degenerate Hopf bifurcations.

In this chapter we will study Hopf bifurcation as a special case of the general results of chapter 6, section 2. It appears that the general theory gives us results of the Hopf bifurcation type when we take for G the group SO(2) of rotations in the plane. We will work with spaces of periodic functions, on which SO(2) can be represented by phase shifts. Since the equations under consideration will be autonomous, we obtain a bifurcation problem which is equivariant with respect to SO(2).

Our approach has originally been motivated by the work of Chafee [36], which considers the following problem. Suppose that one has an "unperturbed" autonomous ordinary differential equation

$$\dot{x} = f_0(x) \tag{1}$$

which satisfies the following conditions : (i) $f_0(0) = 0$; (ii) $\pm i$ are simple eigenvalues of the Jacobian matrix $Df_0(x)$; (iii) none of the other eigenvalues is an entire multiple of $\pm i$. These are the usual hypotheses imposed on the unperturbed equation in the classical Hopf bifurcation problem. Now consider the "perturbed equation"

$$\dot{x} = f(x) \tag{2}$$

where f is supposed to be near f_0 in an appropriate function space. The problem is to determine all periodic solutions of (2) near the origin and with period near 2π. In [36] Chafee shows that such solutions are uniquely determined by the solutions of a scalar equation

$$g(\rho,f) = 0 , \tag{3}$$

where $g(\rho,f)$ is defined and sufficiently smooth for any (ρ,f) in a small neighbourhood of $(0,f_0)$; also g is odd in ρ. To the solution $\rho = 0$ of (3) there corresponds a stationary solution of (2); to each sufficiently small solution $\rho > 0$ of (3) there corresponds a family of non-constant periodic solutions of (2); different elements of such family are related to each other by phase shifts. The method also determines the period, which depends on (ρ,f) and is near to 2π. An earlier version of these results, restricted to the two-dimensional case, can be found in [242].

We will show that these results follow from the general theory of chapter 6; for example, the oddness of $g(\rho,f)$ in ρ expresses the equivariance of the bifurcation problem with respect to the group SO(2). In section 2 we start by considering general bifurcation problems which are equivariant with respect to SO(2); we obtain the bifurcation equation and discuss its solutions under a further nondegeneracy condition. In section 3 we make a further analysis of this condition, and show that it is generically satisfied. Section 4 contains the application on the Hopf bifurcation problem; we give a unified treatment for ordinary as well as functional differential equations. Finally, in section 5 we use some of our results to prove the Liapunov Center Theorem.

7.2. BIFURCATION PROBLEMS WITH SO(2)-SYMMETRY

In this section we apply the reduction method of section 6.2 to equations which are equivariant with respect to the group SO(2). Since SO(2) is a major subgroup of O(2), one might expect some results similar to those of section 6.4. However, since SO(2) contains no reflections, the condition (6.4.6) is no longer satisfied. This gives an additional scalar bifurcation equation, which is identically satisfied when the problem has O(2)-symmetry.

7.2.1. The hypotheses. Let X, Z and Λ be real Banach spaces, and $M : \Omega \subset X \times \Lambda \times \mathbb{R} \to Z$ a mapping, defined and of class C^2 in a neighbourhood Ω of the origin in $X \times \Lambda \times \mathbb{R}$. We will consider the equation

$$M(x,\lambda,\sigma) = 0 . \tag{1}$$

In view of remark 6.2.9 and the hypotheses which we will make further on, we will assume that :

$$M(0,\lambda,\sigma) = 0 \qquad , \qquad \forall(\lambda,\sigma) . \tag{2}$$

The main hypotheses on M are the following :

(H1) M is equivariant with respect to $(SO(2),\Gamma,\tilde{\Gamma})$, where $\Gamma : SO(2) \to L(X)$ and $\tilde{\Gamma} : SO(2) \to L(Z)$ are representations of SO(2) over X, respectively Z.

(H2) $L = D_xM(0,0,0)$ is a Fredholm operator with zero index; dim ker L = 2, and ker L has a basis $\{u_1,u_2\}$ such that :

$$\left.\begin{aligned} \Gamma(\alpha)u_1 &= \cos\alpha.u_1 - \sin\alpha.u_2 , \\ \Gamma(\alpha)u_2 &= \sin\alpha.u_1 + \cos\alpha.u_2 , \end{aligned}\right\} \quad \forall\alpha \in \mathbb{R} . \tag{3}$$

(H3) $D_\sigma D_x M(0,0,0).u \notin R(L)$ for some $u \in \ker L \setminus \{0\}$.

7.2.2. Remarks. It follows from theorem 2.6.17 that, if the representation induced by Γ on ker L is irreducible, then either dim ker L = 1 and $\Gamma(\alpha)u = u$, $\forall\alpha \in \mathbb{R}$, $\forall u \in$ ker L, or dim ker L = 2 and ker L has a basis $\{u_1,u_2\}$ such that :

$$\begin{aligned} \Gamma(\alpha)u_1 &= \cos k\alpha . u_1 - \sin k\alpha . u_2 , \\ \Gamma(\alpha)u_2 &= \sin k\alpha . u_1 + \cos k\alpha . u_2 , \end{aligned} \qquad \forall\alpha \in \mathbb{R} , \tag{4}$$

for some $k \in \mathbb{N}\setminus \{0\}$. The hypothesis (H2) rules out the case dim ker L = 1, which can be handled as in section 6.3; all bifurcating solutions will in this case satisfy $\Gamma(\alpha)x = x$, $\forall\alpha \in \mathbb{R}$.

(H2) says that dim ker L = 2 and that we have k = 1 in (4); the following argument shows that this gives no loss of generality. Assume that (H1)-(H3) are satisfied, with (3) replaced by (4), for some general $k \neq 0$. Then :

$$\Gamma(\tfrac{2\pi}{k}p)u = u \qquad , \qquad \forall u \in \ker L \ , \ p = 0,1,\dots,k-1 . \tag{5}$$

It follows from section 3.4 that each solution (x,λ,σ) of (1) sufficiently near the origin will have the same invariance property as the elements of ker L :

$$\Gamma(\tfrac{2\pi}{k}p)x = x \qquad , \qquad p = 0,1,\dots,k-1 . \tag{6}$$

Define

$$X_k = \{x \in X \mid \Gamma(\tfrac{2\pi}{k}p)x = x,\ p = 0,1,\dots,k-1\}$$

and

$$Z_k = \{z \in Z \mid \tilde{\Gamma}(\tfrac{2\pi}{k}p)z = z,\ p = 0,1,\dots,k-1\} .$$

Let M_k be the restriction of M to dom $M_k = \Omega_k = \Omega \cap (X_k \times \Lambda \times \mathbb{R})$; then M_k maps Ω_k into Z_k, and our problem (1) is equivalent to the equation :

$$M_k(x,\lambda,\sigma) = 0 . \tag{7}$$

We will show that (7) satisfies (H1)-(H3).

Define representations $\Gamma_k : SO(2) \to L(X_k)$ and $\tilde{\Gamma}_k : SO(2) \to L(Z_k)$ as follows :

$$\Gamma_k(\alpha) = \Gamma(\tfrac{\alpha}{k})|_{X_k} \quad , \quad \tilde{\Gamma}_k(\alpha) = \tilde{\Gamma}(\tfrac{\alpha}{k})|_{Z_k} \quad , \quad \forall\alpha \in \mathbb{R} . \tag{8}$$

It is clear that M_k is equivariant with respect to $(SO(2),\Gamma_k,\tilde{\Gamma}_k)$, so that (7) satisfies (H1). Let $L_k = D_x M_k(0,0,0)$; then, since $\ker L \subset X_k$ by (4), we have $\ker L_k = \ker L$, and (3) is satisfied if we replace Γ and $\tilde{\Gamma}$ by Γ_k and $\tilde{\Gamma}_k$. Also, $R(L_k) = R(L) \cap Z_k$. Indeed, it is clear that $R(L_k) \subset R(L) \cap Z_k$, since $L_k = L|_{X_k}$ and since L commutes with $\Gamma(\alpha)$. Conversely, if $z \in R(L) \cap Z_k$, then $z = Lx$ for some $x \in X$, and, since $z \in Z_k$, we have :

$$z = \frac{1}{k} \sum_{p=0}^{k-1} \tilde{\Gamma}(\tfrac{2\pi}{k}p)z = L(\frac{1}{k} \sum_{p=0}^{k-1} \Gamma(\tfrac{2\pi}{k}p)x) ,$$

where $\frac{1}{k} \sum_{p=0}^{k-1} \Gamma(\frac{2\pi}{k}p)x \in X_k$; so $z \in R(L_k)$. If $Q \in L(Z)$ is a projection such that $\ker Q = R(L)$ and $\tilde{\Gamma}(\alpha)Q = Q\tilde{\Gamma}(\alpha)$, $\forall\alpha \in \mathbb{R}$, then lemma 6.2.4 shows that $R(Q) \subset Z_k$. Then it follows from $R(L_k) = R(L) \cap Z_k = \{z \in Z_k \mid Qz = 0\}$ that L_k is a Fredholm operator with zero index. Finally (7) also satisfies (H3), as can easily be seen from the foregoing and $D_\sigma D_x M_k(0,0,0) = D_\sigma D_x M(0,0,0)|_{X_k}$.

We conclude that there is no loss of generality in restricting to the case $k = 1$. In the application on differential equations given in section 4, the foregoing reduction can be performed by a simple time rescale.

7.2.3. It is clear that the hypotheses (H1)-(H3) above form a particular case of the general hypotheses (H1)-(H3) of section 6.2, so that we can apply the results of that section, and in particular theorem 6.2.17. Also lemma 6.2.16 applies, since (3) shows that $\Gamma(\pi)u = -u$ for each $u \in \ker L$. By theorem 6.2.17 we have to find solutions (ρ,θ,λ) of the bifurcation equation

$$G(\rho,\theta,\lambda) = 0 , \tag{9}$$

where (ρ,λ) are near the origin, $\rho > 0$ and $\theta \in S = \{\beta_1u_1+\beta_2u_2 \mid \beta_1^2+\beta_2^2 = 1\}$. The mapping G is of class C^1, and satisfies (6.2.47)-(6.2.50) and (6.2.52).

Since each $\theta \in S$ has the form $\theta = \Gamma(\alpha)u_1$ for some $\alpha \in \mathbb{R}$, and $G(\rho,\Gamma(\alpha)u_1,\lambda) = \Gamma(\alpha)G(\rho,u_1,\lambda)$, it follows that (ρ,θ,λ) solves (9) if and only if (ρ,λ) is a solution of

$$G(\rho,u_1,\lambda) = 0 . \tag{10}$$

Since $\langle u_1,G(\rho,u_1,\lambda)\rangle = 0$ we have $G(\rho,u_1,\lambda) = g(\rho,\lambda)u_2$, where $g(\rho,\lambda)$ is a scalar function, defined and of class C^1 in a neighbourhood of the origin, and such that $g(0,0)= 0$ and $g(-\rho,\lambda) = g(\rho,\lambda)$. Our problem reduces to that of solving the scalar bifurcation equation :

$$g(\rho,\lambda) = 0 . \tag{11}$$

The following theorem summarizes the relation between the equations (1) and (11).

7.2.4. <u>Theorem</u>. Let M satisfy (2) and (H1)-(H3). Let $P \in L(X)$ be a projection such that $R(P) = \ker L$ and $\Gamma(\alpha)P = P\Gamma(\alpha)$, $\forall\alpha \in \mathbb{R}$.

Then there exist a neighbourhood Ω_1 of the origin in $X\times\Lambda\times\mathbb{R}$, a neighbourhood ω of the origin in $\mathbb{R}\times\Lambda$, and mappings $\tilde{x} : \omega \to X$, $\tilde{\sigma} : \omega \to \mathbb{R}$ and $g : \omega \to \mathbb{R}$, all of class C^1, such that :

$$\{(x,\lambda,\sigma)\in\Omega_1 \mid M(x,\lambda,\sigma) = 0 \text{ and } x\neq 0\}$$

$$= \{(\Gamma(\alpha)\tilde{x}(\rho,\lambda),\lambda,\tilde{\sigma}(\rho,\lambda)) \mid \alpha\in\mathbb{R},\ (\rho,\lambda)\in\omega,\ \rho>0 \text{ and } g(\rho,\lambda) = 0\} .$$

The mappings $\tilde{x}$, $\tilde{\sigma}$ and g have the following properties, for all $(\rho,\lambda) \in \omega$:

(i) $\tilde{x}(0,\lambda) = 0$, $P\tilde{x}(\rho,\lambda) = \rho u_1$, and

$$\tilde{x}(-\rho,\lambda) = \Gamma(\pi)\tilde{x}(\rho,\lambda) . \tag{12}$$

(ii) $\tilde{\sigma}(0,0) = 0$ and

$$\tilde{\sigma}(-\rho,\lambda) = \tilde{\sigma}(\rho,\lambda) . \tag{13}$$

(iii) $g(0,0)$ and

$$g(-\rho,\lambda) = g(\rho,\lambda) \ . \quad \square \tag{14}$$

7.2.5. For further use, let us recall here how the mapping $\tilde{x}$, $\tilde{\sigma}$ and g can be obtained. Let $Q \in L(Z)$ be a projection such that $\ker Q = R(L)$ and $Q\tilde{\Gamma}(\alpha) = \tilde{\Gamma}(\alpha)Q$, $\forall\alpha \in \mathbb{R}$. Let $v^*(u,\lambda,\sigma)$ be the unique solution $v \in \ker P$ of the equation

$$(I-Q)M(u+v,\lambda,\sigma) = 0 \ . \tag{15}$$

Let $B = QD_\sigma D_x M(0,0,0)|_{\ker L}$ and

$$F(u,\lambda,\sigma) = B^{-1}QM(u+v^*(u,\lambda,\sigma),\lambda,\sigma) \ . \tag{16}$$

Then

$$F(\rho u_1,\lambda,\sigma) = f_1(\rho,\lambda,\sigma)u_1 + f_2(\rho,\lambda,\sigma)u_2 \ , \tag{17}$$

where the scalar functions $f_i(\rho,\lambda,\sigma)$ $(i = 1,2)$ can also be obtained as follows. Define $w_i \in R(Q)$ $(i = 1,2)$ by :

$$w_i = QD_\sigma D_x M(0,0,0).u_i \qquad (i = 1,2) \ . \tag{18}$$

Then there are functionals $Q_i : Z \to \mathbb{R}$ such that

$$Qz = Q_1(z)w_1 + Q_2(z)w_2 \quad , \quad \forall z \in Z \ . \tag{19}$$

The functions $f_i(\rho,\lambda,\sigma)$ are given by :

$$f_i(\rho,\lambda,\sigma) = Q_i M(\rho u_1+v^*(\rho u_1,\lambda,\sigma),\lambda,\sigma) \quad , \quad i = 1,2 \ ; \tag{20}$$

they are odd in ρ, and can be written in the form

$$f_i(\rho,\lambda,\sigma) = \rho h_i(\rho,\lambda,\sigma) \quad , \quad i = 1,2 \ , \tag{21}$$

where

$$h_i(\rho,\lambda,\sigma) = \int_0^1 D_\rho f_i(t\rho,\lambda,\sigma)dt \quad , \quad i = 1,2 . \tag{22}$$

The function $\tilde{\sigma}(\rho,\lambda)$ is the unique solution of the equation

$$h_1(\rho,\lambda,\sigma) = 0 , \tag{23}$$

while

$$g(\rho,\lambda) = h_2(\rho,\lambda,\tilde{\sigma}(\rho,\lambda)) . \tag{24}$$

Finally, $\tilde{x}(\rho,\lambda)$ is given by :

$$\tilde{x}(\rho,\lambda) = \rho u_1 + v^*(\rho u_1,\lambda,\tilde{\sigma}(\rho,\lambda)) . \tag{25}$$

7.2.6. SO(2)-symmetry versus O(2)-symmetry. Suppose that the equation (1) is equivariant with respect to some representation of the group O(2), while also (H2) and (H3) are satisfied. (This is the situation studied in section 6.4). Then of course the foregoing results remain valid. But now one can choose $\{u_1,u_2\}$ such that $\Gamma_\tau u_1 = u_1$ and $\Gamma_\tau u_2 = -u_2$, and in addition to the foregoing properties we also have $\Gamma_\tau F(u,\lambda,\sigma) = F(\Gamma_\tau u,\lambda,\sigma)$. This implies that $F(\rho u_1,\lambda,\sigma) = F(\rho\Gamma_\tau u_1,\lambda,\sigma) = \Gamma_\tau F(\rho u_1,\lambda,\sigma)$, and consequently $f_2(\rho,\lambda,\sigma) = 0$ for all (ρ,λ,σ). Then we see from (24) that also $g(\rho,\lambda) = 0$ for all (ρ,λ), i.e. the bifurcation equation (11) is identically satisfied. As a consequence, $(\tilde{x}(\rho,\lambda),\lambda,\tilde{\sigma}(\rho,\lambda))$ will be a solution of (1) for *arbitrary* $(\rho,\lambda) \in \omega$.

In case M is only equivariant with respect to SO(2), then this argument fails, and there remains the bifurcation equation (11) to solve : $(\tilde{x}(\rho,\lambda),\lambda, \tilde{\sigma}(\rho,\lambda))$ will only be a solution of (1) if (ρ,λ) satisfies (11). In his recent book [247], Sattinger came to a similar conclusion using a rather different approach.

Now there are different ways to attack the equation (11). We consider first the classical approach, which consists in solving (11) for λ as a function of ρ; of course, in order to be able to do so, we have to restrict to one single scalar parameter $\lambda \in \mathbb{R}$.

7.2.7. Theorem. Let M satisfy (2) and (H1)-(H3), with $\Lambda = \mathbb{R}$. Assume that

$$Q_2 D_x D_\lambda M(0,0,0).u_1 \neq 0 . \tag{26}$$

Then there is a neighbourhood Ω_1 of the origin in $X \times \mathbb{R} \times \mathbb{R}$, a $\rho_0 > 0$ and C^1-mappings $\bar{x} :]-\rho_0,\rho_0[\to X$, $\bar{\lambda} :]-\rho_0,\rho_0[\to \mathbb{R}$ and $\bar{\sigma} :]-\rho_0,\rho_0[\to \mathbb{R}$ such that the nontrivial solutions $(x,\lambda,\sigma) \in \Omega_1$ of (1) are given by :

$$\{(\Gamma(\alpha)\bar{x}(\rho),\bar{\lambda}(\rho),\bar{\sigma}(\rho)) \mid \alpha \in \mathbb{R},\ 0<\rho<\rho_0\} .$$

Furthermore $\bar{x}(0) = 0$, $\bar{\lambda}(0) = 0$, $\bar{\sigma}(0) = 0$, $\bar{\lambda}(-\rho) = \bar{\lambda}(\rho)$, $\bar{\sigma}(-\rho) = \bar{\sigma}(\rho)$, $\bar{x}(-\rho) = \Gamma(\pi)\bar{x}(\rho)$ and $P\bar{x}(\rho) = \rho u_1$.

P r o o f. From the formulas of subsection 5 one obtains the following expression for $g(0,\lambda)$:

$$g(0,\lambda) = Q_2 D_x M(0,\lambda,\tilde{\sigma}(0,\lambda)).(u_1+D_u v^*(0,\lambda,\tilde{\sigma}(0,\lambda)).u_1) , \tag{27}$$

from which we obtain :

$$D_\lambda g(0,0) = Q_2 D_x D_\lambda M(0,0,0).u_1 . \tag{28}$$

By assumption this is different from zero. Using the implicit function theorem, equation (11) has a unique solution $\lambda = \bar{\lambda}(\rho)$, with $\bar{\lambda}(0) = 0$ and $\bar{\lambda}(-\rho) = \bar{\lambda}(\rho)$, because of (14). Then the conclusion follows from theorem 4, by defining $\bar{x}(\rho) = \tilde{x}(\rho,\bar{\lambda}(\rho))$ and $\bar{\sigma}(\rho) = \tilde{\sigma}(\rho,\bar{\lambda}(\rho))$. □

7.2.8. In case Λ is a general Banach space one can try to split off a scalar parameter and apply theorem 7; the resulting solutions will then depend on ρ and the remaining parameters.

One can also apply the approach of chapter 5 to the bifurcation equation (11). We will do this under a "generic condition" which can be described as follows. Assume that M in (1) is of class C^3; then $g(\rho,\lambda)$ will be of class C^2. Since g is even in ρ, we have $D_\rho g(0,\lambda) = 0$ for all λ. The condition will be that $a = D_\rho^2 g(0,0) \neq 0$; in the next section we will show that in some sense this condition is generically satisfied.

In order to obtain a more explicit expression for a, remark that

$$F(\rho u_1,\lambda,\tilde{\sigma}(\rho,\lambda)) = \rho g(\rho,\lambda)u_2 \ ;$$

replace F by the expression given in (16), differentiate three times in ρ, and put $(\rho,\lambda) = (0,0)$; using $D_u v^*(0,0,0) = 0$, $D_\sigma v^*(0,0,0) = 0$ (this follows from (2)) and $D_\rho\tilde{\sigma}(0,0) = 0$ (see (13)) one obtains the following result :

$$\begin{aligned} 3D^2 g(0,0)u_2 &= B^{-1}QD_x^3M(0,0,0).(u_1,u_1,u_1) \\ &+ 3B^{-1}QD_x^2M(0,0,0).(u_1,D_u^2v^*(0,0,0).(u_1,u_1)) \\ &+ 3D_\rho^2\tilde{\sigma}(0,0)u_1 \ , \end{aligned} \tag{29}$$

where

$$D_u^2v^*(0,0,0).(u_1,u_1) = -K(I-Q)D_x^2M(0,0,0).(u_1,u_1) \ . \tag{30}$$

Putting the coefficients of u_1 and u_2 in (29) equal to zero, we find :

$$\begin{aligned} D_\rho^2\tilde{\sigma}(0,0) &= -\tfrac{1}{3}Q_1D_x^3M(0,0,0).(u_1,u_1,u_1) \\ &- Q_1D_x^2M(0,0,0).(u_1,D_u^2v^*(0,0,0).(u_1,u_1)) \end{aligned} \tag{31}$$

and

$$\begin{aligned} a &= \tfrac{1}{3}Q_2D_x^3M(0,0,0).(u_1,u_1,u_1) \\ &+ Q_2D_x^2M(0,0,0).(u_1,D_u^2v^*(0,0,0).(u_1,u_1)) \ . \end{aligned} \tag{32}$$

7.2.9. <u>Theorem</u>. Let $M \in C^3(\Omega;Z)$ satisfy (2),(H1),(H2) and (H3). Suppose that

$$a \neq 0 \ , \tag{33}$$

where a is given by (32) and (30). Let $\gamma(\lambda) = g(0,\lambda)$ be given by (27).

Then there is a neighbourhood U of the origin in $X \times \mathbb{R}$, and a neighbourhood W of the origin in W, such that for each $\lambda \in W$ the following holds :

(i) if $a\gamma(\lambda) < 0$, then (1) has exactly one family of nontrivial solutions in U, of the form :

$$\{(\Gamma(\alpha)\bar{x}(\lambda), \bar{\sigma}(\lambda)) \mid \alpha \in \mathbb{R}\} ,$$

with $\bar{x}(\lambda) = O(|\gamma(\lambda)|^{1/2})$;

(ii) if $a\gamma(\lambda) \geqslant 0$, then (1) has no nontrivial solutions $(x,\sigma) \in U$.

P r o o f. We use an argument similar to the one used in subsection 5.5.7. Let $p(r,\lambda) = g(|r|^{1/2},\lambda)$; since $g(\rho,\lambda) = \gamma(\lambda) + \rho^2 g_1(\rho,\lambda)$, with $\gamma(0) = 0$ and $g_1(0,0) = a/2$, we see that $p(r,\lambda)$ is of class C^1, with $p(0,0) = 0$ and $D_r p(0,0) = a/2 \neq 0$. So the equation $p(r,\lambda) = 0$ has a unique solution $r = \bar{r}(\lambda)$ for (r,λ) in a neighbourhood of the origin. Moreover, it is easily seen that $\bar{r}(\lambda) = -\gamma(\lambda)\bar{q}(\lambda)$, where $\bar{q}(\lambda)$ is continuous and $\bar{q}(0) = 2a^{-1}$.

Now (ρ,λ) is a solution of (11) if and only if $(r,\lambda) = (\rho^2,\lambda)$ is a solution of $p(r,\lambda) = 0$, i.e. if and only if $\rho^2 = -\gamma(\lambda)\bar{q}(\lambda)$. The result now follows from theorem 4 and the fact that $\text{sgn}\, \bar{q}(\lambda) = \text{sgn}\, a$ for all λ in a sufficiently small neighbourhood of the origin. If $a\gamma(\lambda) < 0$, then one defines $\bar{x}(\lambda) = \tilde{x}((-\gamma(\lambda)\bar{q}(\lambda))^{1/2},\lambda)$ and $\bar{\sigma}(\lambda) = \tilde{\sigma}((-\gamma(\lambda)\bar{q}(\lambda))^{1/2},\lambda)$. □

7.2.10. <u>Remark</u>. In the general case that

$$D_\rho^{2k} g(0,0) \neq 0 \ , \ D_\rho^j g(0,0) = 0 \ , \ j = 0,1,\ldots,2k-1 \ , \qquad (34)$$

an argument as the one used in the proof of the previous theorem, possibly complemented with a rescaling method (see e.g. Chow, Hale and Mallet-Paret [40], Vanderbauwhede [228]) could in principle work, but becomes very complicated for higher k. When all mappings under consideration are C^∞, then one can use the division theorem in Banach spaces (Michor [164]) to show that under the hypothesis (34) the equation (11) is, for (ρ,λ) sufficiently small, equivalent to an equation of the form :

$$\rho^{2k} + \gamma_{k-1}(\lambda)\rho^{2k-2} + \ldots + \gamma_1(\lambda)\rho + \gamma_0(\lambda) = 0 \ , \qquad (35)$$

where the $\gamma_i(\lambda)$ are smooth functionals with $\gamma_i(0) = 0$. Such approach has, for the case of a finite number of parameters, been used by Hale [82] and Takens [213]. For each λ the number of solutions $\rho > 0$ of (35) can be determined; it depends on the values of $\gamma_0(\lambda),\ldots,\gamma_{k-1}(\lambda)$. This gives a partition of a neighbourhood of the origin in Λ into subregions, such that (35) has a finite and constant number of solutions $\rho > 0$ in each of these subregions; the number of solutions changes (i.e. there is bifurcation) as λ goes from one subregion to another. It is interesting to work this out for small values of k, and to interpret the results of Golubitsky and Langford [250] as a classification of inequivalent one-dimensional paths through such picture.

7.2.11. Remark. It is clear from its definition and from the expression (32) that the number a in the condition (33) will depend on our choice of the projections P and Q, and on the basis $\{u_1,u_2\}$ we have taken for ker L. Let us investigate this dependence in more detail.

It follows from (32) that

$$a = \frac{1}{3}D_\rho^3 Q_2 M(x(\rho),0,0)\big|_{\rho=0}\ , \tag{36}$$

where $x(\rho) = \rho u_1+v^*(\rho u_1,0,0)$ is the unique solution of $(I-Q)M(x,0,0) = 0$ such that $Px = \rho u_1$; we have $D_\rho x(0) = u_1$ and $x(-\rho) = \Gamma(\pi)x(\rho)$.

Now replace P by some other projection P' satisfying all the requirements. Then we have to replace $x(\rho)$ in (36) by $x'(\rho)$, the unique solution of $(I-Q)M(x,0,0) = 0$ such that $P'x = \rho u_1$; we still have $D_\rho x'(0) = u_1$ and $x'(-\rho) = \Gamma(\pi)x'(\rho)$. If we let $u'(\rho) = Px'(\rho)$, then $x'(\rho) = u'(\rho)+v^*(u'(\rho),0,0)$, while $u'(-\rho) = -u'(\rho)$ and $D_\rho u'(0) = u_1$. A direct calculation shows that replacing $x(\rho)$ by $x'(\rho)$ in (36) does not change the result; so a is independent of our choice of the projection P.

Also, replacing Q by some other appropriate projection Q' does not change a. Indeed, we will have $Q' = Q'Q$ and the basic vectors $\{w_1',w_2'\}$ for $R(Q')$ take the form $w_i' = Q'D_\sigma D_x M(0,0,0).u_i = Q'w_i$, $(i = 1,2)$. It follows that $Q_i' = Q_i$ $(i = 1,2)$, and the expression (36) does not change.

Finally, let us replace the basis $\{u_1,u_2\}$ of ker L by some other basis $\{u_1',u_2'\}$ transforming under $\Gamma(\alpha)$ according to (3). We will have $u_1' = \rho_1\Gamma(\alpha_1)u_1$ for some $\rho_1 > 0$ and $\alpha_1 \in \mathbb{R}$. Then it follows from (3) that $u_2' = \Gamma(-\pi/2)u_1' = \rho_1\Gamma(\alpha_1)\Gamma(-\pi/2)u_1 = \rho_1\Gamma(\alpha_1)u_2$. We also have to replace w_i by $w_i' = \rho_1\tilde{\Gamma}(\alpha_1)w_i$ $(i=1,2)$. This implies that $Q_i' = \rho_1^{-1}Q_i\tilde{\Gamma}(-\alpha_1)$. Bringing this in (32) shows that the new constant a' will be given by $a' = \rho_1^2 a$. So $a \neq 0$ if and only if $a' \neq 0$, and even sgn a' = sgn a.

The foregoing proves that the conditions formulated in theorem 9 do not depend on the particular choice of the projections and basic vectors.

7.3. GENERIC BIFURCATION UNDER SO(2)-SYMMETRY

7.3.1. Introduction. In this section we will study bifurcation problems with SO(2)-symmetry from a generic point of view. The problem, as we will state it below, forms an abstraction of the problem discussed by Chafee [36] for periodic solutions of autonomous ordinary differential equations. At the same time our approach will allow us to give a precise meaning to the statement that the condition $a \neq 0$ of theorem 2.9 is generically satisfied.

Let X and Z be real Banach spaces, and let $\Gamma : SO(2) \to L(X)$ and $\tilde{\Gamma} : SO(2) \to L(Z)$ be given representations of SO(2) over X, respectively Z. Let Ω_0 be a neighbourhood of the origin in X such that $\Gamma(\alpha)(\Omega_0) = \Omega_0$ for all $\alpha \in \mathbb{R}$. Let $r \geq 3$. We will consider equations of the type :

$$m(x,\sigma) = 0 \ , \qquad (1)$$

where $m : \Omega_0 \times \mathbb{R} \to Z$ is of class C^3, equivariant with respect to $(SO(2),\Gamma,\tilde{\Gamma})$:

$$m(\Gamma(\alpha)x,\sigma) = \tilde{\Gamma}(\alpha)m(x,\sigma) \quad , \quad \forall \alpha \in \mathbb{R} \ , \ \forall (x,\sigma) \ , \qquad (2)$$

and such that

$$m(0,\sigma) = 0 \quad , \quad \forall \sigma \in \mathbb{R} \ . \qquad (3)$$

Let m_0 be a mapping of the class just described, and let $\sigma_0 \in \mathbb{R}$ be such that (m_0,σ_0) satisfies the following hypotheses :

(H). (i) $L_0 = D_x m_0(0,\sigma_0)$ is a Fredholm operator with zero index, dim ker L_0 = 2, and ker L_0 has a basis $\{u_1,u_2\}$ such that

$$\begin{aligned} \Gamma(\alpha)u_1 &= \cos\alpha . u_1 - \sin\alpha . u_2 \ , \\ \Gamma(\alpha)u_2 &= \sin\alpha . u_1 + \cos\alpha . u_2 \ , \end{aligned} \qquad \forall \alpha \in \mathbb{R} \ . \qquad (4)$$

(ii) $D_\sigma D_x m_0(0,\sigma_0).u \notin R(L_0)$ for some $u \in \ker L_0 \setminus \{0\}$.

The problem is to describe the nontrivial solutions of (1) near $(0,\sigma_0)$ for each m near m_0 (in a sense to be made precise further on). In particular, we want to determine for each m near m_0 the number of families of nontrivial solutions of (1) near $(0,\sigma_0)$.

As a first step we bring this problem in the form (2.1).

7.3.2. <u>The abstract setting</u>. Let Λ_0 be the space of all mapping $m : \Omega_0 \times \mathbb{R} \to Z$ of class C^r, such that m satisfies (2) and (3), and such that

$$|m|_r = \sup_{(x,\sigma) \in \Omega_0 \times \mathbb{R}} (\|m(x,\sigma)\| + \|Dm(x,\sigma)\| + \ldots + \|D^r m(x,\sigma)\|) < \infty . \qquad (5)$$

Using $|.|_r$ as a norm, Λ_0 becomes a Banach space. Define :

$$\Sigma_0 = \{(m_0,\sigma_0) \in \Lambda_0 \times \mathbb{R} \mid (H) \text{ is satisfied at } (m_0,\sigma_0)\} . \qquad (6)$$

Also, let $\Omega = \Omega_0 \times \Lambda_0 \times \mathbb{R}$ and define $M : \Omega \to Z$ by :

$$M(x,m,\sigma) = m(x,\sigma) \qquad , \qquad \forall (x,m,\sigma) \in \Omega . \qquad (7)$$

Then our problem can be reformulated as follows : given $(m_0,\sigma_0) \in \Sigma_0$, describe the nontrivial solutions (x,m,σ) near $(0,m_0,\sigma_0)$ in $X \times \Lambda_0 \times \mathbb{R}$ of the equation :

$$M(x,m,\sigma) = 0 . \qquad (8)$$

It is easily verified that the hypotheses (H1)-(H3) of section 2 are satisfied at the point $(0,m_0,\sigma_0)$ (instead of $(0,0,0)$, as formulated in section 2). So we can apply the results of section 2. Before doing so we first study the structure of the set Σ_0 in more detail. Since the condition on m_0 in (H) is a condition on the linearization of m_0 at $x = 0$, we start by considering a linear problem.

7.3.3. <u>A linear problem</u>. Let Λ be a real Banach space, ω_1 a neighbourhood of (λ_0,σ_0) in $\Lambda \times \mathbb{R}$, and $L : \omega_1 \to L(X,Z)$ a C^1-map. Assume the following :

$(H)_L$ (i) $L(\lambda,\sigma)\Gamma(\alpha) = \tilde{\Gamma}(\alpha)L(\lambda,\sigma)$, $\forall\alpha \in \mathbb{R}$, $\forall(\lambda,\sigma) \in \omega_1$; (9)

(ii) $L_0 = L(\lambda_0,\sigma_0)$ is a Fredholm operator with zero index, dim ker L_0 = 2, and ker L_0 has a basis $\{u_1,u_2\}$ such that (4) holds;

(iii) $D_\sigma L(\lambda_0,\sigma_0).u \notin R(L_0)$ for some $u \in \ker L_0 \setminus \{0\}$.

We wish to determine ker $L(\lambda,\sigma)$ and $R(L(\lambda,\sigma))$ for each (λ,σ) near (λ_0,σ_0).

7.3.4. Determination of ker $L(\lambda,\sigma)$ and $R(L(\lambda,\sigma))$. Consider the linear equation

$$L(\lambda,\sigma).x = z \ , \tag{10}$$

where $z \in Z$ is given, and we want to solve for $x \in X$. As usual, let P and Q be projections onto ker L_0 and onto a complement of $R(L_0)$, respectively, both commuting with the symmetry operators. Write $x \in X$ as $u+v$, with $u = Px \in \ker L_0$ and $v = (I-P)x \in \ker P$. Consider the equation :

$$(I-Q)L(\lambda,\sigma).(u+v) = (I-Q)z \ . \tag{11}$$

Since $L_0 = (I-Q)L(\lambda_0,\sigma_0)$ is an isomorphism between ker P and $R(L_0)$, the same holds for $(I-Q)L(\lambda,\sigma)$, if (λ,σ) is sufficiently close to (λ_0,σ_0). So we can solve (11) for v; we obtain :

$$v = V^*(\lambda,\sigma).u + W^*(\lambda,\sigma).z \ , \tag{12}$$

where for each (λ,σ) near (λ_0,σ_0), we have $V^*(\lambda,\sigma) \in L(\ker L_0,\ker P)$ and $W^*(\lambda,\sigma) \in L(Z,\ker P)$. Also $V^*(\lambda_0,\sigma_0) = 0$, and

$$V^*(\lambda,\sigma)\Gamma(\alpha) = \Gamma(\alpha)V^*(\lambda,\sigma) \quad , \quad W^*(\lambda,\sigma)\tilde{\Gamma}(\alpha) = \Gamma(\alpha)W^*(\lambda,\sigma) \ . \tag{13}$$

Equation (10) reduces to

$$QL(\lambda,\sigma).(u+V^*(\lambda,\sigma).u) = Q(\lambda,\sigma).z \ , \tag{14}$$

where

$$Q(\lambda,\sigma).z = Qz - QL(\lambda,\sigma).W^*(\lambda,\sigma).z \quad , \quad \forall z \in Z . \tag{15}$$

Since $W^*(\lambda,\sigma).z$ has the form $\tilde{W}^*(\lambda,\sigma).(I-Q)z$, it is easily seen that $Q(\lambda,\sigma)$ is a projection in Z onto $R(Q)$ and commuting with the $\tilde{\Gamma}(\alpha)$.

Using the condition $(H)_L$(iii) the argument of lemma 6.2.4 shows that $\tilde{\Gamma}$ induces on $R(Q)$ an irreducible representation of $SO(2)$ equivalent to the representation (4). An appropriate basis for $R(Q)$ is given by $\{w_1,w_2\}$, where $w_i = QD_\sigma L(\lambda_0,\sigma_0).u_i$ $(i=1,2)$.

Consider now the operator $C(\lambda,\sigma) \in L(\ker L_0, R(Q))$ defined by :

$$C(\lambda,\sigma) = QL(\lambda,\sigma)(I_{\ker L_0} + V^*(\lambda,\sigma)) . \tag{16}$$

Since $C(\lambda,\sigma)\Gamma(\alpha) = \tilde{\Gamma}(\alpha)C(\lambda,\sigma)$, $\forall \alpha \in \mathbb{R}$, it follows from Schur's lemma (see section 2.6) that either $C(\lambda,\sigma)$ is an isomorphism, or $C(\lambda,\sigma) = 0$. In case $C(\lambda,\sigma)$ is an isomorphism equation (14) has a unique solution u for each $z \in Z$; in particular, taking $z = 0$, we see that $\ker L(\lambda,\sigma) = \{0\}$. In case $C(\lambda,\sigma) = 0$ (14) (with $z = 0$) shows that $\ker L(\lambda,\sigma) = \{u+V^*(\lambda,\sigma).u \mid u \in \ker L_0\}$; since the mapping $u \mapsto u+V^*(\lambda,\sigma).u$ is an isomorphism (we have $V^*(\lambda_0,\sigma_0) = 0$), we see that $\dim \ker L(\lambda,\sigma) = 2$, and $\ker L(\lambda,\sigma)$ has a basis $\{u_1+V^*(\lambda,\sigma).u_1, u_2+V^*(\lambda,\sigma).u_2\}$ transforming under the symmetry operators $\Gamma(\alpha)$ according to (4). Also, (14) shows that $z \in R(L(\lambda,\sigma))$ if and only if $Q(\lambda,\sigma)z = 0$.

Finally, it follows that $C(\lambda,\sigma) = 0$ if and only if $C(\lambda,\sigma).u_1 = 0$. Now $C(\lambda,\sigma).u_1 = h_1(\lambda,\sigma)w_1 + h_2(\lambda,\sigma)w_2$, and it is easily seen from the definitions of $C(\lambda,\sigma)$ and w_i that $h_i(\lambda_0,\sigma_0) = 0$ $(i=1,2)$ and $D_\sigma h_1(\lambda_0,\sigma_0) = 1$. It follows that $h_1(\lambda,\sigma) = 0$ has a unique solution $\sigma = \tilde{\sigma}_1(\lambda)$ in a neighbourhood of (λ_0,σ_0); $\tilde{\sigma}_1$ is a C^1-function with $\tilde{\sigma}_1(\lambda_0) = \sigma_0$. Defining

$$\gamma(\lambda) = h_2(\lambda,\tilde{\sigma}_1(\lambda)) \tag{17}$$

we obtain the following result.

7.3.5. <u>Lemma</u>. Suppose that $L : \omega_1 \to L(X,Z)$ satisfies $(H)_L$. Then there is a neighbourhood ω of (λ_0,σ_0) in $\Lambda \times \mathbb{R}$ such that, for each $(\lambda,\sigma) \in \omega$, $L(\lambda,\sigma)$ is a Fredholm operator with zero index. For such (λ,σ) there are two possibilities : either

(i) $\ker L(\lambda,\sigma) = \{0\}$ and $R(L(\lambda,\sigma)) = Z$;

or

(ii) $\dim \ker L(\lambda,\sigma) = 2$, $\ker L(\lambda,\sigma)$ has a basis $\{u_1+V^*(\lambda,\sigma).u_1, u_2+ V^*(\lambda,\sigma)u_2\}$ transforming under $\Gamma(\alpha)$ according to (4), and

$$R(L(\lambda,\sigma)) = \{z \in Z \mid Q(\lambda,\sigma).z = 0\} ,$$

where $Q(\lambda,\sigma)$ is a projection in Z onto $R(Q)$, commuting with $\tilde{\Gamma}(\alpha)$ and such that $Q(\lambda_0,\sigma_0) = Q$.

One has the possibility (ii) if and only if

$$\sigma = \tilde{\sigma}_1(\lambda) \quad \text{and} \quad \gamma(\lambda) = 0 , \tag{18}$$

where $\tilde{\sigma}_1(\lambda)$ and $\gamma(\lambda)$ are C^1-functions with $\tilde{\sigma}_1(\lambda_0) = \sigma_0$ and $\gamma(\lambda_0) = 0$. □

7.3.6. Relation with the nonlinear problem. There is a relation between the linear problem just considered and the nonlinear problem of section 2, as follows. Let $M(x,\lambda,\sigma)$ be as in section 2, and define $L(\lambda,\sigma)$ by :

$$L(\lambda,\sigma) = D_x M(0,\lambda,\sigma) . \tag{19}$$

Then L satisfies the hypotheses $(H)_L$, and it is an easy exercise to establish the following relations between the mappings v^*, $\tilde{\sigma}$ and g of section 2, from one side, and the mappings V^*, $\tilde{\sigma}_1$ and γ connected with the linear problem, from the other side :

$$. \quad V^*(\lambda,\sigma) = D_u v^*(0,\lambda,\sigma) ; \tag{20}$$

$$\tilde{\sigma}_1(\lambda) = \tilde{\sigma}(0,\lambda) , \tag{21}$$

and

$$\gamma(\lambda) = g(0,\lambda) . \tag{22}$$

In particular, (22) shows that the function $\gamma(\lambda)$ given by (17) coincides with the function $\gamma(\lambda)$ used in section 2, in case $L(\lambda,\sigma)$ is given by (19).

An application of the foregoing results on the mapping M defined by (7) gives the following information on Σ_0.

7.3.7. <u>Theorem</u>. Suppose $r \geq 2$ and $(m_0,\sigma_0) \in \Sigma_0$. Then there exists a neighbourhood ω of (m_0,σ_0) in $\Lambda_0 \times \mathbb{R}$ such that

$$\Sigma_0 \cap \omega = \{(m,\sigma) \in \omega \mid \sigma = \tilde{\sigma}(0,m) \text{ and } \gamma(m) = 0\} , \qquad (23)$$

where $\sigma(\rho,m)$ is as in theorem 2.4 and $\gamma(m)$ as in theorem 2.9. So Σ_0 is locally a C^{r-1}-submanifold of $\Lambda_0 \times \mathbb{R}$ with codimension two.

P r o o f. Let $L(m,\sigma) = D_x m(0,\sigma)$; by lemma 5 $L(m,\sigma)$ is a Fredholm operator with zero index for each (m,σ) near (m_0,σ_0). A necessary and sufficient condition for $L(m,\sigma)$ to have a nontrivial kernel is that $\sigma = \tilde{\sigma}_1(m) = \tilde{\sigma}(0,m)$ and $\gamma(m) = 0$. When these conditions are satisfied then lemma 5 implies that (H)(i) is satisfied at (m,σ). Under the same conditions we have $u_1+V^*(m,\sigma).u_1 \in \ker L(m,\sigma)$, while

$$Q(m,\sigma).D_\sigma D_x m(0,\sigma).(u_1+V^*(m,\sigma).u_1)$$

is a continuous function of (m,σ), which for $(m,\sigma) = (m_0,\sigma_0)$ reduces to $QD_\sigma D_x m_0(0,\sigma_0).u_1 \neq 0$, since $(m_0,\sigma_0) \in \Sigma_0$. So this expression will be different from zero for all (m,σ) near (m_0,σ_0), and (H)(ii) is satisfied at (m,σ). This proves (23).

The last statement of the theorem follows from (23) and the fact that

$$D_m\gamma(m_0).\tilde{m} = Q_2 D_x \tilde{m}(0,\sigma_0) \quad , \quad \forall \tilde{m} \in \Lambda_0 , \qquad (24)$$

(see (2.28)). Indeed, the expression (24) is different from zero for an appropriate $\tilde{m} \in \Lambda_0$ (for example, take $\tilde{m}(\beta_1 u_1+\beta_2 u_2+v) = \beta_1 w_2-\beta_2 w_1$, for $\beta_1,\beta_2 \in \mathbb{R}$, $v \in \ker P$, and w_1,w_2 as in section 2). □

7.3.8. The generic condition $a \neq 0$. Let $r \geq 3$. For each point $(m,\sigma) \in \Sigma_0$ we can examine whether or not the condition $a \neq 0$ of theorem 2.9 is satisfied. According to remark 2.11 it is sufficient to choose appropriate projections and basic vectors, and to calculate a from (2.32) and (2.30). This will give us a mapping $a : \Sigma_0 \to \mathbb{R}$, $(m,\sigma) \mapsto a(m,\sigma)$. Let

$$\Sigma_1 = \{(m,\sigma) \in \Sigma_0 \mid a(m,\sigma) \neq 0\} \ . \tag{25}$$

Then the statement that the condition $a \neq 0$ is generically satisfied means that the points of Σ_1 are generic in Σ_0; more precisely :

7.3.9. Theorem. The subset Σ_1 is open and dense in Σ_0.

P r o o f. Let $(m_0,\sigma_0) \in \Sigma_1$, and let ω be a neighbourhood of (m_0,σ_0) in $\Lambda_0 \times \mathbb{R}$ such that the results of lemma 5 hold in ω. Using the notation of subsection 4 and lemma 5, define, for $(m,\sigma) \in \omega$:

$$P(m,\sigma) = P + V^*(m,\sigma)P \tag{26}$$

and

$$Q(m,\sigma) = Q - QD_x m(0,\sigma).W^*(m,\sigma) \ , \tag{27}$$

(see (15)). Then $P(m,\sigma)$ and $Q(m,\sigma)$ are projections in X, respectively Z, commuting with the symmetry operators $\Gamma(\alpha)$, respectively $\tilde{\Gamma}(\alpha)$. Lemma 5 shows that if $(m,\sigma) \in \Sigma_0 \cap \omega$, then $R(P(m,\sigma)) = \ker D_x m(0,\sigma)$ and $\ker Q(m,\sigma) = R(D_x m(0,\sigma))$. Also $P(m_0,\sigma_0) = P$ and $Q(m_0,\sigma_0) = Q$. It is easily seen that the pseudo-inverse $K(m,\sigma)$ of $D_x m(0,\sigma)$ is given by the restriction of $W^*(m,\sigma)$ to $R(D_x m(0,\sigma))$.

For the basis $\{u_1(m,\sigma), u_2(m,\sigma)\}$ of $\ker D_x m(0,\sigma)$ we choose

$$u_i(m,\sigma) = u_i + V^*(m,\sigma).u_i \quad , \quad i = 1,2 \ ;$$

the corresponding vectors $w_i(m,\sigma)$ $(i = 1,2)$ are given by

$$w_i(m,\sigma) = Q(m,\sigma)D_\sigma D_x m(0,\sigma).u_i(m,\sigma) \quad , \quad i = 1,2 \ .$$

Since both $Q(m,\sigma)$ and $w_i(m,\sigma)$ depend continuously on (m,σ), it follows that also the functionals $Q_i(m,\sigma) \in L(Z,\mathbb{R})$ depend continuously on (m,σ).

Now we define :

$$a(m,\sigma) = \frac{1}{3}Q_2(m,\sigma)D_x^3m(0,\sigma).(u_1(m,\sigma),u_1(m,\sigma),u_1(m,\sigma))$$
$$- Q_2(m,\sigma)D_x^2m(0,\sigma). \tag{28}$$
$$(u_1(m,\sigma),W^*(m,\sigma)(I-Q(m,\sigma))D_x^2m(0,\sigma).(u_1(m,\sigma),u_1(m,\sigma))).$$

This is a continuous function of (m,σ). Since $(m_0,\sigma_0) \in \Sigma_1$, we have $a(m_0,\sigma_0) \neq 0$, and consequently $a(m,\sigma) \neq 0$ for all (m,σ) near (m_0,σ_0). This shows that Σ_1 is open in Σ_0.

To show that Σ_1 is dense in Σ_0, suppose that $(m_0,\sigma_0) \in \Sigma_0 \setminus \Sigma_1$. Define $\bar{m} : X \to Z$ by

$$\bar{m}(\beta_1u_1+\beta_2u_2+v) = (\beta_1^2+\beta_2^2)(\beta_1w_2-\beta_2w_1) \quad , \quad \forall\beta_1,\beta_2 \in \mathbb{R} \ , \ \forall v \in \ker P \ .$$

It is easy to see that $\bar{m}(\Gamma(\alpha)x) = \Gamma(\alpha)\bar{m}(x)$ for all $\alpha \in \mathbb{R}$ and all $x \in X$. Define, for $\varepsilon \in \mathbb{R}$:

$$m_\varepsilon(x,\sigma) = m_0(x,\sigma) + \varepsilon\bar{m}(x) \ .$$

Since $D_xm_\varepsilon(0,\sigma) = D_xm_0(0,\sigma)$ and $D_\sigma D_xm_\varepsilon(0,\sigma) = D_\sigma D_xm_0(0,\sigma)$, it follows that $(m_\varepsilon,\sigma_0) \in \Sigma_0$ for all ε. A direct calculation shows that $a_\varepsilon = 2\varepsilon$; so $(m_\varepsilon,\sigma_0) \in \Sigma_1$ for all $\varepsilon \neq 0$. This proves that Σ_1 is dense in Σ_0. □

Application of theorem 2.9 to equation (8) gives the following result.

7.3.10. Theorem. Let $r \geq 3$ and $(m_0,\sigma_0) \in \Sigma_1$. Then there exists a neighbourhood ω of $(0,\sigma_0)$ in $X \times \mathbb{R}$, and a neighbourhood U of m_0 in Λ_0, such that for each $m \in U$ the equation (1) has a nontrivial solution $(x,\sigma) \in \omega$ $(x \neq 0)$ if and only if

$$a_0\gamma(m) < 0 \ , \tag{29}$$

where $a_0 = a(m_0,\sigma_0)$.

If the condition (29) is satisfied, then the set of nontrivial solutions of (1) in ω has the form $\{(\Gamma(\alpha)\bar{x}(m),\bar{\sigma}(m)) \mid \alpha\in\mathbb{R}\}$, where $P\bar{x}(m) = \bar{\rho}(m)u_1$, for some $\bar{\rho}(m) > 0$. One has $\bar{\rho}(m) = O(|\gamma(m)|^{1/2})$.

The set

$$B_0 = \{m\in U \mid \gamma(m) = 0\} \tag{30}$$

is a C^{r-1}-submanifold of Λ_0, with codimension one, containing m_0, and dividing U in two subregions : in one of these subregions (1) has no nontrivial solutions in ω, while for m in the other subregion (1) has exactly one family of nontrivial solutions as described above.

P r o o f. Apply theorem 2.9. The fact that B_0 is a submanifold of codimension one was already proved in theorem 7. The fact that $\bar{x}(m) = O(|\gamma(m)|^{1/2})$ shows that B_0 is indeed the (local) bifurcation set of the problem : if m approaches B_0 from the subregion where (29) is satisfied, then the corresponding family of nontrival solutions approaches $x = 0$. □

7.3.11. Remark. When considering a one-parameter family of equations ($\lambda \in \mathbb{R}$), such as in theorem 2.7, then we consider in fact a one-dimensional path in Λ_0 through $m_0(.,.) = M(.,0,.)$. The condition (2.26) means that this path is *transversal* to the submanifold B_0 at m_0 (see (2.28)) : the path crosses the bifurcation manifold B_0, and bifurcation takes place.

In the paper [268] we briefly discussed a situation where the condition $a \neq 0$ is not satisfied. This (nongeneric) situation is interesting from the point of view of imperfect bifurcation (see Golubitsky and Schaeffer [74] and [75], and Golubitsky and Langford [250]).

7.4. HOPF BIFURCATION

In this section we consider a particular application of the results of section 2. We suppose that M is a mapping of a specified form between spaces of periodic functions. As a special case we obtain the classical Hopf bifurcation for ordinary differential equations as well as for functional differential equations. Our general setting gives a unified approach for both cases.

7.4.1. The problem. Let Z be the space of all continuous 2π-periodic functions $z : \mathbb{R} \to \mathbb{R}^n$, let X be the subspace of all continuously differentiable functions in Z, and let

$$M(x,\lambda,\sigma) = -\frac{dx}{dt} + F(x,\lambda,\sigma) . \qquad (1)$$

Here $x \in X$, $\lambda \in \Lambda$ and $\sigma \in \mathbb{R}$, while we suppose that $F : Z \times \Lambda \times \mathbb{R} \to Z$ is a C^2-map such that

$$F(0,\lambda,\sigma) = 0 \quad , \quad \forall(\lambda,\sigma) \in \Lambda \times \mathbb{R} . \qquad (2)$$

The equation $M(x,\lambda,\sigma) = 0$ takes the form :

$$\frac{dx}{dt} = F(x,\lambda,\sigma) . \qquad (3)$$

This is a functional differential equation for the 2π-periodic function $x = x(t)$.

First we examine what conditions should be imposed on F such that M is equivariant with respect to SO(2). This leads to the concept of an *autonomous* equation, as follows.

7.4.2. The symmetry condition. There is an obvious representation of SO(2) over Z, defined by

$$(\Gamma(\alpha)z)(t) = z(t+\alpha) \quad , \quad \forall t \in \mathbb{R} , \forall \alpha \in \mathbb{R} , \forall z \in Z . \qquad (4)$$

The restriction of $\Gamma(\alpha)$ to X gives a corresponding representation over X. The mapping M defined by (1) will be equivariant with respect to $(SO(2),\Gamma)$ (i.e. $M(\Gamma(\alpha)x,\lambda,\sigma) = \Gamma(\alpha)M(x,\lambda,\sigma)$, $\forall\alpha \in \mathbb{R}$, $\forall(x,\lambda,\sigma) \in X \times \Lambda \times \mathbb{R}$) if and only if

$$F(\Gamma(\alpha)z,\lambda,\sigma) = \Gamma(\alpha)F(z,\lambda,\sigma) \quad , \quad \forall\alpha \in \mathbb{R} , \forall(z,\lambda,\sigma) \in Z \times \Lambda \times \mathbb{R} . \qquad (5)$$

This follows from the continuity of F and the fact that X is dense in Z. Condition (5) says that equation (3) is *autonomous*, that is invariant for time translations. An equivalent form for the condition (5) is

$$F(z,\lambda,\sigma)(t) = F(\Gamma(t)z,\lambda,\sigma)(0) \quad , \quad \forall t \in \mathbb{R}, \ \forall (z,\lambda,\sigma) \in Z \times \Lambda \times \mathbb{R} . \tag{6}$$

When this is satisfied, let us define $f : Z \times \Lambda \times \mathbb{R} \to \mathbb{R}^n$ by

$$f(z,\lambda,\sigma) = F(z,\lambda,\sigma)(0) . \tag{7}$$

We also denote $\Gamma(t)z$ by z_t, i.e. :

$$z_t(s) = (\Gamma(t)z)(s) = z(t+s) \quad , \quad \forall s,t \in \mathbb{R} , \ \forall z \in Z . \tag{8}$$

So $z_t(s)$ is 2π-periodic, both in s and t. This notation has been inspired by a similar notation introduced by Hale in the theory of functional differential equations (see e.g. [79]). However, its meaning here is slightly different, since we restrict to periodic functions, and allow s to take values in all $\mathbb{R}$ (and not only in an interval on the negative axis).

Using (6) and the notation (8) we can rewrite (3) in the form :

$$\frac{dx}{dt}(t) = f(x_t,\lambda,\sigma) \quad , \quad \forall t \in \mathbb{R} . \tag{9}$$

M is equivariant with respect to $(SO(2),\Gamma)$ if and only if (3) can be rewritten in the form (9), for some C^2-map $f : Z \times \Lambda \times \mathbb{R} \to \mathbb{R}^n$.

In order to translate the hypotheses (H2) and (H3) of section 2 in terms of our particular M, we have to study the linearization of M.

7.4.3. The linearization. Fix some $(\lambda_0,\sigma_0) \in \Lambda \times \mathbb{R}$. We define, for each $\sigma \in \mathbb{R}$:

$$A(\sigma) = D_z f(0,\lambda_0,\sigma) \in L(Z,\mathbb{R}^n) . \tag{10}$$

Let $A_0 = A(\sigma_0)$ and $L = D_x M(0,\lambda_0,\sigma_0)$. Then L is explicitly given by :

$$(Lx)(t) = -\frac{dx}{dt}(t) + A_0 x_t \quad , \quad \forall x \in X , \ \forall t \in \mathbb{R} . \tag{11}$$

We want to determine ker L and R(L). To do so we have to study the linear equation

$$-\frac{dx}{dt}(t) + A_0 x_t = z(t) \quad , \quad \forall t \in \mathbb{R} , \tag{12}$$

where $z \in Z$ is given, and we want to solve for $x \in X$.

We will use Fourier series to determine the solutions of (12). For each $k \in \mathbb{Z}$ we denote by $c_k[z]$ the k-th Fourier coefficient of $z \in Z$:

$$c_k[z] = \frac{1}{2\pi}\int_0^{2\pi} e^{-ikt} z(t)dt \ ; \tag{13}$$

since z is real valued, we have $c_{-k}[z] = \bar{c}_k[z]$. For each $N \in \mathbb{N}$ we denote by $P_N z$ the truncated Fourier series of z :

$$(P_N z)(t) = \sum_{|k| \leqslant N} c_k[z]e^{ikt} \ . \tag{14}$$

On Z_c, the complexification of Z, we use the bilinear form :

$$\langle z^*, z\rangle = \frac{1}{2\pi}\int_0^{2\pi} (z^*(t), z(t))dt \quad , \quad \forall z, z^* \in Z \ , \tag{15}$$

where $(.,.)$ is the usual inner product on $\mathbb{C}^n$: $(a,b) = \sum_{i=1}^{n} \bar{a}_i b_i$, $\forall a,b \in \mathbb{C}^n$.

We need the following facts about the Fourier expansion of $z \in Z$: we have $\sum_{k\in\mathbb{Z}} \| c_k[z] \|^2 < \infty$, and $P_N z$ converges in $L_2((0,2\pi),\mathbb{R}^n)$ to z; consequently, we have for each $z, z^* \in Z$:

$$\lim_{N\to\infty} \langle z^*, P_N z\rangle = \langle z^*, z\rangle \ . \tag{16}$$

If $x \in X$, then $c_k[\dot{x}] = ikc_k[x]$, and $\sum_{k\in\mathbb{Z}} (1+k^2)\| c_k[x] \|^2 < \infty$. It follows that

$$\Big(\sum_{k\in\mathbb{Z}} \| c_k[x] \|\Big)^2 \leqslant \sum_{k\in\mathbb{Z}} (1+k^2)^{-1} \sum_{k\in\mathbb{Z}} (1+k^2)\| c_k[x] \|^2 < \infty \ ,$$

which implies that $x(t) = \lim_{N\to\infty} (P_N x)(t)$, uniformly for $t \in \mathbb{R}$. The same holds for x_t, and since A_0 is continuous from Z into $\mathbb{R}^n$, it follows that

$$A_0 x_t = \sum_{k\in\mathbb{Z}} e^{ikt} A_0(c_k[x]e^{iks}) \ , \tag{17}$$

uniformly for $t \in \mathbb{R}$, and for each $x \in X$. (In (17) we have to use the complexification of A_0, which we also denote by A_0).

Now suppose that $x \in X$ and $z \in Z$ satisfy (12). Taking Fourier coefficients we find : $\Delta_k . c_k[x] = c_k[z]$, $\forall k \in \mathbb{Z}$, (18)

where $\Delta_k \in L(\mathbb{C}^n)$ is defined by :

$$\Delta_k . c = -ikc + A_0(ce^{iks}) \quad , \quad \forall c \in \mathbb{C}^n . \tag{19}$$

It is immediately verified that $\Delta_{-k}.\bar{c} = \overline{\Delta_k . c}$ for all $c \in \mathbb{C}^n$ and all $k \in \mathbb{Z}$. We denote by Δ_k^* the adjoint of Δ_k :

$$(\Delta_k^* . a, b) = (a, \Delta_k . b) \quad , \quad \forall a,b \in \mathbb{C}^n . \tag{20}$$

Finally, we denote by n_k the dimension of the (complex) subspace $\ker \Delta_k$ of $\mathbb{C}^n$; we have $n_{-k} = n_k$.

7.4.4. Lemma. There exists some $k_0 \in \mathbb{N}$ such that $n_k = 0$ if $|k| > k_0$. Then

$$\dim \ker L = \sum_{|k| \leq k_0} n_k < \infty \tag{21}$$

and

$$\ker L = \{ \sum_{k=0}^{k_0} \mathrm{Re}(c_k e^{ikt}) \mid c_k \in \ker \Delta_k, \forall k \} . \tag{22}$$

P r o o f. For $k \neq 0$ we can write Δ_k in the form :

$$\Delta_k = -ikI + D_k = -ik(I-(ik)^{-1}D_k) ,$$

where $D_k . c = A_0(ce^{iks})$, for each $c \in \mathbb{C}^n$. Since A_0 is bounded, there exists a uniform bound for $\|D_k\|$, and consequently :

$$\|(ik)^{-1}D_k\| \leq 1-\eta < 1 \quad \text{and} \quad \|(I-(ik)^{-1}D_k)^{-1}\| \leq \eta^{-1} ,$$

for some $\eta > 0$ and for all $|k| > k_0$, with k_0 sufficiently large. It follows that for $|k| > k_0$ the operator Δ_k is invertible, $n_k = 0$ and $\|\Delta_k^{-1}\| \leq C|k|^{-1}$ for some $C > 0$.

Now notice that $\{\mathrm{Re}(c_k e^{ikt}) \mid c_k \in \ker \Delta_k\}$ is a $2n_k$-dimensional (real) subspace of X if $k \neq 0$, and a n_0-dimensional subspace if $k = 0$. Then, using

$n_{-k} = n_k$, (21) is an easy consequence of (22). So it remains to prove (22).

Let $x \in \ker L$. Taking $z = 0$ in (18), we see that $c_k[x] \in \ker \Delta_k$ for all $k \in \mathbb{Z}$, and consequently $c_k[x] = 0$ for $|k| > k_0$. But then :

$$x(t) = \sum_{|k| \leqslant k_0} c_k[x] e^{ikt} ,$$

with $c_k[x] \in \ker \Delta_k$ and $c_{-k}[x] = \overline{c_k[x]}$, so that x belongs to the set at the r.h.s. of (22). Conversely, a direct calculation shows that the function $\mathrm{Re}(c_k e^{ikt})$ belongs to ker L if $c_k \in \ker \Delta_k$. This completes the proof. □

7.4.5. Lemma. R(L) is closed and dim ker L = codim R(L), i.e. L is a Fredholm operator with zero index. Also :

$$R(L) = \{z \in Z \mid (c_k^*, c_k[z]) = 0 \quad , \ \forall c_k^* \in \ker \Delta_k^* \ , \ 0 \leqslant k \leqslant k_0\} . \tag{23}$$

P r o o f. Since $\dim \ker \Delta_k^* = \dim \ker \Delta_k = n_k$, the first part of the lemma is a consequence of (23) and lemma 4. So it is sufficient to prove (23).

Let $z \in R(L)$; then (12) and (18) hold for some $x \in X$, i.e. $c_k[z] \in R(\Delta_k)$ for all $k \in \mathbb{Z}$. This is equivalent to :

$$(c_k^*, c_k[z]) = 0 \quad , \quad \forall c_k^* \in \ker \Delta_k^* \ , \forall k \in \mathbb{Z} . \tag{24}$$

Since $\ker \Delta_k^*$ is trivial for $|k| > k_0$, and since $c_k^* \in \ker \Delta_k^*$ if and only if $\bar{c}_k^* \in \ker \Delta_{-k}^*$, it follows that z belongs to the set at the r.h.s. of (23).

Conversely, if z is an element of the set at the r.h.s. of (23), then $z \in Z$ and (24) holds. This implies that the equation

$$\Delta_k . c_k = c_k[z] \tag*{$(25)_k$}$$

has at least one solution $c_k \in \mathbb{C}^n$ for each $k \in \mathbb{Z}$. If c_k is a solution of $(25)_k$, then $\bar{c}_k$ is a solution of $(25)_{-k}$. Let now $\{c_k \mid k \in \mathbb{Z}\}$ be any solution set of (25), such that $c_{-k} = \bar{c}_k$. Consider the Fourier series $\sum c_k e^{ikt}$; we will show that this series converges to some $x \in X$, and that $Lx = z$; so $z \in R(L)$, and the lemma will be proved.

If $|k| > k_0$, then $(23)_k$ and lemma 4 imply that $c_k = \Delta_k^{-1}.c_k[z]$ and $\|c_k\| \leqslant C|k|^{-1}\|c_k[z]\|$. So we have :

$$(\sum_{|k|>k_0} \|c_k\|)^2 \leq C^2(\sum_{|k|>k_0} |k|^{-2}) . \sum_{|k|>k_0} \|c_k[z]\|^2 < \infty .$$

This shows that $\sum c_k e^{ikt}$ is uniformly convergent, and that its limit x(t) is continuous, real-valued and 2π-periodic. Define $y : \mathbb{R} \to \mathbb{R}^n$ by :

$$y(t) = x(0) + \int_0^t (A_0 x_s - z(s))ds . \tag{26}$$

The integrandum in (26) is continuous and 2π-periodic, and $(23)_0$ implies that its mean value is zero. Consequently $y \in X$ and $y(0) = x(0)$. If we can show that $x(t) = y(t)$ for all $t \in \mathbb{R}$, then $x \in X$ and $Lx = z$, by (26).

From (25) and (26) it follows that $P_N\dot{y} = \frac{d}{dt}P_N x$. Let now $\psi : \mathbb{R} \to \mathbb{R}^n$ be any C^1-function. Then :

$$\begin{aligned}
\langle\dot{\psi},x\rangle &= \lim_{N\to\infty} \langle\dot{\psi},P_N x\rangle \\
&= \lim_{N\to\infty} (\psi(t),(P_N x)(t))\Big|_{t=0}^{t=2\pi} - \lim_{N\to\infty} \langle\psi,\frac{d}{dt}P_N x\rangle \\
&= (\psi(t),x(t))\Big|_{t=0}^{t=2\pi} - \lim_{N\to\infty} \langle\psi,P_N\dot{y}\rangle \\
&= (\psi(t),y(t))\Big|_{t=0}^{t=2\pi} - \langle\psi,\dot{y}\rangle = \langle\dot{\psi},y\rangle .
\end{aligned}$$

By a standard argument this implies that $x(t) = y(t)$ for all $t \in \mathbb{R}$. □

7.4.6. We want to find conditions on $f(z,\lambda,\sigma)$ such that $M(z,\lambda,\sigma)$ satisfies the hypotheses (H2) and (H3) of section 2. To formulate such conditions we will need the operator $\Delta_k(\sigma) \in L(\mathbb{C}^n)$ defined by :

$$\Delta_k(\sigma).c = -ikc + A(\sigma)(ce^{iks}) \quad , \quad \forall c \in \mathbb{C}^n , \tag{27}$$

where $A(\sigma)$ is given by (10). We denote by $\Delta_k'(\sigma)$ the derivative of $\Delta_k(\sigma)$ in the variable σ. Also, let :

$$H_k(\sigma) = \det \Delta_k(\sigma) . \tag{28}$$

Finally, we remind the definition 6.3.1 of a B-simple eigenvalue of a linear operator A.

7.4.7. <u>Theorem</u>. Let $(\lambda_0,\sigma_0) \in \Lambda\times\mathbb{R}$, and let $f : Z\times\Lambda\times\mathbb{R} \to \mathbb{R}^n$ be a C^2-map such that $f(0,\lambda,\sigma) = 0$, $\forall(\lambda,\sigma) \in \Lambda\times\mathbb{R}$. Let M be defined by (1), where $F : Z\times\Lambda\times\mathbb{R} \to Z$ is defined by $F(z,\lambda,\sigma)(t) = f(z_t,\lambda,\sigma)$. Let L, $A(\sigma)$, $\Delta_k(\sigma)$ and $H_k(\sigma)$ be given by (11), (10), (27) and (28) respectively.

Then the following statements are equivalent :

(i) dim ker L = 2, ker L has a basis $\{u_1,u_2\}$ such that (2.3) holds, and $D_\sigma D_x M(0,\lambda_0,\sigma_0).u \notin R(L)$ for some $u \in \ker L\setminus\{0\}$;

(ii) zero is a $\Delta_1'(\sigma_0)$-simple eigenvalue of $\Delta_1(\sigma_0)$, and $\Delta_k(\sigma_0)$ is invertible for all $k \neq \pm 1$;

(iii) σ_0 is a simple zero of H_1 (i.e. $H_1(\sigma_0) = 0$ and $H_1'(\sigma_0) \neq 0$), and $H_k(\sigma_0) \neq 0$ for all $k \neq \pm 1$.

P r o o f. Suppose that (i) is satisfied. Let $c_1 = u_1(0) + iu_2(0)$, and $\zeta(t) = c_1e^{it}$. It follows from (2.3) that $u_1(t) = \mathrm{Re}(\zeta(t))$, $u_2(t) = \mathrm{Im}(\zeta(t))$ and $\ker L = \{\mathrm{Re}(z\zeta) \mid z\in\mathbb{C}\}$. If $u \in \ker L$, then $c_k[u] = 0$ for all $k \neq \pm 1$. By lemma 4 this implies that $n_k = 0$ for $k \neq \pm 1$, and that $\Delta_k = \Delta_k(\sigma_0)$ is invertible for such k. Since dim ker L = 2, (21) also shows that $n_1 = n_{-1} = 1$. Indeed, we have $\ker \Delta_1(\sigma_0) = \mathrm{span}\{c_1\}$. Now, by definition of $\Delta_1(\sigma)$ and c_1 :

$$(D_xM(0,\lambda_0,\sigma).u_1)(t) = \mathrm{Re}(\Delta_1(\sigma).c_1e^{it})$$

and

$$(D_\sigma D_xM(0,\lambda_0,\sigma_0).u_1)(t) = \mathrm{Re}(\Delta_1'(\sigma_0).c_1e^{it}) .$$

Since this last expression does not belong to R(L) (by (i) and the general theory), it follows from lemma 5 that $(c_1^*,\Delta_1'(\sigma_0).c_1) \neq 0$, where $c_1^* \in \mathbb{C}^n$ is such that $\ker \Delta_1^*(\sigma_0) = \mathrm{span}\{c_1^*\}$. But this means precisely that $\Delta_1'(\sigma_0).c_1 \notin R(\Delta_1(\sigma_0))$, i.e. zero is a $\Delta_1'(\sigma_0)$-simple eigenvalue of $\Delta_1(\sigma_0)$.

Conversely, suppose that (ii) holds. Then lemma 4 shows that dim ker L = 2 and $\ker L = \{\mathrm{Re}(z\zeta) \mid z\in\mathbb{C}\}$, where $\zeta(t) = c_1e^{it}$ and c_1 is defined by

$\ker \Delta_1(\sigma_0) = \mathrm{span}\{c_1\}$. It is clear that the basis $\{u_1, u_2\}$ of $\ker L$, given by $u_1 = \mathrm{Re}\ \zeta$ and $u_2 = \mathrm{Im}\ \zeta$ has the required transformation properties. Since also $(c_1^*, \Delta_1'(\sigma_0).c_1) \neq 0$, the calculations above show that $D_\sigma D_x M(0,\lambda_0,\sigma_0).u_1 \notin R(L)$. We conclude that (i) and (ii) are equivalent.

As for the equivalence of (ii) and (iii), it is clear that $\Delta_k(\sigma_0)$ will be invertible if and only if $H_k(\sigma_0) \neq 0$. The remaining part of the proof follows from the following general result. □

7.4.8. Lemma. Let $\Delta : \mathbb{R} \to L(\mathbb{C}^n)$ be defined and of class C^1 in a neighbourhood of $\sigma = \sigma_0$. Let $H(\sigma) = \det \Delta(\sigma)$. Then zero is a $\Delta'(\sigma_0)$-simple eigenvalue of $\Delta(\sigma_0)$ if and only if σ_0 is a simple zero of H, i.e. if and only if $H(\sigma_0) = 0$ and $H'(\sigma_0) \neq 0$.

P r o o f. A proof using the canonical Jordan form of $\Delta(\sigma_0)$ can be found in Hale and de Oliveira [91],[170]. Here we give a proof based on exterior forms (see [238]). We start with the following identity, which can be considered as the definition of $H(\sigma)$:

$$H(\sigma)c_1 \wedge c_2 \wedge \ldots \wedge c_n = \Delta(\sigma).c_1 \wedge \Delta(\sigma).c_2 \wedge \ldots \wedge \Delta(\sigma).c_n , \quad \forall\sigma \in \mathbb{R} , \tag{29}$$

where $\{c_1,\ldots,c_n\}$ is any basis of $\mathbb{C}^n$.

Suppose zero is a $\Delta'(\sigma_0)$-simple eigenvalue of $\Delta(\sigma_0)$. Choose the basis of $\mathbb{C}^n$ such that $\Delta(\sigma_0).c_1 = 0$. Then $\{\Delta(\sigma_0).c_j \mid j = 2,3,\ldots,n\}$ will form a basis for $R(\Delta(\sigma_0))$. Putting $\sigma = \sigma_0$ in (29) gives $H(\sigma_0) = 0$, since $c_1 \wedge \ldots \wedge c_n \neq 0$. Differentiation of (29) at $\sigma = \sigma_0$ gives then :

$$H'(\sigma_0)c_1 \wedge \ldots \wedge c_n = \Delta'(\sigma_0).c_1 \wedge \Delta(\sigma_0).c_2 \wedge \ldots \wedge \Delta(\sigma_0).c_n . \tag{30}$$

Since $\Delta'(\sigma_0).c_1 \notin R(L)$, the right hand side of (30) is different from zero, and we conclude that $H'(\sigma_0) \neq 0$.

Conversely, suppose that $H(\sigma_0) = 0$ and $H'(\sigma_0) \neq 0$. Then (29), for $\sigma = \sigma_0$ and any basis of $\mathbb{C}^n$, shows that the vectors $\{\Delta(\sigma_0).c_j \mid j = 1,\ldots,n\}$ are linearly dependent, and consequently $\dim \ker \Delta(\sigma_0) = k \geq 1$. Let now $\{c_1,\ldots,c_n\}$ be a basis of $\mathbb{C}^n$ such that $\Delta(\sigma_0).c_j = 0$ for $1 \leq j \leq k$. Then (30) holds, and if $k > 1$, it follows that $H'(\sigma_0) = 0$, which contradicts the hypothesis. So dim ker $\Delta(\sigma_0) = 1$. Furthermore, (30) and $H'(\sigma_0) \neq 0$ imply that $\Delta'(\sigma_0).c_1$ does not belong to $\text{span}\{\Delta(\sigma_0).c_j \mid j = 2,\ldots,n\} = R(\Delta(\sigma_0))$. So zero is a $\Delta'(\sigma_0)$-simple eigenvalue of $\Delta(\sigma_0)$. □

7.4.9. The projection operators. When the equivalent conditions (i), (ii) and (iii) of theorem 7 are satisfied, then we can apply the general results of section 2 to the equation (9). In order to find a more explicit expression for the bifurcation function $g(\rho,\lambda)$ we need to introduce appropriate projections P and Q.

Fix $c_1 \in \ker \Delta_1(\sigma_0)$ and $c_1^* \in \ker \Delta_1^*(\sigma_0)$ such that

$$(c_1^*, \Delta_1'(\sigma_0).c_1) = 2 \ . \qquad (31)$$

Let $\zeta(t) = c_1 e^{it}$, $\zeta^*(t) = c_1^* e^{it}$, $u_1(t) = \text{Re}\ \zeta(t)$ and $u_2(t) = \text{Im}\ \zeta(t)$. Then $\ker L = \text{span}\{u_1,u_2\} = \{\text{Re}(z\zeta) \mid z \in \mathbb{C}\}$ and $R(L) = \{z \in Z \mid \langle\zeta^*, z\rangle = 0\}$, as can easily be seen from lemma 4 and lemma 5.

Define $P \in L(X)$ by :

$$Px = \text{Re}(\langle\zeta^*, D_\sigma A(\sigma_0).x_t\rangle\zeta) \qquad , \qquad \forall x \in X \ ; \qquad (32)$$

since $D_\sigma A(\sigma_0) \in L(Z,\mathbb{R}^n)$ we have :

$$\begin{aligned}\langle\zeta^*, D_\sigma A(\sigma_0).x_t\rangle &= \frac{1}{2\pi}\int_0^{2\pi} (\zeta^*(t), D_\sigma A(\sigma_0).x_t)dt \\ &= (c_1^*, D_\sigma A(\sigma_0).(\frac{1}{2\pi}\int_0^{2\pi} e^{-it}x(t+s)dt)) \\ &= (c_1^*, \Delta_1'(\sigma_0).c_1[x]) \ .\end{aligned}$$

From this it follows easily that P is a projection onto ker L, such that $P\Gamma(\alpha) = \Gamma(\alpha)P$, for all $\alpha \in \mathbb{R}$.

Similarly, we define $Q \in L(Z)$ by

$$Qz = \text{Re}(\langle\zeta^*, z\rangle\chi) \qquad , \qquad \forall z \in Z \ , \qquad (33)$$

where $\chi(t) = \Delta_1'(\sigma_0).c_1 e^{it} = D_\sigma A(\sigma_0).\zeta_t$. Again, Q is a projection commuting with the symmetry operators, and such that $\ker Q = R(L)$. If we let w_i =

$QD_\sigma D_x M(0,\lambda_0,\sigma_0).u_i = QD_\sigma A(\sigma_0).u_{i,t}$ $(i=1,2)$, then $w_1 = \mathrm{Re}(\chi)$ and $w_2 = \mathrm{Im}(\chi)$. From this it follows that :

$$Q_1(z) = \mathrm{Re}\langle\zeta^*,z\rangle \quad , \quad Q_2(z) = -\mathrm{Im}\langle\zeta^*,z\rangle \quad , \quad \forall z \in Z . \tag{34}$$

Using these projections one can, for example, explicitly work out the condition (2.26) of theorem 2.7. We will do this for a few particular case.

7.4.10. Hopf bifurcation for functional differential equations. Let $r \geq 0$, $C = C^0([-r,0],\mathbb{R}^n)$ and $h : C \times \Lambda \to \mathbb{R}^n$ a C^2-mapping such that $h(0,\lambda) = 0$ for all $\lambda \in \Lambda$. Let $x : [t_0-r,t_0+\alpha[\to \mathbb{R}^n$ be a continuous function, for some $\alpha > 0$; then we define, for each $t \in [t_0,t_0+\alpha[$ an element $x_t \in C$ by :

$$x_t(s) = x(t+s) \quad , \quad \forall s \in [-r,0] . \tag{35}$$

We want to find periodic solutions of the equation :

$$\dot{y}(t) = h(y_t,\lambda) \; ; \tag{36}$$

these are continuously differentiable functions $y : \mathbb{R} \to \mathbb{R}^n$, satisfying (36) for all $t \in \mathbb{R}$, and such that there is some $T > 0$ for which $y(t+T) = y(t)$, $\forall t \in \mathbb{R}$. Since the equation (36) is autonomous, the period T of such solution remains an unknown of the problem. We will rescale the independent variable t in such a way that the rescaled period has a fixed value, say 2π; at the same time this will bring the problem in the form (9).

Let $y(t)$ be a T-periodic solution of (36), with $T = 2\pi\sigma$ $(\sigma > 0)$. Define $x : \mathbb{R} \to \mathbb{R}^n$ by

$$x(t) = y(\sigma t) \quad , \quad \forall t \in \mathbb{R} . \tag{37}$$

Then $x(t)$ is 2π-periodic (i.e. $x \in X$) and

$$\dot{x}(t) = \sigma\dot{y}(\sigma t) = \sigma h(y_{\sigma t},\lambda) \quad , \quad \forall t \in \mathbb{R} .$$

Now we have for each $s \in [-r,0]$:

$$y_{\sigma t}(s) = y(\sigma t+s) = y(\sigma(t+\frac{s}{\sigma})) = x(t+\frac{s}{\sigma}) = x_{t,\sigma}(s) \ ,$$

where, for each $z \in Z$, $t \in \mathbb{R}$ and $\sigma > 0$ we define $z_{t,\sigma} \in C$ by

$$z_{t,\sigma}(s) = z(t+\frac{s}{\sigma}) \quad , \quad \forall s \in [-r,0] \ . \tag{38}$$

It follows that $x(t)$ is a 2π-periodic solution of the equation

$$\dot{x}(t) = \sigma h(x_{t,\sigma},\lambda) \ . \tag{39}$$

Conversely, if $x(t)$ is a 2π-periodic solution of (39) for some $\sigma > 0$, then $y(t) = x(\frac{t}{\sigma})$ is a $2\pi\sigma$-periodic solution of (36). So we have to determine 2π-periodic solutions of (39).

To bring this in the form (9), define $f : Z \times \Lambda \times]0,\infty[\to \mathbb{R}^n$ by :

$$f(z,\lambda,\sigma) = \sigma h(z_{0,\sigma},\lambda) \ . \tag{40}$$

Now remark that for each $s \in [-r,0]$:

$$(z_t)_{0,\sigma}(s) = (\Gamma(t)z)_{0,\sigma}(s) = z(t+\frac{s}{\sigma}) = z_{t,\sigma}(s) \ .$$

It follows that (9), with f defined by (40), coincides with the equation (39). Remark in this context that σ has to be considered as an unknown of the problem : given $\lambda \in \Lambda$ one has to adjust σ such that (39) has a 2π-periodic solution; if such σ can be found, then (36) has a corresponding $2\pi\sigma$-periodic solution.

For the critical value (λ_0,σ_0) we will take $\sigma_0 = 1$; the general case can be reduced to this particular one by a preliminary time rescale in the equation (36), and an appropriate redefinition of the mapping h. So we will look for solutions $(x,\lambda,\sigma) \in X \times \Lambda \times]0,\infty[$ of (39) near $(0,\lambda_0,1)$.

7.4.11. Remark. The mapping f defined by (40) is only continuous in σ, but not continuously differentiable : this is due to the presence of the term $z_{0,\sigma}$. But when we consider the restriction of f to $X \times \Lambda \times]0,\infty[$, then f is of class C^1 in σ, and of class C^2 in (x,λ). This is sufficient for the theory of the previous sections to remain valid.

In a number of cases a similar remark can be made about differentiability in the parameter λ. For example, equation (36) may contain some delays; considering these delays as parameters, f will only be continuous in these parameters. Again, the restriction of f to X will be continuously differentiable (see Hale [86]). A general remark in this context is that when we differentiate in λ and σ, this usually only happens after replacing x by solutions of certain equations, and such solutions may have more smoothness than general x.

Now define $A : \Lambda \to L(C,\mathbb{R}^n)$, $\Delta : \Lambda \times \mathbb{C} \to L(\mathbb{C}^n)$ and $H : \Lambda \times \mathbb{C} \to \mathbb{C}$ by :

$$A(\lambda) = D_z h(0,\lambda) , \tag{41}$$

$$\Delta(\lambda,\mu).c = -\mu c + A(\lambda).(ce^{\mu s}) \quad , \quad \forall c \in \mathbb{C}^n , \tag{42}$$

and

$$H(\lambda,\mu) = \det \Delta(\lambda,\mu) . \tag{43}$$

It is easily seen that these mappings are of class C^1; in particular, $H(\lambda,\mu)$ is analytic in μ. We let $H'(\lambda,\mu) = D_\mu H(\lambda,\mu)$ and $\Delta'(\lambda,\mu) = D_\mu \Delta(\lambda,\mu)$.

7.4.12. Lemma. For the equation (39) and $\sigma_0 = 1$, the statements (i), (ii) and (iii) of theorem 7 are equivalent with :

(iv) $H(\lambda_0,i) = 0$, $H'(\lambda_0,i) \neq 0$ and $H(\lambda_0,ik) \neq 0$, for all $k \in \mathbb{Z}$, $k \neq \pm 1$.

P r o o f. From the definition (27) of $\Delta_k(\sigma)$ we find for the case under consideration :

$$\begin{aligned} \Delta_k(\sigma).c &= -ikc + \sigma A(\lambda_0).(ce^{iks})_{0,\sigma} \\ &= -ikc + \sigma A(\lambda_0).(ce^{iks/\sigma}) \\ &= \sigma\Delta(\lambda_0,\tfrac{ik}{\sigma}).c \quad , \quad \forall c \in \mathbb{C}^n . \end{aligned}$$

Consequently :

$$\Delta_k(\sigma) = \sigma\Delta(\lambda_0,\frac{ik}{\sigma}) \quad , \quad \forall\sigma > 0 \ , \ \forall k \in \mathbb{Z} \ , \tag{44}$$

and

$$H_k(\sigma) = \sigma^n H(\lambda_0,\frac{ik}{\sigma}) \ . \tag{45}$$

It follows that $H_k(1) = H(\lambda_0,ik)$ and $H_1'(1) = nH(\lambda_0,i) - iH'(\lambda_0,i)$. The result is now obvious. □

Under the condition (iv) of lemma 12 we can apply the general theory of section 2 to equation (39). In particular, the application of theorem 2.7 to (39) will give us the classical theorem on Hopf bifurcation for functional differential equations (see e.g. Hale [79]). The following lemmas give us a more explicit form for the condition (2.26) of that theorem.

7.4.13. Lemma. Let $\Delta : \Lambda \times \mathbb{C} \to L(\mathbb{C}^n)$ be continuously differentiable, $(\lambda_0,\mu_0) \in \Lambda \times \mathbb{C}$, and zero a $\Delta'(\lambda_0,\mu_0)$-simple eigenvalue of $\Delta(\lambda_0,\mu_0)$. Let $\ker \Delta(\lambda_0,\mu_0) = \text{span}\{c_1\}$.

Then there exist a neighbourhood U of λ_0 in Λ, a neighbourhood V of μ_0 in $\mathbb{C}$, and continuously differentiable functions $\mu : U \to \mathbb{C}$ and $c_1 : U \to \mathbb{C}^n$ such that the following holds :

(i) for each $(\lambda,\mu) \in U \times V$, $\ker \Delta(\lambda,\mu)$ is nontrivial if and only if $\mu = \mu(\lambda)$;

(ii) $\ker \Delta(\lambda,\mu(\lambda)) = \text{span}\{c_1(\lambda)\}$, $\forall\lambda \in U$;

(iii) $\mu(\lambda_0) = \mu_0$, $c_1(\lambda_0) = c_1$.

P r o o f. We have to solve the equation $\Delta(\lambda,\mu).c = 0$ for $c \in \mathbb{C}^n$. Since zero is a $\Delta'(\lambda_0,\mu_0)$-simple eigenvalue of $\Delta(\lambda_0,\mu_0)$, we can find some $c_1^* \in \ker \Delta^*(\lambda_0,\mu_0)$ such that $(c_1^*,\Delta'(\lambda_0,\mu_0).c_1) = 1$. We define the following projections in $\mathbb{C}^n$:

$$P_0.c = (c_1^*,\Delta'(\lambda_0,\mu_0).c)c_1 \ , \ Q_0.c = (c_1^*,c)\Delta'(\lambda_0,\mu_0).c_1 \quad , \quad \forall c \in \mathbb{C}^n \ . \tag{46}$$

Each $c \in \mathbb{C}^n$ can be written as $c = \alpha c_1+d$, with $\alpha \in \mathbb{C}$ and $d \in \ker P_0$. Now $(I-Q_0)\Delta(\lambda_0,\mu_0)$ is an isomorphism between $\ker P_0$ and $R(\Delta(\lambda_0,\mu_0)) = \ker Q_0$; the same holds for $(I-Q_0)\Delta(\lambda,\mu)$ if (λ,μ) is sufficiently near to (λ_0,μ_0). The equation :

$$(I-Q_0)\Delta(\lambda,\mu).(\alpha c_1+d) = 0$$

has a unique solution $d = \alpha d^*(\lambda,\mu)$, where $d^*(\lambda,\mu)$ is uniquely defined and continuously differentiable for (λ,μ) in a neighbourhood of (λ_0,μ_0); moreover $d^*(\lambda_0,\mu_0) = 0$.

Now ker $\Delta(\lambda,\mu)$ will be nontrivial if and only if the equation

$$\alpha Q_0\Delta(\lambda,\mu).(c_1+d^*(\lambda,\mu)) = 0$$

has a solution $\alpha \neq 0$, i.e. if and only if :

$$\beta(\mu,\lambda) \equiv (c_1^*,\Delta(\mu,\lambda).(c_1+d^*(\lambda,\mu))) = 0 .$$

Now $\beta(\lambda_0,\mu_0) = 0$ and $D_\mu\beta(\lambda_0,\mu_0) = (c_1^*,\Delta'(\lambda_0,\mu_0).c_1) = 1$, since $(c_1^*,\Delta(\lambda_0,\mu_0).c) = 0$ for all $c \in \mathbb{C}^n$. So we can uniquely solve the equation $\beta(\lambda,\mu) = 0$ for $\mu = \mu(\lambda)$. If we let $c_1(\lambda) = c_1+d^*(\lambda,\mu(\lambda))$, then this proves the lemma. □

7.4.14. Lemma. Suppose that equation (39) satisfies the condition (iv) of lemma 12. Let $\mu(\lambda)$ and $c_1(\lambda)$ be the mappings given by lemma 13, such that $\mu(\lambda_0) = i$, $c_1(\lambda_0) = c_1$ and

$$\Delta(\lambda,\mu(\lambda)).c_1(\lambda) = 0 \qquad , \qquad \forall\lambda \in U . \tag{46}$$

Let $g(\rho,\lambda)$ be the bifurcation function for the equation (39), as given by theorem 2.4. Then :

$$D_\lambda g(0,\lambda_0).\tilde{\lambda} = \text{Re}\ D_\lambda\mu(\lambda_0).\tilde{\lambda} \qquad , \qquad \forall\lambda \in \Lambda . \tag{47}$$

P r o o f. Using (44), the normalization condition (31) for $c_1 \in \ker \Delta(\lambda_0,i)$ and $c_1^* \in \ker \Delta^*(\lambda_0,i)$ takes the form :

$$(c_1^*,\Delta'(\lambda_0,i).c_1) = 2i . \tag{48}$$

Differentiation of (46) at $\lambda = \lambda_0$, and taking the inner product with c_1^* gives :

$$(c_1^*, (D_\lambda \Delta(\lambda_0, i).\lambda).c_1) = -2iD_\lambda \mu(\lambda_0).\tilde{\lambda} \quad , \quad \forall \tilde{\lambda} \in \Lambda , \tag{49}$$

where we have used (48) and the fact that $(c_1^*, \Delta(\lambda_0, i).c) = 0$ for all $c \in \mathbb{C}^n$.

From (2.28) we have :

$$D_\lambda g(0,\lambda_0).\tilde{\lambda} = Q_2 D_x D_\lambda M(0,\lambda_0,1).(u_1,\tilde{\lambda}) ,$$

where $u_1(t) = \text{Re}(c_1 e^{it})$. For the equation (39) we have $M(x,\lambda,1)(t) = -\dot{x}(t) + h(x_t,\lambda)$. This implies

$$(D_x M(0,\lambda,1).u_1)(t) = \text{Re}(\Delta(\lambda,i).c_1 e^{it}) ,$$

and, using (34) where $\zeta^* = c_1^* e^{it}$:

$$D_\lambda g(0,\lambda_0).\tilde{\lambda} = -\frac{1}{2} \text{Im}(c_1^*, (D_\lambda \Delta(\lambda_0,i).\tilde{\lambda}).c_1) . \tag{50}$$

Combining with (49) gives (47). □

Using lemma 14, theorem 2.7 takes for the equation (36) the following form.

7.4.15. Theorem. Consider the equation (36) with $\Lambda = \mathbb{R}$. Define $H(\lambda,\mu)$ by (43) and let $\lambda_0 \in \mathbb{R}$. Suppose the following :

(i) $H(\lambda_0, i) = 0$, $H'(\lambda_0, i) \neq 0$ and $H(\lambda_0, ik) \neq 0$ for all $k \in \mathbb{Z}$, $k \neq \pm 1$;

(ii) if $\mu(\lambda)$ is the solution branch of $H(\lambda,\mu) = 0$ such that $\mu(\lambda_0) = i$, then

$$\text{Re}\, D_\lambda \mu(\lambda_0) \neq 0 . \tag{51}$$

Then there exist C^1-functions $\bar{\lambda}(\rho)$ and $\bar{\sigma}(\rho)$, both even in ρ and with $\bar{\lambda}(0) = \lambda_0$ and $\bar{\sigma}(0) = 1$, such that for each sufficiently small $\rho > 0$ the equation (36) has for $\lambda = \bar{\lambda}(\rho)$ a family of $2\pi\bar{\sigma}(\rho)$-periodic solutions, of the form $\{\Gamma(\alpha)\bar{x}(\rho) \mid \alpha\in\mathbb{R}\}$, with $\bar{x}(\rho) = O(\rho)$. For λ near λ_0, these are the only non-trivial solutions of (36) near $x = 0$ and with a period near 2π. □

Theorem 15 is the classical Hopf bifurcation result for functional differential equations. It contains as a special case the Hopf bifurcation for ordinary differential equations, which we describe now.

7.4.16. <u>Hopf bifurcation for ordinary differential equations</u>. Let $h : \mathbb{R}^n \times \Lambda \to \mathbb{R}^n$ be a C^2-function, with $h(0,\lambda) = 0$ for all λ. The ordinary differential equation

$$\dot{y} = h(y,\lambda) \tag{52}$$

forms a special case of (36) : it is sufficient to take $r = 0$. Then the space C coincides with $\mathbb{R}^n$. All results for (36), and in particular theorem 15, are also valid for (52). There are even a few simplifications.

The operator $A(\lambda)$ defined by (41) belongs to $L(\mathbb{R}^n)$, and $\Delta(\lambda,\mu) = -\mu I + A(\lambda)$ for all $\mu \in \mathbb{C}$. The condition (iv) of lemma 12 takes the form :

(v) i is a simple eigenvalue of $A(\lambda_0)$, i.e. $\dim \ker (A(\lambda_0)-iI) = 1$, and if $c_1 \in \mathbb{C}^n$ is a corresponding eigenvector, then $c_1 \notin R(A(\lambda_0)-iI)$; moreover, if $k \in \mathbb{Z}$, $k \neq \pm 1$, then ik is not an eigenvalue of $A(\lambda_0)$.

The function $\mu(\lambda)$ is the eigenvalue branch of $A(\lambda)$ such that $\mu(\lambda_0) = i$. Using these modifications theorem 15 becomes the classical Hopf bifurcation theorem (Hopf [98], Marsden and McCracken [156], Crandall and Rabinowitz [52]). Of course we can also apply our other results (such as theorem 2.9) to the equations (36) and (52). To conclude this section, let us calculate for (52) the constant a appearing in theorem 2.9.

7.4.17. For the Hopf bifurcation problem for (52), the operator $M : X \times \Lambda \times \mathbb{R} \to Z$ is given by :

$$M(x,\lambda,\sigma)(t) = -\dot{x}(t) + \sigma h(x(t),\lambda) \quad , \quad \forall t \in \mathbb{R} . \tag{53}$$

Let us assume (v) above. Let $A(\lambda) = D_x h(0,\lambda)$. Then $\Delta(\lambda,\mu) = -\mu I + A(\lambda)$, the vectors $c_1 \in \mathbb{C}^n$ and $c_1^* \in \mathbb{C}^n$ are such that

$$A(\lambda_0)c_1 = ic_1 \quad , \quad A^T(\lambda_0)c_1^* = -ic_1 \ , \tag{54}$$

while the normalization condition (48) takes the form :

$$(c_1^*,c_1) = -2i \ . \tag{55}$$

Putting, as before, $\zeta(t) = c_1 e^{it}$ and $\zeta^*(t) = c_1^* e^{it}$, it follows that

$$\dot{\zeta}(t) = i\zeta(t) = A(\lambda_0)\zeta(t) \ , \ -\dot{\zeta}^*(t) = -i\zeta^*(t) = A^T(\lambda_0)\zeta^*(t) \ . \tag{56}$$

Also $\Delta_1'(1) = \Delta(\lambda_0,i)-i\Delta'(\lambda_0,i) = A(\lambda_0)$, such that in (33) we have $\chi = A(\lambda_0)\zeta = i\zeta$. Then it follows from (32) that for each $z \in Z$:

$$Pz = \mathrm{Re}(<\zeta^*,A(\lambda_0)z>\zeta) = \mathrm{Re}(<A^T(\lambda_0)\zeta^*,z>\zeta)$$

$$= \mathrm{Re}(i<\zeta^*,z>\zeta) = Qz \ . \tag{57}$$

Writing $\zeta = u_1+iu_2$ and $\zeta^* = -u_2^*+iu_1^*$ this takes the form :

$$Pz = <u_1^*,z>u_1 + <u_2^*,z>u_2 \quad , \quad \forall z \in Z \ . \tag{58}$$

Since $<\zeta^*,\zeta> = -2i$ (from (55)) and $<\zeta^*,\bar{\zeta}> = 0$, it follows easily that

$$<u_i^*,u_j> = \delta_{ij} \qquad i,j = 1,2 \ . \tag{59}$$

Finally, we have from (34) :

$$Q_1(z) = \mathrm{Re}<\zeta^*,z> = -<u_2^*,z> \quad , \quad \forall z \in Z \ . \tag{60}$$

and

$$Q_2(z) = -\mathrm{Im}<\zeta^*,z> = <u_1^*,z> \quad , \quad \forall z \in Z \ . \tag{61}$$

Now we use the formulas (2.32) and (2.30) to determine the number a. First consider $v_2 = D_u^2 v^*(0,0,0).(u_1,u_1)$. From (53) and the definition of u_1

it follows immediately that $QD_x^2M(0,\lambda_0,1)(u_1,u_1) = 0$. So, by (2.30), v_2 is the unique solution of

$$-\dot{v}_2 + A(\lambda_0)v_2 = -D_x^2h(0,\lambda_0).(u_1,u_1)$$

such that $Pv_2 = 0$, i.e. such that $\langle\zeta^*,v_2\rangle = 0$ (by (57)). A direct calculation shows that

$$\begin{aligned} v_2 = -\tfrac{1}{4}\{&(-2i+A(\lambda_0))^{-1}.D_x^2h(0,\lambda_0).(c_1,c_1)e^{2it} \\ &+ 2A(\lambda_0)^{-1}D_x^2h(0,\lambda_0).(c_1,\bar{c}_1) \\ &+ (2i+A(\lambda_0))^{-1}.D_x^2h(0,\lambda_0).(\bar{c}_1,\bar{c}_1)e^{-2it}\} .\end{aligned}$$

Using (61) and (2.32) an easy, straightforward calculation shows that a is given by :

$$\begin{aligned} a = -\tfrac{1}{8}\mathrm{Im}\{&(c_1^*,D_x^3h(0,\lambda_0).(c_1,c_1,\bar{c}_1)) \\ &-(c_1^*,D_x^2h(0,\lambda_0).(\bar{c}_1,(-2i+A(\lambda_0))^{-1}D_x^2h(0,\lambda_0).(c_1,c_1))) \\ &-2(c_1^*,D_x^2h(0,\lambda_0).(c_1,A(\lambda_0)^{-1}D_x^2h(0,\lambda_0).(c_1,\bar{c}_1)))\} . \end{aligned} \tag{62}$$

This constant a is not only important in the application of theorem 2.9, but also in the situation of theorem 15 (applied to equation (52)) it will give further information on the mapping $\bar{\lambda}(\rho)$ (i.e. on the form of the bifurcation diagram). Indeed, it follows from lemma 14 and the definition of a that we have (take $\Lambda = \mathbb{R}$) :

$$g(\rho,\lambda) = (\mathrm{Re}\ D_\lambda\mu(\lambda_0))\lambda + \tfrac{1}{2}a\rho^2 + o(|\lambda|+\rho^2) . \tag{63}$$

From this it follows that $\bar{\lambda}(\rho)$ is approximately given by :

$$\bar{\lambda}(\rho) = -\tfrac{a}{2}(\mathrm{Re}\ D_\lambda\mu(\lambda_0))^{-1}\rho^2 + o(\rho^2) . \tag{64}$$

Depending on the sign of the coefficient of ρ^2 in (64) we will have subcritical or supercritical bifurcation; this is important for the determination of the stability properties of the bifurcating periodic solutions (see Marsden and McCracken [156], Iooss and Joseph [255], Hassard, Kazarinoff and Wan [253]).

7.5. DEGENERATE HOPF BIFURCATION

In this section we briefly discuss two examples of degenerate Hopf bifurcation; by this we mean bifurcation problems of the form discussed in section 4, but such that $g(\rho,\lambda_0) = 0$ for all ρ. Both examples are concerned with autonomous ordinary differential equations of the form (4.52). Two kind of assumptions lead to degenerate Hopf bifurcation : the first assumption is that for $\lambda = \lambda_0$ the equation is reversible, the second that for $\lambda = \lambda_0$ the equation has a first integral. This second assumption will give us the Liapunov Center Theorem which we prove here under a somewhat weaker condition than is usually done (see e.g. Alexander and Yorke [4], Kirchgraber [256], Schmidt [201]). A direct treatment of the results of this section has been given in [271].

We start by stating some general results.

7.5.1. Theorem. Under the conditions of theorem 7.2.4, and using the notation of that theorem, suppose that for some λ near λ_0 we have $g(\rho,\lambda) = 0$ for all ρ near 0. Then the equation $M(x,\lambda,\sigma) = 0$ has for this value of the parameter λ a 2-parameter family of solutions given by :

$$\{(\Gamma(\alpha)\tilde{x}(\rho,\lambda),\lambda,\tilde{\sigma}(\rho,\lambda)) \mid \alpha\in\mathbb{R},\ |\rho|<\rho_0\}\ ,$$

(for some $\rho_0 > 0$).

Conversely, if for some λ near λ_0 and for all ρ near 0 we have $M(\tilde{x}(\rho,\lambda),\lambda,\tilde{\sigma}(\rho,\lambda)) = 0$, then $g(\rho,\lambda) = 0$ for all ρ.

Proof. This follows immediately from theorem 7.2.4. □

7.5.2. Theorem. Under the conditions of theorem 7.2.4, assume that $\Lambda = \mathbb{R}$, $g(\rho,\lambda_0) = 0$ for all ρ, and $D_\lambda g(0,\lambda_0) \neq 0$.

Then the equation $M(x,\lambda,\sigma) = 0$ has for all λ near λ_0 only nontrivial solutions (x,σ) near $(0,\sigma_0)$ if and only if $\lambda = \lambda_0$. For $\lambda = \lambda_0$ the equation has a 2-parameter family of solutions given by :

$$\{(\Gamma(\alpha)x(\rho,\lambda_0),\sigma(\rho,\lambda_0)) \mid \alpha\in\mathbb{R},\ |\rho|<\rho_0\} .$$

P r o o f. Since $\Lambda = \mathbb{R}$ and $D_\lambda g(0,\lambda_0) \neq 0$ we can apply theorem 2.7. But $g(\rho,\lambda_0) = 0$ for all ρ, such that we will have $\bar{\lambda}(\rho) = \lambda_0$ for all ρ. Then the result follows from theorem 2.7. □

7.5.3. Periodic solutions of reversible systems. Consider the autonomous equation

$$\dot{y} = h(y,\lambda) , \tag{1}$$

with $h : \mathbb{R}^n \times \Lambda \to \mathbb{R}^n$ of class C^2, and such that $h(0,\lambda) = 0$ for all λ. As in section 4, the problem of finding periodic solutions of (1) with period near 2π can be brought in the form

$$M(x,\lambda,\sigma) = 0 \tag{2}$$

when we define $M : X \times \Lambda \times \mathbb{R} \to Z$ by :

$$M(x,\lambda,\sigma)(t) = -\dot{x}(t) + \sigma h(x(t),\lambda) \quad , \quad \forall t \in \mathbb{R} . \tag{3}$$

As before, we let

$$A(\lambda) = D_x h(0,\lambda) . \tag{4}$$

Now assume that the equation (1) is reversible (see section 3.5) : there exists some symmetric $S \in L(\mathbb{R}^n)$ with $S^2 = I$ and such that

$$h(Sy,\lambda) = -Sh(y,\lambda) \quad , \quad \forall(y,\lambda) \in \mathbb{R}^n \times \Lambda . \tag{5}$$

Using representations $\Gamma : O(2) \to L(Z)$ and $\tilde{\Gamma} : O(2) \to L(Z)$ defined by :

$$\begin{aligned} (\Gamma(\alpha)z)(t) &= (\tilde{\Gamma}(\alpha)z)(t) = z(t+\alpha) \quad , \quad \forall\alpha \in \mathbb{R} \\ (\Gamma_\tau z)(t) &= Sz(-t) \quad , \quad (\tilde{\Gamma}_\tau z)(t) = -Sz(-t) , \end{aligned} \tag{6}$$

we see that M, as given by (3), is equivariant with respect to $(O(2),\Gamma,\tilde{\Gamma})$.

Now suppose that also the other hypotheses of section 4 are satisfied, i.e. :

(i) i is a simple eigenvalue of $A(\lambda_0)$;

(ii) ik is not an eigenvalue of $A(\lambda_0)$ for all $k \neq \pm 1$.

From (5) it follows that $A(\lambda)S = -SA(\lambda)$, and if $c_1 \in \mathbb{C}^n$ is an eigenvector of $A(\lambda_0)$ corresponding to the eigenvalue $+i$, then so is $S\bar{c}_1$. Replacing c_1 by $(c_1+S\bar{c}_1)$ it follows easily that c_1 can be chosen such that $\Gamma_\tau u_1 = u_1$ and $\Gamma_\tau u_2 = -u_2$. Then the theory of section 6.4 is applicable. In particular it follows from (6.4.7), (2.17), (2.21) and (2.24) that $g(\rho,\lambda) = 0$ for all (ρ,λ). If (1) is only reversible for $\lambda = \lambda_0$, then $g(\rho,\lambda_0) = 0$ for all ρ. Then theorems 1 and 2 and the theory of section 6.4 give the following result.

7.5.4. Theorem. Consider the equation (1), and let $\lambda_0 \in \Lambda$ be such that hypotheses (i)-(ii) above are satisfied.

(a) If (1) is reversible for all λ near λ_0, then for all such λ (2) has a 2-parameter family of solutions, given by $\{(\Gamma(\alpha)\tilde{x}(\rho,\lambda),\tilde{\sigma}(\rho,\lambda)) \mid \alpha\in\mathbb{R}, |\rho|<\rho_0\}$, where $\tilde{x}(\rho,\lambda)$ and $\tilde{\sigma}(\rho,\lambda)$ have the properties given in theorem 2.4. Moreover $\Gamma_\tau\tilde{x}(\rho,\lambda) = \tilde{x}(\rho,\lambda)$.

(b) If $\Lambda = \mathbb{R}$, (1) is only reversible for $\lambda = \lambda_0$, and $D_\lambda\mu(\lambda_0) \neq 0$, where $\mu(\lambda)$ is the eigenvalue branch of $A(\lambda)$ such that $\mu(\lambda_0) = i$, then (2) has for $\lambda = \lambda_0$ a 2-parameter family of solutions, given by $\{(\Gamma(\alpha)\tilde{x}(\rho,\lambda_0),\tilde{\sigma}(\rho,\lambda_0)) \mid \alpha\in\mathbb{R}, |\rho|<\rho_0\}$, where also $\Gamma_\tau\tilde{x}(\rho,\lambda_0) = \tilde{x}(\rho,\lambda_0)$. For $\lambda \neq \lambda_0$ and near λ_0 the equation (2) has no nontrivial solutions (x,σ) near $(0,1)$. □

Translating part (a) of this theorem back to the equation (1), we obtain a result which was already given by Palmer in [263].

7.5.5. Equations with a first integral. Now suppose that the equation (1) has a first integral; more precisely, assume that there exists a C^2-function $I : \mathbb{R}^n \times \Lambda \to \mathbb{R}$ such that

$$(D_yI(y,\lambda),h(y,\lambda)) = 0 \quad , \quad \forall(y,\lambda) . \tag{7}$$

In (7) we consider $D_y I$ as a map from $\mathbb{R}^n \times \Lambda$ into $\mathbb{R}^n$, i.e. we take $D_y I(y,\lambda) = \operatorname{grad} I(y,\lambda)$. Differentiation of (7) at $y = 0$ gives :

$$(D_y I(0,\lambda), A(\lambda).x) = 0 \quad , \quad \forall (x,\lambda) . \tag{8}$$

If also the hypotheses (i) and (ii) hold, then $A(\lambda)$ is invertible for all λ near λ_0, and (8) implies :

$$D_y I(0,\lambda) = 0 \quad , \quad \forall \lambda . \tag{9}$$

7.5.6. Under the hypotheses (i), (ii) and (7) we will prove a result similar to that of theorem 4. The main point in the proof will be the following observation.

After solving the auxiliary equation $(I-Q)M(u+v,\lambda,\sigma) = 0$ and using the equivariance of M with respect to $SO(2)$, equation (2) reduces to that of solving :

$$M(\rho u_1 + v^*(\rho u_1,\lambda,\sigma),\lambda,\sigma) = 0 \tag{10}$$

for (ρ,λ,σ) near $(0,\lambda_0,1)$. For each such (ρ,λ,σ) the left hand side of (10) belongs to $R(Q)$. By the results at the end of section 4 we have $R(Q) = R(P) = \ker L = \{\operatorname{Re}(z\zeta) \mid z \in \mathbb{C}\}$. From (2.17), (2.20) and (4.34) it follows that the bifurcation function $F(\rho u_1,\lambda,\sigma)$ used in the general theory is given by :

$$F(\rho u_1,\lambda,\sigma) = \operatorname{Re}(<\zeta^*, M(\rho u_1 + v^*(\rho u_1,\lambda,\sigma),\lambda,\sigma)>\zeta) . \tag{11}$$

That means, we obtain the bifurcation equation $F(\rho u_1,\lambda,\sigma) = 0$ from (10) by application of the isomorphism $u \mapsto \operatorname{Re}(<\zeta^*,u>\zeta)$ from $\ker L$ onto itself; in fact, this isomorphism is nothing else than the operator B^{-1}. Identifying $\operatorname{Re}(z\zeta) \in \ker L$ with $z \in \mathbb{C}$, this isomorphism takes the form

$$z \mapsto <\zeta^*, \operatorname{Re}(z\zeta)> = -iz , \tag{12}$$

where we have used (4.55).

Now the idea is to replace ζ^* in (12) by $\psi^*(\rho,\lambda,\sigma)$ where $\psi^* : \mathbb{R} \times \Lambda \times \mathbb{R} \to Z_C$ is defined and continuous for (ρ,λ,σ) near $(0,\lambda_0,1)$. For fixed (ρ,λ,σ) and considering $\mathbb{C}$ as a two-dimensional *real* vectorspace, the mapping

$$z \mapsto \langle \psi^*(\rho,\lambda,\sigma), \mathrm{Re}(z\zeta)\rangle \tag{13}$$

is a linear map from $\mathbb{C}$ into itself. It will be an isomorphism for all (ρ,λ,σ) near $(0,\lambda_0,1)$ if it is an isomorphism for $(\rho,\lambda,\sigma) = (0,\lambda_0,1)$. When this is the case then the equation (11) is equivalent with the modified bifurcation equation

$$\tilde{F}(\rho,\lambda,\sigma) \equiv \langle \psi^*(\rho,\lambda,\sigma) \quad , \quad M(\rho u_1 + v^*(\rho u_1,\lambda,\sigma),\lambda,\sigma)\rangle = 0 . \tag{14}$$

We will take ψ^* of the form :

$$\psi^*(\rho,\lambda,\sigma) = \mathrm{Re}\ \zeta^* + i\xi^*(\rho,\lambda,\sigma) , \tag{15}$$

for some continuous $\xi^* : \mathbb{R}\times\Lambda\times\mathbb{R} \to Z$. Then the condition for (13) to be a linear isomorphism for all (ρ,λ,σ) near $(0,\lambda_0,1)$ becomes :

$$\langle \xi^*(0,\lambda_0,1),\mathrm{Re}\zeta\rangle = \mathrm{Re}\langle \xi^*(0,\lambda_0,1),\zeta\rangle \neq 0 . \tag{16}$$

7.5.7. Liapunov Center Theorem. Let (1) satisfy (i) and (ii). Let $c_1 \in \mathbb{C}^n$ be an eigenvector of $A(\lambda_0)$ with eigenvalue i. Suppose that (1) has a first integral $I(y,\lambda)$ for all λ near λ_0, and such that

$$(D_y^2 I(0,\lambda_0).c_1,c_1) \neq 0 . \tag{17}$$

Then the equation (2) has for each λ near λ_0 a 2-parameter family of solutions given by $\{(\Gamma(\alpha)\tilde{x}(\rho,\lambda),\tilde{\sigma}(\rho,\lambda)) \mid \alpha\in\mathbb{R}, |\rho| < \rho_0\}$, where $\tilde{x}(\rho,\lambda)$ and $\tilde{\sigma}(\rho,\lambda)$ are as in theorem 2.4. Correspondingly the equation (1) has for each λ near λ_0 a 2-parameter family of periodic solutions near $y = 0$, with period near 2π and depending on the amplitude ρ and the parameter λ.

P r o o f. Define $\beta : X\times\Lambda \to Z$ by :

$$\beta(x,\lambda)(t) = D_y I(x(t),\lambda) \qquad , \qquad \forall t \in \mathbb{R} . \tag{18}$$

Using (7) and the fact that $I(x(t),\lambda)$ is 2π-periodic in t (if $x \in X$), it follows easily that :

$$<\beta(x,\lambda),M(x,\lambda,\sigma)> = 0 \quad , \quad \forall(x,\lambda,\sigma) . \tag{19}$$

Let $\gamma(\rho,\lambda,\sigma) = \beta(\rho u_1+v^*(\rho u_1,\lambda,\sigma),\lambda)$. From (9) and (18) it follows that $\gamma(0,\lambda,\sigma) = 0$, and consequently $\gamma(\rho,\lambda,\sigma) = \rho\gamma_1(\rho,\lambda,\sigma)$, where :

$$\gamma_1(\rho,\lambda,\sigma) = \int_0^1 D_x\beta(\tau\rho u_1+v^*(\tau\rho u_1,\lambda,\sigma),\lambda).(u_1+D_u v^*(\tau\rho u_1,\lambda,\sigma).u_1)d\tau \ ; \tag{20}$$

in particular :

$$\gamma_1(0,\lambda_0,1)(t) = D_y^2 I(0,\lambda_0).u_1(t) \quad , \quad \forall t \in \mathbb{R} . \tag{21}$$

From (19) we have :

$$<\gamma_1(\rho,\lambda,\sigma),M(\rho u_1+v^*(\rho u_1,\lambda,\sigma),\lambda,\sigma)> = 0 \quad , \quad \forall(\rho,\lambda,\sigma) . \tag{22}$$

Now we take $\xi^*(\rho,\lambda,\sigma) = \gamma_1(\rho,\lambda,\sigma)$ in (15); this is allowed since by (21) :

$$<\gamma_1(0,\lambda_0,1),\zeta> = \frac{1}{2\pi}\int_0^{2\pi} (D_y^2 I(0,\lambda_0).u_1(t),\zeta(t))dt$$
$$= \frac{1}{2}(D_y^2 I(0,\lambda_0).c_1,c_1) \neq 0$$

by (17). Since $D_y^2 I(0,\lambda_0) \in L(\mathbb{R}^n)$ is real and symmetric, this expression is also real, and consequently the condition (16) is satisfied.

With this choice for $\xi^*(\rho,\lambda,\sigma)$, and using (22) and (4.60) the modified bifurcation equation (14) takes the form :

$$Q_1 M(\rho u_1+v^*(\rho u_1,\lambda,\sigma),\lambda,\sigma) = 0 . \tag{23}$$

From the general theory of section 2 we know that this equation can, for $\rho \neq 0$, be solved uniquely for $\sigma = \tilde{\sigma}(\rho,\lambda)$. This proves the theorem. □

7.5.8. Remark. Our version of the Liapunov Center Theorem requires the condition (17), while usually one finds the stronger condition that $D_y^2 I(0,\lambda_0) \in$

$L(\mathbb{R}^n)$ is nonsingular. For another proof of the theorem, see [271].

The main part of the argument, as explained in subsection 6, can be abstracted into a kind of nonlinear Liapunov-Schmidt method. This was done in Loud and Vanderbauwhede [260], where also an application to boundary value problems is given.

Finally, if (1) has a first integral only for $\lambda = \lambda_0$, then the arguments used above may be adapted to prove a result similar to part (b) of theorem 4. We leave it to the reader to work this out explicitly.

8 Symmetry and bifurcation near families of solutions

8.1. INTRODUCTION

In the previous chapters we studied local bifurcation for equations of the type

$$M(x,\lambda) = 0 \; ; \tag{1}$$

each time we restricted our attention to a neighbourhood of the origin in $X \times \Lambda$, which was supposed to be a solution of (1). In practical examples, however, one may want to study (1) near a solution (x_0,λ_0) which is not necessarily at the origin. By a simple translation this problem can be reduced to the case $(x_0,\lambda_0) = (0,0)$. However, such translation may disturb the symmetry properties of (1). Indeed, if M is equivariant with respect to some group G, then the translated equation will only remain equivariant with respect to G if x_0 is invariant under the action of G, i.e. if $\Gamma(s)x_0 = x_0$, $\forall s \in G$.

In case x_0 is not invariant under G, then for $\lambda = \lambda_0$ (1) will have a compact family of solutions, given by

$$\gamma_0 = \{\Gamma(s)x_0 \mid s \in G\} \; . \tag{2}$$

But then there is no reason to restrict our attention to a neighbourhood of x_0; a more reasonable problem is to consider the bifurcation problem in a neighbourhood of the compact family γ_0. Remark also that it is sufficient that M is equivariant for $\lambda = \lambda_0$ in order to obtain a solution family such as γ_0. The study of bifurcation near a compact family of solutions of the unperturbed problem was initiated in recent years by Hale and Taboas ([81],[89]), who studied a specific example, a generalization of which will be treated in section 5 of this chapter. Other contributions were made by Dancer ([247], [248]), Schmitt and Mazzanti [200] and Vanderbauwhede ([230],[231]).

In section 2 we show how the Liapunov-Schmidt procedure can be adapted to the problem of bifurcation near a compact family of solutions of the unperturbed equation; it is assumed that this family is generated by the symmetry of the unperturbed equation. The main point is the construction of a symmetry invariant tubular neighbourhood of the compact family γ_0. In section 3 we show that if the perturbed equation maintains some symmetry, then the perturbed equation may have certain branches of solutions near γ_0 which are induced by this symmetry. An application to periodic perturbations of conservative oscillation equations is given in section 4. Finally, in section 5 we discuss the bifurcation equation for generic two-parameter perturbations of the same conservative oscillation equation.

8.2. REDUCTION TO A FINITE-DIMENSIONAL PROBLEM

8.2.1. The problem. Let X, Z and Λ be real Banach spaces, and $M : X \times \Lambda \to Z$ a continuous map, continuously differentiable with respect to the variable $x \in X$. Consider the equation :

$$M(x,\lambda) = 0 . \tag{1}$$

Assume that $M(x_0,0) = 0$, for some $x_0 \in X$, and that for $\lambda = 0$ equation (1) is equivariant with respect to some compact group action. Then the orbit $\gamma_0 = \{\Gamma(s)x_0 \mid s \in G\}$ of x_0 forms a compact family of solutions of the equation $M(x,0) = 0$. We want to describe all solutions (x,λ) of (1) near $\gamma_0 \times \{0\}$ in $X \times \Lambda$. We will show how this problem can be reduced to a finite-dimensional one by an adapted version of the Liapunov-Schmidt method. We begin with a definition.

8.2.2. Definition. A *Lie group* is a finite-dimensional differentiable manifold G on which there is defined a smooth mapping $G \times G \to G$, $(s,t) \mapsto s.t$ for which G forms a group. The dimension of the manifold is called the *dimension* of the Lie group.

Every Lie group is a topological group. For fixed $s \in G$ the mapping $L_s : G \to G$, $t \mapsto L_s t = s.t$ is a diffeomorphism of G onto itself. By the inverse function theorem 2.1.17 this implies that the tangent map $T_t L_s : T_t G \to T_{s.t} G$ is an isomorphism, for each $s,t \in G$. Consider now the equation $L_s t = e$, where e is the identity element of G. For $s = s_0$ this equation has the solution $t = s_0^{-1}$; since $T_{s_0^{-1}} L_{s_0}$ is an isomorphism, it follows from the implicit function theorem that the equation has a unique solution $t = s^{-1}$ which depends smoothly on s near s_0. Hence, also the mapping $s \mapsto s^{-1}$ is of class C^∞.

8.2.3. Let now G be a compact k-dimensional Lie group, and $\Gamma : G \to L(X)$ a representation of G over X. Such a representation induces an *action* of G over X, given by $\Phi : G \times X \to X$, $(s,x) \mapsto \Phi(s,x) = \Gamma(s)x$. For fixed $x \in X$ we denote by Φ_x the partial mapping $\Phi_x : G \to X$, $s \mapsto \Phi_x(s) = \Phi(s,x)$. We call $\gamma(x) = R(\Phi_x) = \{\Gamma(s)x \mid s \in G\}$ the *orbit* of x under the action Φ; $\gamma(x)$ is a compact subset of X. Finally, we denote by $G_x = \Phi_x^{-1}(x) = \{s \in G \mid \Gamma(s)x = x\}$ the *isotropy subgroup* of x; it is a compact subgroup of G.

8.2.4. Lemma. For some fixed $x_0 \in X$, let $\Phi_0 = \Phi_{x_0}$, $\gamma_0 = \gamma(x_0)$ and $G_0 = G_{x_0}$.
If Φ_0 is of class C^1, then γ_0 is a C^1-submanifold of X, and G_0 is a C^1-submanifold of X.

P r o o f. We made already the remark that the mapping $L_s : G \to G$, $t \mapsto s.t$ is a diffeomorphism for each $s \in G$. Since $\Phi_0 \circ L_s = \Gamma(s) \circ \Phi_0$, $\forall s \in G$, it follows that Φ_0 has constant rank on G. Then it follows from the rank theorem 2.1.18 and its corollary 2.1.20 that each $s \in G$ has a neighbourhood U_s in G such that $\Phi_0(U_s)$ is a submanifold of X, while also for each $s \in G$ the set $\Phi_0^{-1}(\Phi_0(s))$ forms a submanifold of G. In particular G_0 is a C^1-submanifold of X.

Fix some $s \in G$, and let $V_s = \Phi_0^{-1}(\Phi_0(U_s)) = \{t.s_0 \mid t \in U_s,\ s_0 \in G_0\}$; V_s is open in G, its complement $G \setminus V_s$ is closed and therefore compact, and consequently also $\Phi_0(G \setminus V_s)$ is compact. Since $\Phi_0(s) \notin \Phi_0(G \setminus V_s)$, there is a neighbourhood W_s of $\Phi_0(s)$ in X such that $W_s \cap \Phi_0(G \setminus V_s) = \phi$. This implies that $\gamma_0 \cap W_s = \Phi_0(U_s) \cap W_s$, which shows that γ_0 is indeed a C^1-submanifold of X.

It follows from the rank theorem that $\dim \gamma_0 = n$ and $\dim G_0 = k-n$, where $n = \operatorname{rank} \Phi_0 = \dim T_e\Phi_0(T_eG) \leq k = \dim G$. □

8.2.5. Theorem. Let $x_0 \in X$ and $P \in L(X)$ be such that the following is satisfied (we use the notation of lemma 4) :

(a) Φ_0 is of class C^1 ;

(b) P is a projection with $R(P) = T_e\Phi_0(T_eG)$;

(c) $P\Gamma(s) = \Gamma(s)P$, $\forall s \in G_0$;

(d) $P\circ\Phi_x : G \to X$ is of class C^1, for each $x \in X$.

Then there exists a neighbourhood Ω of the origin in ker P such that the following holds :

(i) $\Gamma(s)(\Omega) = \Omega$, $\forall s \in G_0$;

(ii) $\{\Gamma(s)(x_0+y) \mid s\in G,\ y\in\Omega\}$ is an invariant open neighbourhood of γ_0 in X ;

(iii) for all $s,s' \in G$ and $y,y' \in \Omega$ one has $\Gamma(s)(x_0+y) = \Gamma(s')(x_0+y')$ if and only if there is some $s_0 \in G_0$ such that $s' = s.s_0$ and $y = \Gamma(s_0)y'$.

P r o o f. Let $\alpha : U\subset G \to \tilde{U} = \alpha(U) \subset \mathbb{R}^k$ be a C^1-chart of G, such that $e \in U$, $\alpha(e) = 0$ and $\alpha(G_0\cap U) = (\{0\}\times\mathbb{R}^{k-n})\cap\tilde{U}$; it is possible to find such α, since, by lemma 4, G_0 is a C^1-submanifold of G. We write the elements of $\mathbb{R}^k$ in the form (a,b), with $a \in \mathbb{R}^n$ and $b \in \mathbb{R}^{k-n}$.

Let $\tilde{\Phi}_0 = \Phi_0\circ\alpha^{-1} : \tilde{U} \to X$ and $\tilde{\Gamma} = \Gamma\circ\alpha^{-1} : \tilde{U} \to L(X)$. Since α is a chart of G, we have $R(D\tilde{\Phi}_0(0)) = T_e\Phi_0(T_eG) = R(P)$. Now $\tilde{\Phi}_0(0,b) = x_0$ for each b near 0 in $\mathbb{R}^{k-n}$, and consequently $D_b\tilde{\Phi}_0(0,0) = 0$. Since dim $R(P) = n$, this implies that $D_a\tilde{\Phi}_0(0,0)$ is an isomorphism between $\mathbb{R}^n$ and $R(P)$.

Now define $g : \tilde{U}\times X \to R(P)$ by :

$$g(a,b,x) = P(\tilde{\Gamma}(a,b)x-x_0) .$$

Then g is of class C^1, $g(0,0,x_0) = 0$ and $D_ag(0,0,x_0) = PD_a\tilde{\Phi}_0(0,0) = D_a\tilde{\Phi}_0(0,0)$. It follows from the implicit function theorem that there is a neighbourhood $A\times B\times W$ of $(0,0,x_0)$ in $\mathbb{R}^n\times\mathbb{R}^{k-n}\times X$, and a C^1-mapping $a^* : B\times W \to A$ such that

$$g^{-1}(0)\cap(A\times B\times W) = \{(a^*(b,x),b,x) \mid (b,x)\in B\times W\} .$$

In particular, if $y \in \ker P$ is such that $x_0+y \in W$, and if $b \in B$, then :

$$P(\tilde{\Gamma}(0,b)(x_0+y)-x_0) = P\tilde{\Gamma}(0,b)y = \tilde{\Gamma}(0,b)Py = 0 .$$

This implies that $a^*(b,x_0+y) = 0$.

Define $\tilde{a} : W \to A$, $\tilde{y} : W \to \ker P$ and $h : A\times\ker P \to X$ by :

$$\tilde{a}(x) = a^*(0,x) \quad , \quad \tilde{y}(x) = \tilde{\Gamma}(\tilde{a}(x),0)x-x_0 \quad , \quad \forall x \in W$$

and

$$h(a,y) = \tilde{\Gamma}^{-1}(a,0)(x_0+y) \quad , \quad \forall (a,y) \in A\times\ker P .$$

Then $\tilde{a}$, $\tilde{y}$ and h are continuous, and we have :

$$(a,y) = (\tilde{a}(h(a,y)),\tilde{y}(h(a,y))) ,$$

for each $(a,y) \in A\times\ker P$ such that $h(a,y) \in W$. This shows that there is a neighbourhood $A_1\times\Omega$ of $(0,0)$ in $\mathbb{R}^n\times\ker P$ such that the restriction of h to $A_1\times\Omega$ is a homeomorphism onto a neighbourhood W_1 of x_0 in X. Since the set of operators $\{\Gamma(s) \mid s\in G_0\}$ is equibounded, we may suppose that $\Gamma(s)(\Omega) = \Omega$ for all $s \in G_0$. (Use the argument used in the proof of theorem 2.5.13). Then

$$\{\Gamma(s)(x_0+y) \mid s\in G,\ y\in\Omega\} = \cup\{\Gamma(s)(W_1) \mid s\in G\}$$

is an open and invariant neighbourhood of γ_0.

Finally we want to show that it is possible to shrink Ω in such a way that also the requirement (iii) of the theorem is satisfied. Assume this is not possible. Then we can find sequences $\{s_j \mid j\in\mathbb{N}\} \subset G$, $\{s_j' \mid j\in\mathbb{N}\} \subset G$, $\{y_j \mid j\in\mathbb{N}\} \subset \ker P$ and $\{y_j' \mid j\in\mathbb{N}\} \subset \ker P$, such that $\lim_{j\to\infty} y_j = 0$, $\lim_{j\to\infty} y_j' = 0$,

$$\Gamma(s_j)(x_0+y_j) = \Gamma(s_j')(x_0+y_j') \quad , \quad \forall j \in \mathbb{N} \tag{2}$$

and

$$s_j' \notin \{s_js_0 \mid s_0 \in G_0\} \quad , \quad \forall j \in \mathbb{N} .$$

Since G is compact we may suppose that $s_j'^{-1}.s_j$ converges to some element $\bar{s} \in G$. Taking the limit in (2) shows that $\bar{s} \in G_0$. Let $\bar{y}_j = \Gamma(\bar{s})y_j$ and $t_j = s_j'^{-1}.s_j.\bar{s}^{-1}$. Then we have :

$$\lim_{j\to\infty} \bar{y}_j = 0 \quad , \quad \lim_{j\to\infty} t_j = e$$

$$P(\Gamma(t_j)(x_0+\bar{y}_j)-x_0) = 0 \qquad , \qquad \forall j \in \mathbb{N} , \tag{3}$$

and

$$e \notin \{t_j.s_0 \mid s_0 \in G_0\} \qquad , \qquad \forall j \in \mathbb{N} . \tag{4}$$

For j sufficiently large we have $\alpha(t_j) = (a_j,b_j) \in A\times B$ and $x_0+\bar{x}_j \in W$, and consequently $a^*(b_j,x_0+\bar{y}_j) = 0$. From (3) we also have $g(a_j,b_j,x_0+\bar{y}_j) = 0$. We conclude that $a_j = 0$ for j sufficiently large, i.e. $t_j \in G_0$. This, however, contradicts (4). □

8.2.6. Remarks. (i) Since $T_e\Phi_0(T_eG)$ is invariant under $\{\Gamma(s) \mid s\in G_0\}$, we can use theorem 2.5.9 to prove the existence of a projection P satisfying the conditions (b) and (c) of theorem 5.

(ii) Denote by $\Gamma^* : G \to L(X^*)$ the adjoint representation of Γ, defined by $\Gamma^*(s) = (\Gamma(s))^*$, $\forall s \in G$. Let $\{u_1,\dots,u_n\}$ be a basis of $T_e\Phi_0(T_eG)$, and let $\{u_1^*,\dots,u_n^*\} \subset X$ be such that $\langle u_i^*,u_j\rangle = \delta_{ij}$ $(i,j=1,\dots,n)$ and such that the mapping $s \mapsto \Gamma^*(s)u_i^*$ from G into X^* is of class C^1, for each $i = 1,\dots,n$. Then it is immediately verified that the operator $P \in L(X)$ defined by

$$Px = \sum_{i=1}^{n} \langle u_i^*,x\rangle u_i \qquad , \qquad \forall x \in X$$

satisfies the requirements (b) and (d) of theorem 5. Using the second part of theorem 2.5.9 one can construct a new P satisfying (b), (c) and (d).

(iii) Using lemma 1 of [248] one can prove that, given n linearly independent vectors $\{u_i \mid 1\leq i\leq n\}$ in X, it is always possible to find $\{u_i^* \mid 1\leq i\leq n\}$ in X^* satisfying the requirements of remark 2. So there exists at least one projection P satisfying (b), (c) and (d).

(iv) Theorem 5 describes a symmetry-invariant tubular neighbourhood of the orbit γ_0. For a general discussion of orbits and tubes one can see Bredon [244].

8.2.7. Let us now return to the equation (1). We make the following hypotheses :

(H1) For $\lambda = 0$, (1) is equivariant with respect to $(G,\Gamma,\tilde{\Gamma})$, where G is a k-dimensional compact Lie group, while $\Gamma : G \to L(X)$ and $\tilde{\Gamma} : G \to L(Z)$ are representations of G over X, respectively Z.

(H2) $x_0 \in X$ is such that :

(i) $M(x_0,0) = 0$;

(ii) $\Phi_0 = \Phi_{x_0}$ is of class C^1 ;

(iii) $D_xM(x_0,0)$ is a Fredholm operator, with $\dim \ker D_xM(x_0,0) = \dim \gamma_0 = k_0$.

(H3) $P \in L(X)$ satisfies the requirements (b), (c) and (d) of theorem 5.

8.2.8. Remark. It follows from (H2)(ii) and lemma 4 that γ_0 is a C^1-submanifold, with $\dim \gamma_0 = k-n = k_0$. From (H1) and (H2)(i) we have that $M(x,0) = 0$ for all $x \in \gamma_0$; this implies that $T_{x_0}\gamma_0 \subset \ker D_xM(x_0,0)$, and consequently $\dim \ker D_xM(x_0,0) \geq k_0$. The hypothesis (H2)(iii) requires that $\ker D_xM(x_0,0)$ coincides with $T_{x_0}\gamma_0 = T_e\Phi_0(T_eG)$.

8.2.9. The Liapunov-Schmidt reduction. Our problem is to find the solutions (x,λ) of (1) in a neighbourhood of $\gamma_0 \times \{0\}$ in $X \times \Lambda$. By theorem 5 this reduces to finding solutions $(s,y,\lambda) \in G \times \ker P \times \Lambda$, with (y,λ) near the origin, of the equation

$$M(\Gamma(s)(x_0+y),\lambda) = 0 \ . \tag{5}$$

By conclusion (iii) of theorem 5, different solutions (s,y,λ) and (t,z,λ) of (5) will correspond to the same solutions of (1) if and only if there is some $s_0 \in G_0$ such that $t = s.s_0$ and $y = \Gamma(s_0)z$.

First we rewrite (5) in the equivalent form :

$$\tilde{M}(s,y,\lambda) \equiv \tilde{\Gamma}(s^{-1})M(\Gamma(s)(x_0+y),\lambda) = 0 \ . \tag{6}$$

Now $R(D_xM(x_0,0))$ is a closed subspace of Z, with a finite codimension, and invariant under $\{\tilde{\Gamma}(s) \mid s \in G_0\}$. Consequently we can find a projection $Q \in L(Z)$ such that

$$\ker Q = R(D_x M(x_0,0)) \quad \text{and} \quad \tilde{\Gamma}(s)Q = Q\tilde{\Gamma}(s) \quad , \quad \forall s \in G_0 \ . \tag{7}$$

Then equation (6) is equivalent to :

$$\begin{aligned} &(a) \quad (I-Q)\tilde{M}(s,y,\lambda) = 0 \ , \\ &(b) \quad Q\,\tilde{M}(s,y,\lambda) = 0 \ . \end{aligned} \tag{8}$$

8.2.10. Lemma. Assume (H1)-(H3). Then there exist a neighbourhood Ω_0 of the origin in ker P, a neighbourhood ω of the origin in Λ and a continuous mapping $y^* : G \times \omega \to \Omega_0$ such that

(i) $\Gamma(s)(\Omega_0) = \Omega_0 \quad , \quad \forall s \in G_0$;

(ii) for each $(s,y,\lambda) \in G \times \Omega_0 \times \omega$ equation (8.a) is satisfied if and only if $y = y^*(s,\lambda)$;

(iii) $y^*(s,0) = 0 \quad , \quad \forall s \in G$;

(iv) $\Gamma(s_0^{-1})y^*(s,\lambda) = y^*(s.s_0,\lambda)$,

$$\forall s \in G \ , \ \forall s_0 \in G_0 \ , \ \forall \lambda \in \omega \ . \tag{9}$$

P r o o f. Fix $s \in G$; then we have (using (H1)) :

$$(I-Q)D_y\tilde{M}(s,0,0) = (I-Q)D_xM(x_0,0) = D_xM(x_0,0) \ ,$$

which is an isomorphism between ker P and $R(D_xM(x_0,0)) = \ker Q$. Then the implicit function theorem and the compactness of G give us the existence of Ω_0, ω and y^* satisfying (i), (ii) and (iii). As for (iv), this follows from the uniqueness of the solution $y^*(s,\lambda)$ and the following identity :

$$\Gamma(s_0^{-1})\tilde{M}(s,y,\lambda) = \tilde{M}(s.s_0,\Gamma(s_0^{-1})y,\lambda) \ , \tag{10}$$

$$\forall s_0 \in G_0, \ \forall (s,y,\lambda) \in G \times \ker P \times \Lambda \ . \qquad \square$$

Bringing the solution $y^*(s,\lambda)$ of (8.a) into (8.b) we obtain the bifurcation equation :

$$F(s,\lambda) \equiv Q\tilde{M}(s,y^*(s,\lambda),\lambda) = 0 \ . \tag{11}$$

The mapping $F : G \times \omega \to R(Q)$ is continuous, $F(s,0) = 0$, $\forall s \in G$, and

$$F(s.s_0,\lambda) = \tilde{\Gamma}(s_0^{-1})F(s,\lambda) \quad , \quad \forall s_0 \in G_0, \forall s \in G, \forall \lambda \in \omega . \tag{12}$$

8.2.11. Theorem. Assume (H1), (H2) and (H3), and let $Q \in L(Z)$ be a projection satisfying (7). Then there exist a neighbourhood U of γ_0 in X, and a neighbourhood ω of the origin in Λ such that each solution $(x,\lambda) \in U \times \omega$ of (1) can be written in the form

$$x = \Gamma(s)(x_0+y^*(s,\lambda)) , \tag{13}$$

where (s,λ) is a solution of (11) and $y^*(s,\lambda)$ is the mapping given by lemma 10.

Conversely, if $(s,\lambda) \in G \times \omega$ is a solution of (11), and x is given by (13), then (x,λ) is a solution of (1).

The representation (13) of solutions of (1) in $U \times \omega$ is unique modulo right multiplication by elements of G_0; that means, two solutions (s,λ) and (t,λ) of (11) give the same solution of (1) if and only if $t = s.s_0$ for some $s_0 \in G_0$. □

This theorem reduces the problem (1) to that of solving the bifurcation equation (11). For a discussion of some aspects of this equation, one can see Vanderbauwhede [230]. In section 5 we will discuss a particular example of such bifurcation equation.

8.3. SYMMETRIC SOLUTIONS

In this section we show how the symmetry properties of the equations may sometimes imply the existence of certain branches of solutions near a compact manifold of solutions of the unperturbed equation. Our presentation is independent from the results of the preceding section, and is based on the ideas in Vanderbauwhede [231] and on a reformulation of these ideas by D. Chillingworth (private communication). Dancer [248] gives a somewhat different approach.

8.3.1. The hypotheses. Let X, Z and Λ be real Banach spaces, and $M : X \times \Lambda \to Z$ a continuous mapping which is continuously differentiable in $x \in X$. Consider the equation

$$M(x,\lambda) = 0 . \qquad (1)$$

We make the following hypothesis :

(H) There is a compact group G, a closed subgroup H, and representations $\Gamma : G \to L(X)$ and $\tilde{\Gamma} : G \to L(Z)$ such that :

$$M(\Gamma(s)x,\lambda) = \tilde{\Gamma}(s)M(x,\lambda) \quad , \quad \forall s \in H \ , \ \forall (x,\lambda) \in X \times \Lambda \qquad (2)$$

and

$$M(\Gamma(s)x,0) = \tilde{\Gamma}(s)M(x,0) \quad , \quad \forall s \in G \ , \ \forall x \in X \ ; \qquad (3)$$

i.e. M is equivariant with respect to $(H,\Gamma,\tilde{\Gamma})$, and for $\lambda = 0$ even equivariant with respect to G.

We will also use the following notation : if $T \subset G$, then we let

$$\mathrm{Fix}_X(T) = \{x \in X \mid \Gamma(s)x = x, \ \forall s \in T\} . \qquad (4)$$

8.3.2. <u>Theorem</u>. Assume (H), and let $x_0 \in X$ be such that :

(i) $M(x_0,0) = 0$;

(ii) $L = D_x M(x_0,0)$ is a Fredholm operator :

(iii) $\mathrm{Fix}_Z(H_0) \subset R(L)$,

where $H_0 = \{s \in H \mid \Gamma(s)x_0 = x_0\}$ is the isotropy subgroup of x_0.

Then there exists a mapping :

$$x^* : \omega \to \mathrm{Fix}_X(H_0) \quad , \quad \lambda \mapsto x^*(\lambda)$$

defined and continuous for λ in a neighbourhood ω of the origin in Λ, and with $x^*(0) = x_0$, such that

$$M(x^*(\lambda),\lambda) = 0 \quad , \quad \forall \lambda \in \omega . \qquad (5)$$

P r o o f. Since $L\Gamma(s) = \tilde{\Gamma}(s)L$, $\forall s \in G_0 = \{t \in G \mid \Gamma(t)x_0 = x_0\}$, there exist projections $P \in L(X)$ and $Q \in L(Z)$ such that

$$R(P) = \ker L \quad , \quad \ker Q = R(L) \; ,$$

and

$$P\Gamma(s) = \Gamma(s)P \quad , \quad Q\tilde{\Gamma}(s) = \tilde{\Gamma}(s)Q \quad , \quad \forall s \in G_0 \; .$$

Now define $F : \ker P \times \Lambda \to R(L)$ by

$$F(y,\lambda) = (I-Q)M(x_0+y,\lambda) \; . \tag{6}$$

We have $F(0,0) = 0$, while $D_yF(0,0) = L|_{\ker P}$ is an isomorphism between ker P and R(L). It follows that the equation $F(y,\lambda) = 0$ has a unique solution $y = y^*(\lambda)$, with $y^* : \omega \subset \Lambda \to \ker P$ continuous and $y^*(0) = 0$.

Since $F(\Gamma(s)y,\lambda) = \tilde{\Gamma}(s)F(y,\lambda)$, for all $s \in H_0$, and all $(y,\lambda) \in \ker P \times \Lambda$, it follows that $\Gamma(s)y^*(\lambda) = y^*(\lambda)$, $\forall s \in H_0$, $\forall \lambda \in \omega$, i.e. $y^*(\lambda) \in \mathrm{Fix}_X(H_0)$, $\forall \lambda \in \omega$. This in turn implies that

$$M(x_0+y^*(\lambda),\lambda) \in \mathrm{Fix}_Z(H_0) \subset R(L) \quad , \quad \forall \lambda \in \omega \; .$$

Defining $x^* : \omega \to X$ by $x^*(\lambda) = x_0+y^*(\lambda)$, (5) then follows from $F(y^*(\lambda),\lambda) = 0$, $\forall \lambda \in \omega$. It is also clear that $x^*(\lambda) \in \mathrm{Fix}_X(H_0)$, $\forall \lambda \in \omega$. □

8.3.3. Let now $x_0 \in X$ satisfy the conditions (i) and (ii) of theorem 2. Then the same conditions are also satisfied at each point $\tilde{x}_0$ of the orbit $\gamma_0 = \{\Gamma(t)x_0 \mid t \in G\}$, since we have $D_xM(\Gamma(t)x_0,0) = \tilde{\Gamma}(t)D_xM(x_0,0)\Gamma(t^{-1})$. So we may ask at which points of γ_0 condition (iii) of theorem 2 is satisfied.

Let $\tilde{x}_0 = \Gamma(t)x_0$ $(t \in G)$, and let $\tilde{H}_0$ be the isotropy subgroup of $\tilde{x}_0$. Then it is not too difficult to see that $\tilde{H}_0 = H \cap t.G_0.t^{-1}$, and consequently

$$\mathrm{Fix}_Z(\tilde{H}_0) = \tilde{\Gamma}(t)(\mathrm{Fix}_Z(G_0 \cap t^{-1}Ht)) \; .$$

Since $R(D_xM(\tilde{x}_0,0)) = \tilde{\Gamma}(t)R(L)$, we have the following result.

8.3.4. Corollary. Assume (H), and let $x_0 \in X$ satisfy the conditions (i) and (ii) of theorem 2. Let $t \in G$ be such that

$$\text{(iii)}' \quad \mathrm{Fix}_Z(G_0 \cap t^{-1}Ht) \subset R(L) \ .$$

Then there exists a mapping

$$x_t^* : \omega \to \mathrm{Fix}_X(t.G_0.t^{-1} \cap H) \quad , \quad \lambda \mapsto x_t^*(\lambda)$$

defined and continuous in a neighbourhood ω of the origin in Λ, and with $x_t^*(0) = \Gamma(t)x_0$, such that

$$M(x_t^*(\lambda),\lambda) = 0 \quad , \quad \forall\lambda \in \omega \ . \quad \square \tag{7}$$

8.3.5. Remark. If the condition (iii)' of corollary 4 is satisfied for some $t \in G$, then it is also satisfied for $t' = s.t.s_0$, where $s \in H$ and $s_0 \in G_0$. We leave it as an exercise to show from the proof of theorem 2 that one will have

$$x_{t'}^*(\lambda) = \Gamma(s)x_t^*(\lambda) \quad , \quad \forall\lambda \in \omega \ .$$

Consequently, all the solutions obtained in this way belong to the orbit of solutions $\{\Gamma(s)x_t^*(\lambda) \mid s \in H\}$ generated by the solution $x_t^*(\lambda)$.

8.3.6. Remark. Suppose M satisfies the hypotheses (H1)-(H3) of section 2, while also (H) is satisfied. Using the notation of section 2, it is then easily seen that $\widetilde{M}(s.t,y,\lambda) = \widetilde{M}(t,y,\lambda)$, for all $s \in H$, $t \in G$ and $(y,\lambda) \in \ker P \times \Lambda$. This implies

$$y^*(s.t,\lambda) = y^*(t,\lambda) \quad , \quad \forall(s,t,\lambda) \in H \times G \times \omega \tag{8}$$

and

$$F(s.t,\lambda) = F(t,\lambda) \quad , \quad \forall(s,t,\lambda) \in H \times G \times \omega \ . \tag{9}$$

If now $s \in G_0 \cap t^{-1}Ht$, then we have from (9) and (2.12) :

$$\tilde{\Gamma}(s^{-1})F(t,\lambda) = F(t.s,\lambda) = F((t.s.t^{-1}).t,\lambda) = F(t,\lambda) \; ; \tag{10}$$

i.e. $F(t,\lambda) \in \mathrm{Fix}_Z(G_0 \cap t^{-1}Ht)$. If also (iii)' holds, then $\mathrm{Fix}_Z(G_0 \cap t^{-1}Ht) \cap R(Q) = \{0\}$, and consequently $F(t,\lambda) = 0$ for all $\lambda \in \omega$. This gives another proof of corollary 4; however, part of the hypotheses (H2)-(H3) are unnecessary to obtain the result, as the direct proof shows.

8.3:7. Corollary. Let X and Z be real Banach spaces, G a compact group, $\Gamma : G \to L(X)$ and $\tilde{\Gamma} : G \to L(Z)$ representations, and $M_0 : X \to Z$ a C^1-mapping such that

$$M_0(\Gamma(t)x) = \tilde{\Gamma}(t)M_0(x) \quad , \quad \forall t \in G \; , \; \forall x \in X \; . \tag{11}$$

Assume the following :

(i) $x_0 \in X$ is such that $M_0(x_0) = 0$ while $L = D_xM_0(x_0)$ is a Fredholm operator ;

(ii) $G_0 = \{t \in G \mid \Gamma(t)x_0 = x_0\}$;

(iii) H is a closed subgroup of G ;

(iv) $t \in G$ is such that

$$\mathrm{Fix}_Z(G_0 \cap t^{-1}Ht) \subset R(L) \; . \tag{12}$$

Then the equation

$$M_0(x) = p \tag{13}$$

has for each sufficiently small $p \in \mathrm{Fix}_Z(H)$ an orbit of solutions $\{\Gamma(s)x_t^*(p) \mid s \in H\}$, where $x_t^*(p)$ depends continuously on p, $x_t^*(0) = \Gamma(t)x_0$, and

$$x_t^*(p) \in \mathrm{Fix}_X(t.G_0.t^{-1} \cap H) \quad , \quad \forall p \; .$$

P r o o f. One can apply corollary 4 with $\Lambda = \mathrm{Fix}_Z(H)$ and $M : X \times \Lambda \to Z$ defined by $M(x,p) = M_0(x) - p$. □

8.4. APPLICATION : PERIODIC PERTURBATIONS OF CONSERVATIVE SYSTEMS

8.4.1. The problem. In this section we will apply theorem 3.2 and corollary 3.7 to scalar equations of the form :

$$\ddot{x} + g(x) = p(t) \tag{1}$$

where $g : \mathbb{R} \to \mathbb{R}$ is of class C^1, and $p : \mathbb{R} \to \mathbb{R}$ is continuous and 2π-periodic, with $\|p\| = \sup\{|p(t)| \mid t \in \mathbb{R}\}$ sufficiently small. We want to find 2π-periodic solutions of (1).

Let $Z = \{z : \mathbb{R} \to \mathbb{R} \mid z$ is continuous and 2π-periodic$\}$ and $X = \{x \in Z \mid x$ is of class $C^2\}$; in Z we use the C^0-sup norm, and in X the C^2-sup norm. In order to simplify our notation further on, we will identify Z with the set of continuous functions $z : S^1 \to \mathbb{R}$, where S^1 is the unit circle in $\mathbb{R}^2$. Define $M_0 : X \to Z$ by

$$M_0(x)(t) = \ddot{x}(t) + g(x(t)) \quad , \quad \forall t \in \mathbb{R} \; , \; \forall x \in X \; . \tag{2}$$

Then our problem takes the form (3.13).

The basic group with which we will work is $O(2) \times \mathbb{Z}_2 = \{(\gamma,\varepsilon) \mid \gamma \in O(2), \varepsilon = \pm 1\}$; this group acts on Z (and X) by :

$$\Gamma(\gamma,\varepsilon)z = \varepsilon z \circ \gamma^{-1} \quad , \quad \forall z \in Z \; , \; \forall (\gamma,\varepsilon) \in O(2) \times \mathbb{Z}_2 \; , \tag{3}$$

(where we identify $\gamma \in O(2)$ with its restriction to S^1). For general $g(x)$, M_0 is equivariant with respect to $O(2) \times \{1\} \cong O(2)$; if $g(x)$ is odd, then M_0 is equivariant with respect to $O(2) \times \mathbb{Z}_2$.

8.4.2. The linearization. Let $M_0(x_0) = 0$ for some $x_0 \in X$, i.e. $x_0(t)$ is a 2π-periodic solution of

$$\ddot{x} + g(x) = 0 \; . \tag{4}$$

We will assume that $x_0(t)$ is non-constant, with least period $2\pi/k$ ($k \in \mathbb{N} \setminus \{0\}$). The operator $L = D_x M_0(x_0) \in L(X,Z)$ is explicitly given by

$$(Lx)(t) = \ddot{x}(t) + g'(x_0(t)).x(t) \quad , \quad \forall t \in \mathbb{R} \ , \ \forall x \in X \ , \tag{5}$$

and has been studied in section 2.2. There we have shown that if h_0 is the total energy of the solution $x_0(t)$, then there exists a function $T(h)$, defined and continuously differentiable for h in a neighbourhood of h_0, such that $T(h_0) = 2\pi/k$, while for each h near h_0, (4) has a periodic solution with least period $T(h)$ and total energy h. The operator L is a Fredholm operator with zero index, and if

$$\frac{dT}{dh}(h_0) \neq 0 \tag{6}$$

then

$$\ker L = \{\mu\dot{x}_0 \mid \mu \in \mathbb{R}\} \tag{7}$$

and

$$R(L) = \{z \in Z \mid \int_{S^1} z(t)\dot{x}_0(t)dt = 0\} \ . \tag{8}$$

8.4.3. Lemma. Assume the following :

(i) x_0 is a non-constant 2π-periodic solution of (4), such that (6) holds :

(ii) $(\gamma,\varepsilon) \in O(2) \times \mathbb{Z}_2$, and γ is orientation reversing ;

(iii) $x_0 \circ \gamma = \varepsilon x_0$.

Then

$$\mathrm{Fix}_Z(\gamma,\varepsilon) \subset R(L) \ . \tag{9}$$

P r o o f. If $z \in \mathrm{Fix}_Z(\gamma,\varepsilon)$, i.e. if $z\circ\gamma = \varepsilon z$, then it is easily seen from a change of integration variable that

$$\int_{S^1} z(t)\dot{x}(t)dt = \int_{\gamma(S^1)} z(t)\dot{x}_0(t)dt \ ;$$

since γ is orientation reversing, we conclude that the integral is zero, and consequently $z \in R(L)$. □

It follows from this lemma that under the condition (6) for x_0 we can apply theorem 3.2 as soon as H_0 contains an element (γ,ε), with γ orientation reversing. An immediate result along these lines is the following.

8.4.4. Theorem. Under the conditions of lemma 3, assume further that $g(\varepsilon x) = \varepsilon g(x)$, $\forall x \in \mathbb{R}$.

Then the equation (1) has for each sufficiently small $p \in \mathrm{Fix}_Z(\gamma,\varepsilon)$ a 2π-periodic solution $x^*(p) \in \mathrm{Fix}_X(\gamma,\varepsilon)$. Here $x^*(p)$ depends continuously on p, and $x^*(0) = 0$.

P r o o f. Apply theorem 3.2, taking for H the (closure of the) subgroup of $O(2) \times \mathbb{Z}_2$ generated by (γ,ε). □

8.4.5. Let us now look in a more systematic way for 2π-periodic solutions of (1) bifurcating from 2π-periodic solutions of (4). Consider first the general case that $g(x)$ is not necessarily odd. Then M_0 is equivariant with respect to $O(2)$. If H is a closed subgroup of $O(2)$, then (1) will be equivariant with respect to H if and only if $p \in \mathrm{Fix}_Z(H)$. Let us restrict ourselves for the moment to subgroups H such that the corresponding class of perturbations $\mathrm{Fix}_Z(H)$ contains functions whose least period is 2π. Then H cannot contain nontrivial pure rotations, and the only nontrivial choice for H is $H = \{\phi(0), \sigma\circ\phi(\alpha_0)\}$, for some fixed $\alpha_0 \in \mathbb{R}$ (we use the notation of subsection 2.6.13). By a simple phase shift in (1) this can be reduced to the case $H = \{\phi(0), \sigma\}$.

8.4.6. Theorem. Let x_0 be a non-constant periodic solution of (4), with least period $2\pi/k$, and such that (6) holds. Then :

(i) there exists an $\alpha_0 \in \mathbb{R}$ such that $x_1(t) = x_0(t+\alpha_0)$ and $x_2(t) = x_0(t+\alpha_0+\pi/k)$ are even periodic solutions of (4) ;

(ii) equation (1) has for each sufficiently small $p \in \mathrm{Fix}_Z(\sigma)$ (i.e. p = even) at least two different 2π-periodic solutions $x_1^*(p)$ and $x_2^*(p)$;

(iii) $x_1^*(p)$ and $x_2^*(p)$ are even functions of t, they depend continuously on p, $x_1^*(0) = x_1$ and $x_2^*(0) = x_2$.

P r o o f. Let $\alpha_0 \in \mathbb{R}$ be such that $x_0(\alpha_0) = \max\{x_0(t) \mid t \in \mathbb{R}\}$; then $\dot{x}_1(0) = \dot{x}_0(\alpha_0) = 0$, and since $x_1(t)$ solves (4), this implies that $x_1(t)$ is even. Since $x_1(t)$ is $2\pi/k$-periodic, also $x_2(t)$ is even. This proves (i). Parts (ii) and (iii) follows from an application of theorem 4 to x_1 and x_2, with $(\gamma,\varepsilon) = (\sigma,1)$. □

8.4.7. If g(x) in (1) is odd, then M_0 is equivariant with respect to $O(2) \times \mathbb{Z}_2$, and H may include elements of the form $(\gamma,-1)$, $\gamma \in O(2)$. Under the same restriction as before, the nontrivial choices for H become now, after appropriate phase shifts :

(i) $H = \{(\phi(0),1),(\sigma,1)\}$;

(ii) $H = \{(\phi(0),1),(\sigma,-1)\}$;

(iii) $H = \{(\phi(0),1),(\sigma,1),(\sigma\circ\phi(\pi),-1),(\phi(\pi),-1)\}$.

Case (i) is included in theorem 6. For case (ii) we have the following result, analogous to theorem 6.

8.4.8. Theorem. Let g be odd, and let x_0 be a periodic solution of (4), with least period $2\pi/k$, taking both positive and negative values, and such that (6) holds. Then :

(i) for some $\beta_0 \in \mathbb{R}$, $\tilde{x}_1(t) = x_0(t+\beta_0)$ and $\tilde{x}_2(t) = x_0(t+\beta_0+\pi/k)$ are odd periodic solutions of (4) ;

(ii) equation (1) has for each sufficiently small $p \in \mathrm{Fix}_Z(\sigma,-1)$ (i.e. p = odd) at least two different 2π-periodic solutions $\tilde{x}_1^*(p)$ and $\tilde{x}_2^*(p)$;

(iii) $\tilde{x}_1^*(p)$ and $\tilde{x}_2^*(p)$ are odd functions of t, they depend continuously on p, $\tilde{x}_1^*(0) = \tilde{x}_1^*$ and $\tilde{x}_2^*(0) = \tilde{x}_2^*$.

P r o o f. One takes β_0 such that $x_0(\beta_0) = 0$, and one applies theorem 4 with $(\gamma,\varepsilon) = (\sigma,-1)$. □

8.4.9. When we take H equal to the subgroup of $O(2) \times \mathbb{Z}_2$ generated by $(\sigma,1)$ and $(\sigma\circ\phi(\pi),-1)$, then we have to distinguish between k even and k odd. Let

x_1 be an even periodic solution of (4), with least period $2\pi/k$, and taking both positive and negative values. If k is odd, then the isotropy group of x_1 will be the whole H, and we can apply theorem 3.2 at x_1 and at $x_2(t) = x_1(t+\pi/k)$. If k is even, let $x_j(t) = x_1(t+(j-1)\pi/2k)$, $j = 1,2,3,4$. Then the isotropy subgroup of x_1 and x_3 will be $\{(\phi(0),1),(\sigma,1)\}$, while the isotropy subgroup of x_2 and x_4 is $\{(\phi(0),1),(\sigma\circ\phi(\pi),-1)\}$. At x_1 and x_3 we apply theorem 4 with $(\gamma,\varepsilon) = (\sigma,1)$, at x_2 and x_4 we apply the theorem 4 with $(\gamma,\varepsilon) = (\sigma\circ\phi(\pi),-1)$. This gives us the following results.

8.4.10. Theorem. Let g be odd, and x_0 as in theorem 8, with k odd. Then :

(i) there exists an $\alpha_0 \in \mathbb{R}$ such that $x_1(t) = x_0(t+\alpha_0)$ and $x_2(t) = x_0(t+\alpha_0+\pi/k)$ are even and odd harmonic periodic solutions of (4);

(ii) equation (1) has for each sufficiently small $p \in Z$, which is even and odd harmonic, at least two different 2π-periodic solutions $x_1^*(p)$ and $x_2^*(p)$;

(iii) $x_1^*(p)$ and $x_2^*(p)$ are even and odd harmonic, they depend continuously on p, $x_1^*(0) = x_1$ and $x_2^*(0) = x_2$. □

8.4.11. Theorem. Let g be odd, and x_0 as in theorem 8, with k even. Then :

(i) there exists an $\alpha_0 \in \mathbb{R}$ such that, if we define $x_j(t) = x_0(t+\alpha_0+(j-1)\pi/2k)$, $j = 1,2,3,4$, then x_1 and x_3 are even, while x_2 and x_4 are odd after a phase shift by $\pi/2$;

(ii) equation (1) has for each sufficiently small $p \in Z$, which is even and odd harmonic, at least four different 2π-periodic solutions $x_j^*(p)$ $(j=1,2,3,4)$, depending continuously on p and with $x_j^*(0) = x_j$, $j = 1,2,3,4$.

(iii) $x_1^*(p)$ and $x_3^*(p)$ are even functions of t, while $x_2^*(p)$ and $x_4^*(p)$ are odd after a phase shift by $\pi/2$. □

8.4.12. Remarks. (1) Some of the foregoing results can also be obtained in a systematic way by using corollary 3.7. When one does not restrict to the subgroup H considered here, this corollary gives a systematic approach to find all symmetry induced solution branches from $\gamma_0 = \{\Gamma(\alpha)x_0 \mid \alpha\in\mathbb{R}\}$, where

$x_0(t)$ is a 2π-periodic solution of (4).

(2) For the Duffing equation ($g(x) = bx+x^3$), the results of theorems 10 and 11 were proved by Schmidt and Mazzanti [200] using phase plane techniques.

(3) As another possible application of the results of section 3, let us mention the following boundary value problem :

$$\begin{aligned} \Delta u + f(u) &= p(x) \quad , \qquad x \in B \\ u(x) &= 0 \qquad\quad , \qquad x \in \partial B \,. \end{aligned} \tag{10}$$

When the domain B has some symmetry (e.g. a ball), and when p is small, then one can use the foregoing approach to obtain certain solution near a solution u_0 of (10) with $p = 0$.

8.5. THE BIFURCATION SET : AN EXAMPLE

8.5.1. Introduction. In this section we return to the bifurcation equation (2.11) which resulted from the reduction of section 2. This equation has the form :

$$F(s,\lambda) = 0 \tag{1}$$

where $F : G \times \Lambda \to R(Q)$ is defined and continuous for (s,λ) near $G \times \{0\}$, with $F(s,0) = 0$ for all $s \in G$. If M in (2.1) is of class C^1, then $F(s,\lambda)$ is continuously differentiable in λ, and we find from the results of section 2 that :

$$D_\lambda F(s,0) = Q\tilde{\Gamma}(s^{-1})D_\lambda M(\Gamma(s)x_0,0) \quad , \qquad \forall s \in G \,. \tag{2}$$

By the implicit function theorem (1) can only have nontrivial solutions (i.e. solutions with $\lambda \neq 0$) near those points $(\bar{s},0)$ where $D_\lambda F(\bar{s},0)$ is not injective. If $\dim \Lambda \leqslant \dim R(Q)$, then generically $D_\lambda F(s,0)$ will be injective, and in general nontrivial solutions of (1) will only branch off $G \times \{0\}$ at isolated points. When to the contrary $\dim \Lambda > \dim R(Q)$ then $D_\lambda F(s,0)$ will have a nontrivial kernel for all $s \in G$, and one expects (1) to have nontrivial solutions for all $s \in G$. In [230] we have given a general discussion of the solution set and the bifurcation points for this case. Since this discussion is rather abstract, we will explain here the essential ideas on a particular example, which is sufficiently simple to allow for an easy visualization of what is going on.

8.5.2. The problem. Let $g : \mathbb{R} \to \mathbb{R}$ and $h : \mathbb{R}\times\mathbb{R}^2\times\mathbb{R}^2 \to \mathbb{R}$ be of class C^3, with $h(t,x,y,\lambda_1,\lambda_2)$ 2π-periodic in the first variable t, and $h(t,x,y,0,0) = 0$ for all (t,x,y). We want to study 2π-periodic solutions of the equation

$$\ddot{x} + g(x) + h(t,x,\dot{x},\lambda) = 0 \tag{3}$$

for λ near the origin in $\mathbb{R}^2$. Special cases of this problem have been studied by Hale and Taboas ([81],[89]). The smoothness conditions on g and h are such that at each step of the discussion below we will have automatically all the smoothness we need; by a more carefull analysis it may be possible to obtain the results under weaker smoothness conditions.

To bring the problem in the form (2.1) we use the same spaces X and Z as in section 4, take $\Lambda = \mathbb{R}^2$ and define $M : X\times\Lambda \to Z$ by :

$$M(x,\lambda)(t) = \ddot{x}(t) + g(x(t))+h(t,x(t),\dot{x}(t),\lambda) , \tag{4}$$
$$\forall t \in \mathbb{R} , \forall x \in X , \forall\lambda \in \mathbb{R}^2 .$$

For $\lambda = 0$ the equation (3) becomes

$$\ddot{x} + g(x) = 0 ; \tag{5}$$

$M(x,0)$ is equivariant with respect to $G = SO(2)$, with the usual representation $(\Gamma(\alpha)z(t) = z(t+\alpha))$.

Let $x_0(t)$ be a non-constant periodic solution of (5), with least period $2\pi/k$ $(k \in \mathbb{N})$, and such that (4.6) holds. Then $L = D_xF(x_0,0)$ is a Fredholm operator, with ker L and R(L) given by (4.7), respectively (4.8). For the projections P and Q we take $P = Q|_X$, with

$$Qz = a_0^{-1}\dot{x}_0 \int_0^{2\pi} z(t)\dot{x}_0(t)dt \qquad , \qquad \forall z \in Z , \tag{6}$$

and

$$a_0 = \int_0^{2\pi} \dot{x}_0^2(t)dt \ . \tag{7}$$

Since $x_0(t)$ is a solution of (5), x_0 is of class C^5, and from this it easily follows that all hypotheses of section 2 are satisfied. The operator $\widetilde{M}$ defined by (2.6) takes the form :

$$\begin{aligned}\widetilde{M}(\alpha,y,\lambda)(t) &= \ddot{x}_0(t) + \ddot{y}(t) + g(x_0(t)+y(t)) \\ &\quad + h(t-\alpha, x_0(t)+y(t), \dot{x}_0(t)+\dot{y}(t), \lambda) \ , \\ &\quad \forall t \in \mathbb{R} \ , \ \forall \alpha \in \mathbb{R} \ , \ \forall (y,\lambda) \in \ker P \times \mathbb{R}^2 \ ,\end{aligned} \tag{8}$$

and is clearly of class C^3. Consequently, also the solution $y^*(\alpha,\lambda)$ of the auxiliary equation (2.8a) and the bifurcation function $F(\alpha,\lambda)$ will be of class C^3. Since $\dim R(Q) = 1$ we may consider F to be real-valued, and explicitly given by :

$$F(\alpha,\lambda) = \int_0^{2\pi} \widetilde{M}(\alpha, y^*(\alpha,\lambda),\lambda)(t)\dot{x}_0(t)dt \ . \tag{9}$$

Also, since $G_0 = \{\phi(\frac{2\pi}{k}j) \mid j \in \mathbb{Z}\}$, it follows from (2.12) that $F(\alpha,\lambda)$ will be $2\pi/k$-periodic in α. Solutions (α,λ) and $(\alpha + \frac{2\pi}{k}j, \lambda)$ of the bifurcation equation

$$F(\alpha,\lambda) = 0 \tag{10}$$

correspond to the same solution of the original problem.

8.5.3. <u>The solution set</u>. In order to find the solution set of (10) we will use polar coordinates in the parameter space $\Lambda = \mathbb{R}^2$. Writing $\lambda = (\lambda_1,\lambda_2) = \rho(\cos\theta, \sin\theta)$, we have

$$F(\alpha, \rho\cos\theta, \rho\sin\theta) = \rho\widetilde{F}(\alpha,\theta,\rho) \tag{11}$$

with

$$\widetilde{F}(\alpha,\theta,\rho) = \int_0^1 D_\lambda F(\alpha, \tau\rho(\cos\theta,\sin\theta)).(\cos\theta,\sin\theta)d\tau \ . \tag{12}$$

The function $\tilde{F}(\alpha,\theta,\rho)$ is of class C^2, $2\pi/k$-periodic in α, 2π-periodic in θ, and

$$\tilde{F}(\alpha,\theta+\pi,-\rho) = -\tilde{F}(\alpha,\theta,\rho) \quad , \quad \forall(\alpha,\theta,\rho) . \tag{13}$$

Using (2) we find

$$\tilde{F}(\alpha,\theta,0) = p_1(\alpha)\cos\theta + p_2(\alpha)\sin\theta \quad , \quad \forall(\alpha,\theta) . \tag{14}$$

with

$$p_i(\alpha) = \int_0^{2\pi} \frac{\partial h}{\partial \lambda_i}(t-\alpha,x_0(t),\dot{x}_0(t),0)\dot{x}_0(t)dt \quad , \quad i = 1,2 . \tag{15}$$

For $\rho \neq 0$ equation (10) is equivalent to :

$$\tilde{F}(\alpha,\theta,\rho) = 0 . \tag{16}$$

Now we make the following hypothesis :

$$\text{(H1)} \quad (p_1(\alpha),p_2(\alpha)) \neq (0,0) \quad , \quad \forall\alpha \in \mathbb{R} . \tag{17}$$

Let $\eta(\alpha) = (p_1^2(\alpha)+p_2^2(\alpha))^{1/2}$; then it is a classical result (see e.g. Knobloch and Kappel [129]) that under the hypothesis (H1) there exists a C^2-function $\zeta : \mathbb{R} \to \mathbb{R}$ such that :

$$(-p_2(\alpha),p_1(\alpha)) = \eta(\alpha)(\cos\zeta(\alpha),\sin\zeta(\alpha)) \quad , \quad \forall\alpha \in \mathbb{R} . \tag{18}$$

Since $p_i(\alpha)$ is $2\pi/k$-periodic, we will have :

$$\zeta(\alpha + 2\pi/k) = \zeta(\alpha) + 2\pi m \quad , \quad \forall\alpha \in \mathbb{R} , \tag{19}$$

for some $m \in \mathbb{Z}$.

Then $\tilde{F}(\alpha,\theta,0) = \eta(\alpha)\sin(\theta-\zeta(\alpha))$, $\tilde{F}(\alpha,\zeta(\alpha),0) = 0$ and $\frac{\partial\tilde{F}}{\partial\theta}(\alpha,\zeta(\alpha),0) \neq 0$. It follows from the implicit function theorem, the periodicity properties of F and (13), that there exist a $\rho_0 > 0$ and a C^2-function $\theta^* : \mathbb{R} \times]-\rho_0,\rho_0[\to \mathbb{R}$ such that :

(i) $\{(\alpha,\theta,\rho) \mid \tilde{F}(\alpha,\theta,\rho) = 0$ and $|\rho| < \rho_0\}$

$$= \{(\alpha,\theta^*(\alpha,\rho)+2j\pi,\rho) \mid \alpha\in\mathbb{R},\ |\rho| < \rho_0,\ j\in\mathbb{Z}\}$$

$$\cup\ \{(\alpha,\theta^*(\alpha,-\rho)+(2j+1)\pi,\rho) \mid \alpha\in\mathbb{R},\ |\rho| < \rho_0,\ j\in\mathbb{Z}\}\ ;$$

(ii) $\theta^*(\alpha,0) = \zeta(\alpha) \quad , \quad \forall\alpha \in \mathbb{R}$; (20)

(iii) $\theta^*(\alpha + 2\pi/k,\rho) = \theta^*(\alpha,\rho) + 2\pi m \quad , \quad \forall(\alpha,\rho)$. (21)

Returning to the equation (10) this gives the following result.

8.5.4. Theorem. Let g, h and x_0 be as described above. Define $p_i(\alpha)$ ($i = 1,2$) by (15), and assume (H1) holds. Then there exist a $\rho_0 > 0$ and a C^2-function $\theta^* : \mathbb{R}\times]-\rho_0,\rho_0[\to \mathbb{R}$, satisfying (20) and (21), such that :

$$S \equiv \{(\alpha,\lambda) \in\mathbb{R}\times\mathbb{R}^2 \mid F(\alpha,\lambda) = 0,\ \|\lambda\| < \rho_0\} \tag{22}$$

$$= \{(\alpha,\rho\cos\theta^*(\alpha,\rho),\rho\sin\theta^*(\alpha,\rho)) \mid \alpha\in\mathbb{R},\ |\rho| < \rho_0\}\ .$$

The solution set S is a 2-dimensional C^3-submanifold of $\mathbb{R}\times\mathbb{R}^2$, $2\pi/k$-periodic in the α-direction, containing $\mathbb{R}\times\{0\}$, and along $\mathbb{R}\times\{0\}$ tangent to the C^2-submanifold :

$$S_0 \equiv \{(\alpha,\lambda) \in\mathbb{R}\times\mathbb{R}^2 \mid D_\lambda F(\alpha,0)\lambda = 0\} \tag{23}$$

$$= \{(\alpha,\rho\cos\zeta(\alpha),\rho\sin\zeta(\alpha)) \mid \alpha\in\mathbb{R},\ \rho\in\mathbb{R}\}\ .$$

P r o o f. The first part follows from the foregoing discussion. The fact that S is a C^3-submanifold can be seen from a slightly different approach, as follows. Fix some $\bar{\alpha} \in \mathbb{R}$, and write $\lambda \in \mathbb{R}^2$ in the form $\lambda = \mu(\cos\zeta(\bar{\alpha}), \sin\zeta(\bar{\alpha})) + \nu(-\sin\zeta(\bar{\alpha}),\cos\zeta(\bar{\alpha}))$. Then the equation $F(\alpha,\lambda) = 0$ for S takes the form $\bar{F}(\alpha,\mu,\nu) = 0$, where $\bar{F}$ is of class C^3, $\bar{F}(\alpha,0,0) = 0$ and $D_\nu\bar{F}(\bar{\alpha},0,0) \neq 0$. This shows that near $(\bar{\alpha},0)$ S is a C^3-submanifold of $\mathbb{R}\times\mathbb{R}^2$ with codimension 1. In a similar way one shows that S_0 is a C^2-submanifold, and that S and S_0 are tangent along $\mathbb{R}\times\{0\}$. □

8.5.5. The bifurcation set. Let now $\lambda = (\rho cos\theta, \rho sin\theta) \in \mathbb{R}^2$ be given, with $0 < \rho < \rho_0$. For this parameter value λ, there is, according to the theory of section 2 and the foregoing analysis of equation (10), a one-to-one correspondance between the 2π-periodic solutions of (3) near $\gamma_0 = \{\Gamma(\alpha)x_0 \mid \alpha\in\mathbb{R}\}$ and the points of the set :

$$A(\lambda) = A(\rho,\theta) = \{\alpha\in[0,2\pi/k[\mid \theta^*(\alpha,\rho) = \theta \pmod{2\pi}$$
$$\text{or } \theta^*(\alpha,-\rho) = \theta+\pi \pmod{2\pi}\} .$$

It follows from (21) that this set contains at least $2|m|$ elements. In the limit for $\rho = 0$ we obtain the set :

$$A_0(\theta) = \{\alpha\in[0,2\pi/k[\mid \zeta(\alpha) = \theta \pmod{\pi}\} .$$

Now we make the following hypothesis :

(H2) If $\zeta'(\alpha) = 0$ for some $\alpha \in \mathbb{R}$, then $\zeta''(\alpha) \neq 0$.

This implies that $A_0(\theta)$ is a finite set; we denote the number of its elements by $N_0(\theta)$. If we change θ, then $N_0(\theta)$ will remain constant, except at those θ for which there is an $\alpha \in A_0(\theta)$ such that $\zeta'(\alpha) = 0$. By (H2) there can only be a finite number of $\alpha \in [0,2\pi/k[$ such that $\zeta'(\alpha) = 0$; denote these by $\alpha_1,\alpha_2,\ldots,\alpha_r$, and let $\zeta_j = \zeta(\alpha_j)$, $j = 1,2,\ldots,r$. We will make the following assumption, which is not essential, but which makes the discussion somewhat easier :

(H3) $\zeta_i \neq \zeta_j \pmod{\pi}$, for all $i,j = 1,2,\ldots,r$; $i \neq j$.

Let $\sigma_j = \operatorname{sgn} \zeta''(\alpha_j)$. It follows from (H2) and (H3) that we can find $\varepsilon_0 > 0$ such that for all $\varepsilon \in]0,\varepsilon_0[$, for all $j = 1,2,\ldots,r$ and for all $\ell \in \mathbb{Z}$ we have :

$$N_0(\zeta_j+\ell\pi-\varepsilon)+(-1)^{\sigma_j} = N_0(\zeta_j+\ell\pi) = N_0(\zeta_j+\ell\pi+\varepsilon)-(-1)^{\sigma_j} . \qquad (24)$$

Moreover, if $\theta \neq \zeta_j \pmod{\pi}$ for all $j = 1,2,\ldots,r$, then $N_0(\theta)$ is even.

Consider now the equation

$$\frac{\partial\theta^*}{\partial\alpha}(\alpha,\rho) = 0 \; . \tag{25}$$

Because of (H2) and $\theta^*(\alpha,0) = \zeta(\alpha)$ there exists some $\rho_0 > 0$ such that the solution set of (25) for $|\rho| < \rho_0$ will have the form

$$\{(\alpha_j^*(\rho) + 2\pi\ell/k,\rho) \mid |\rho| < \rho_0,\ \ell\in\mathbb{Z},\ j = 1,2,\ldots,r\} \; ,$$

where $\alpha_j^*(\rho)$ is of class C^1, with $\alpha_j^*(0) = \alpha_j$. Let

$$\theta_j(\rho) = \theta^*(\alpha_j^*(\rho),\rho) \qquad , \qquad j = 1,2,\ldots,r \; , \; |\rho| < \rho_0 \; ;$$

then $\theta_j(0) = \zeta_j$, and by (H3) we can take ρ_0 sufficiently small such that $\theta_i(\rho) \neq \theta_j(\rho')$ (mod π) if $i \neq j$, for all $|\rho|,|\rho'| < \rho_0$. Also

$$\operatorname{sgn} \frac{\partial^2\theta^*}{\partial\alpha^2}(\alpha_j^*(\rho),\rho) = \operatorname{sgn} \zeta''(\alpha_j) = \sigma_j \quad , \quad \forall|\rho| < \rho_0,\ j = 1,2,\ldots,r \; . \tag{26}$$

It follows again that for all (ρ,θ) with $|\rho| < \rho_0$ the set $A(\rho,\theta)$ will be finite; let $N(\rho,\theta)$ denote the number of its elements. For fixed $\rho > 0$, the function $\theta \mapsto N(\rho,\theta)$ is constant and even-valued, except at the points $\theta = \theta_j(\rho)$ (mod 2π) and $\theta = \theta_j(-\rho) + \pi$ (mod 2π), where a relation such as (24) holds.

Summarizing, we obtain our final result, which describes the bifurcation of 2π-periodic solutions of (3) near γ_0.

8.5.6. <u>Theorem</u>. Let g, h, x_0 and γ_0 be as described above, and assume (H1), (H2) and (H3). Let $N_0(\theta)$, ζ_j and σ_j ($\theta\in\mathbb{R}$, $j = 1,2,\ldots,r$) be as defined above.

Then there exist a $\rho_0 > 0$ and r different C^1-functions $\theta_j :]-\rho_0,\rho_0[\to \mathbb{R}$, with $\theta_j(0) = \zeta_j$ ($j = 1,2,\ldots,r$), such that the following holds :

(i) for each $\lambda = (\rho\cos\theta,\rho\sin\theta)$, with $0 < \rho < \rho_0$, equation (3) has a finite number of different 2π-periodic solutions, near γ_0; we denote this number by $N(\lambda) = N(\rho,\theta)$;

(ii) $N(\rho,\theta) \geq 2|m|$, $\forall(\rho,\theta)$, where $m \in \mathbb{Z}$ is as in (21) ;

(iii) if $\theta \neq \zeta_j$ (mod π), $j = 1,2,\ldots,r$, then $\lim_{\rho\to 0} N(\rho,\theta) = N_0(\theta)$;

(iv) the curves :

$$C_j = \{(\rho\cos\theta_j(\rho), \rho\sin\theta_j(\rho)) \mid |\rho| < \rho_0\} , \quad j = 1,2,\ldots,r$$

divide the neighbourhood $\{\lambda \in \mathbb{R}^2 \mid \|\lambda\| < \rho_0\}$ of the origin in $\mathbb{R}^2$ into 2r cone-like regions; in the interior of each of these regions $N(\lambda)$ is constant and even ;

(v) if λ crosses one of the curves C_j, in the direction of increasing θ, then $N(\lambda)$ increases by two if $\sigma_j = +1$, and decreases by two if $\sigma_j = -1$;

(vi) on $C_j \setminus \{0\}$ $N(\lambda)$ is constant, odd-valued, and equal to $N_0(\zeta_j)$;

(vii) the curve C_j is tangent to the line $\{(\rho\cos\zeta_j, \rho\sin\zeta_j) \mid \rho \in \mathbb{R}\}$; the union of the curves C_j form the bifurcation set for the problem. □

In [90] Hale and Taboas discuss the bifurcation of 2π-periodic solutions of a special case of (3) near a family γ_0 generated by a solution x_0 of (5) for which the condition (4.6) is not satisfied.

References

[1] Adams, R.A., Sobolev Spaces. Academic Press, N.Y., 1975.

[2] Agmon, S., A. Douglis and L. Nirenberg, Estimates near the boundary for solutions of elliptic partial differential equations satisfying general boundary conditions. I. Comm. Pure Appl. Math. 12 (1959) 623-727; II. Comm. Pure Appl. Math. 17 (1964) 35-92.

[3] Agmon, S., Lectures on Elliptic Boundary Value Problems. Van Nostrand, Princeton, N.J., 1965.

[4] Alexander, J.C. and J.A. Yorke, Global bifurcation of periodic orbits. Ann. of Math. 100 (1978) 263-292.

[5] Amann, H., Fixed point equations and nonlinear eigenvalue problems in ordered Banach spaces. SIAM Rev. 18 (1976) 620-709.

[6] Ambrosetti, A., Some remarks on the buckling problem for a thin clamped shell. Ricerche di Math. 23 (1974) 161-170.

[7] Ambrosetti, A. and G. Prodi, On the inversion of some differentiable mappings with singularities between Banach spaces. Ann. Mat. Pura Appl. 93 (1972) 231-246.

[8] Antman, S., Buckled states of nonlinearly elastic plates. Arch. Rat. Mech. Anal. 67 (1978) 111-149.

[9] Arnold, V.I., Lectures on bifurcation in versal families. Russ. Math. Surveys 27 (1972) 54-123.

[10] Arnold, V.I., Ordinary Differential Equations. MIT Press, Cambridge, Mass., 1973.

[11] Bancroft, S., J.K. Hale and D. Sweet, Alternative problems for nonlinear functional equations. J. Diff. Eqns. 4 (1968) 40-56.

[12] Bauer, L., H.B. Keller and E.L. Reiss, Axisymmetric buckling of hollow spheres and hemispheres. Comm. Pure Appl. Math. 23 (1970) 529-568.

[13] Bauer, L., H.B. Keller and E.L. Reiss, Multiple eigenvalues lead to secondary bifurcation. SIAM J. Appl. Math. 17 (1975) 101-122.

[14] Bauer, L. and E.L. Reiss, Nonlinear buckling of rectangular plates. SIAM J. 13 (1965) 603-626.

[15] Berger, M.S. and M.S. Berger, Perspectives in Nonlinearity.

Benjamin, N.Y., 1968.

[16] Berger, M.S. and P.C. Fife, On von Kàrmàn's equations and the buckling of a thin elastic plate. Bull. A.M.S. 72 (1966) 1006-1011.

[17] Berger, M.S., On von Kàrmàn's equations and the bunckling of a thin elastic plate. I. The clamped plate. Comm. Pure Appl. Math. 20 (1967) 687-719.

[18] Berger, M.S. and P.C. Fife, Von Kàrmàn's equations and the buckling of a thin elastic plate. II. Plate with general edge conditions. Comm. Pure Appl. Math. 21 (1968) 227-241.

[19] Berger, M.S., A bifurcation theory for real solutions of nonlinear elliptic partial differential equations. In [114], p. 113-216.

[20] Berger, M.S., On the existence of equilibrium state of thin elastic shells (I). Indiana Math. J. 20 (1971) 591-602.

[21] Berger, M.S., Nonlinearity and Functional Analysis. Academic Press, N.Y., 1977.

[22] Bers, L., F. John and M. Schechter, Partial Differential Equations. Interscience, N.Y., 1964.

[23] Böhme R., Die Lösung der Verzweigungsgleichungen für Nichtlineare Eigenwertprobleme. Math. Z. 127 (1972) 105-126.

[24] Brickell, F. and R.S. Clark, Differentiable Manifolds. Van Nostrand Reinhold, London, 1970.

[25] Bröcker, Th., Differentierbare Abbildungen. Lecture Notes, Univ. of Regensburg, 1973.

[26] Bröcker, Th. and L. Lander, Differentiable Germs and Catastrophes. London Math. Soc. Lecture Notes 17, Cambridge Univ. Press, Cambridge, 1975.

[27] Cesari, L., Functional analysis and periodic solutions of nonlinear differential equations. In : Contributions to Differential Equations 1 (1963) 149-187.

[28] Cesari, L., Functional analysis and Galerkin's method. Mich. Math. J. 11 (1964) 385-418.

[29] Cesari, L., Nonlinear analysis. Lecture Notes C.I.M.E. (1972), Ed. Cremonese, Roma, 1973.

[30] Cesari, L., Alternative methods in nonlinear analysis. Int. Conf. Diff. Eqns., Los Angeles, 1974, p.95-198, Acad. Press., N.Y.

[31] Cesari, L., Functional analysis, nonlinear differential equations, and the alternative method. In [33], p. 1-197.

[32] Cesari, L., J.K. Hale and J.P. La Salle, Dynamical Systems - An international symposium, Vol I and II. Academic Press, N.Y., 1976.

[33] Cesari, L., R. Kannan and J. Schuur, Nonlinear Functional Analysis and its Applications. Dekker, N.Y., 1976.

[34] Chafee, N., The bifurcation of one or more closed orbits from an equilibrium point of an autonomous differential equation. J. Diff. Eqns. 4 (1968) 661-679.

[35] Chafee, N., A bifurcation problem for a functional differential equation of finitely retarded type. J. Math. Anal. Appl. 35 (1971) 312-348.

[36] Chafee, N., Generalized Hopf bifurcation and perturbation in a full neighborhood of a given vector field. Indiana Univ. Math. J. 27 (1978) 173-194.

[37] Chillingworth, D.R.J., Differential topology with a view to applications. Research Notes in Math., 9, Pitman, London, 1976.

[38] Chow, S.N. and J.K. Hale, Periodic solutions of autonomous equations. J. Math. Anal. Appl. 66 (1978) 495-506.

[39] Chow, S.N., J.K. Hale and J. Mallet-Paret, Application of generic bifurcation. I. Arch. Rat. Mech. Anal. 59 (1975) 159-188.

[40] Chow, S.N., J.K. Hale and J. Mallet-Paret, Application of generic bifurcation. II. Arch. Rat. Mech. Anal. 62 (1976) 209-236.

[41] Chow, S.N. and J. Mallet-Paret, Hopf bifurcation and the method of averaging. In [156], p. 151-162.

[42] Chow, S.N. and J. Mallet-Paret, Applications of generic bifurcation. In [33], p. 199-207.

[43] Chow, S.N. and J. Mallet-Paret, Integral averaging and bifurcation. J. Diff. Eqns. 26 (1977) 112-159.

[44] Chow, S.N. and J. Mallet-Paret, The Fuller index and global Hopf bifurcation. J. Diff. Eqns. 29 (1978) 66-85.

[45] Chow, S.N., J. Mallet-Paret and J. Yorke, Global Hopf bifurcation from a multiple eigenvalue. Nonlin. Anal., Th., Meth., Appl. 2 (1978) 753-763.

[46] Coddington, E.A. and N. Levinson, Theory of Ordinary Differential Equations, McGraw Hill, N.Y., 1955.

[47] Cohen, D.S., Multiple solutions of nonlinear partial differential equations. Lecture Notes in Math. 322, p.15-77. Springer-Verlag, Berlin 1973.

[48] Courant, R. and D. Hilbert, Methods of Mathematical Physics. Vol. I. Interscience, N.Y. 1953.

[49] Courant, R., and D. Hilbert, Methods of Mathematical Physics.Vol. II. Interscience, N.Y., 1962.

[50] Crandall, M.G. and P.H. Rabinowitz, Bifurcation from simple eigenvalue. J. Funct. Anal. 8 (1971) 321-340.

[51] Crandall, M.G. and P.H. Rabinowitz, Bifurcation, perturbation of simple eigenvalues and linearized stability. Arch. Rat. Mech. Anal. 52 (1973) 161-180.

[52] Crandall, M.G. and P.H. Rabinowitz, The Hopf bifurcation theorem in infinite dimensions. Arch. Rat. Mech. Anal. 67 (1978) 53-72.

[53] Cronin, J., Bifurcation of periodic solutions. J. Math. Anal. Appl. 68 (1979) 130-151.

[54] Cushing, J.M., Nontrivial periodic solutions of some Volterra integral equations. To appear in : Proc. of the Helsinki Conf. on Int. Eqns., Lecture Notes in Math., Springer-Verlag, Berlin.

[55] Dancer, E.N., Bifurcation theory in real Banach spaces. J. London Math. Soc. 23 (1971) 699-734.

[56] Dancer, E.N., Bifurcation theory for analytic operators. Proc. London Math. Soc. 26 (1973) 359-384.

[57] Dancer, E.N., On the existence of bifurcating solutions in the presence of symmetries. Proc. Royal Soc. Edinburgh 85A (1980) 321-336.

[58] Dickey, R.W., Bifurcation problems in nonlinear elasticity. Research Notes in Math. 3, Pitman, London, 1976.

[59] Dunford, N. and J.J. Schwartz, Linear Operators, Part I. Interscience, N.Y., 1958.

[60] Duvaut, G. and J.L. Lions, Inequalities in mechanics and physics. Springer-Verlag, Berlin, 1976.

[61] Fichera, G., Linear elliptic differential systems and eigenvalue problems. Lectures Notes in Math. 8, Springer-Verlag, Berlin, 1965.

[62] Fife, P.C., Nonlinear deflection of thin elastic plates under tension. Comm. Pure Appl. Math. 14 (1961) 81-112.

[63] Flockerzi, D., Bifurcation of periodic solutions from an equilibrium point. Dissertation, Würzburg, 1979.

[64] Folland, G.B., Introduction to partial differential equations. Mathematical Notes 17. Princeton Univ. Press, Princeton, N.J., 1976.

[65] Friedman, A., Partial differential equations of parabolic type.

Prentice Hall, N.J., 1964.

[66] Friedman, A., Partial differential equations. Holt-Rinehart & Winston, 1969.

[67] Friedrichs, K.O., Advanced ordinary differential equations. Gordon and Breach, N.Y., 1965.

[68] Friedrichs, K.O. and J.J. Stoker, The non-linear boundary value problem of the buckled state. Amer. J. Math. 63 (1941) 839-888.

[69] Fucik, S., J. Necas and V. Soucek, Spectral analysis of nonlinear operators. Lecture Notes in Math. 343, Springer-Verlag, Berlin, 1973.

[70] Fujita, H. and T. Kato, On the Navier-Stokes initial value problem. Arch. Rat. Mech. Anal. 16 (1964) 269-315.

[71] Gilbarg, D. and N.S. Trudinger, Elliptic partial differential equations of second order. Springer-Verlag, Berlin, 1977.

[72] Golubitsky, M. and V. Guillemin, Stable mappings and their singularities. Springer-Verlag, Berlin, 1973.

[73] Golubitsky, M., An introduction to catastrophe theory and its applications. SIAM Rev. 28 (1978) 352-387.

[74] Golubitsky, M. and D. Schaeffer, A theory for imperfect bifurcation via singularity theory. Comm. Pure Appl. Math. 32 (1979) 21-98.

[75] Golubitsky, M. and D. Schaeffer, Imperfect bifurcation in the presence of symmetry. Comm. Math. Phys. 67 (1979) 205-232.

[76] Hale, J.K., Oscillations in nonlinear systems. McGraw-Hill, 1963.

[77] Hale, J.K., Ordinary differential equations. Wiley-Interscience, N.Y., 1969.

[78] Hale, J.K., Applications of alternative problems. Lecture Notes , Brown University, 1971.

[79] Hale, J.K., Theory of functional differential equations. Springer-Verlag, Berlin, 1977.

[80] Hale, J.K., Generic bifurcation with applications. In [130], p.59-157.

[81] Hale, J.K., Bifurcation near families of solutions. Proc. Int. Conf. on Diff. Eqns. Uppsala 1977, p. 91-100.

[82] Hale, J.K., Bifurcation with several parameters. VII. Internat. Konf. Nichtlin. Schwing., Berlin (1975), Band I,1, 279-288. Abh. Akad. Wiss. D.D.R., Akademie-Verlag, Berlin, 1977.

[83] Hale, J.K., Bifurcation and alternative problems. Preprint.

[84] Hale, J.K., Restricted generic bifurcation. Nonlinear Analysis (Collection of papers in honour of E.H. Rothe) Academic Press, N.Y., 1978, p.83-98.

[85] Hale, J.K., Bifurcation from simple eigenvalues for several parameter values. Nonlin. Anal., Th., Meth., Appl. 2 (1978) 491-497.

[86] Hale, J.K., Nonlinear oscillations in equations with delays. In : F.C. Hoppensteadt (Ed.), Nonlinear Oscillations in Biology, Lectures in Applied Mathematics, Vol. 17, A.M.S., Providence, 1979, p.157-185.

[87] Hale, J.K., Topics in local bifurcation theory. Annals of the N.Y. Acad. of Sciences 316 (1979) 605-607.

[88] Hale, J.K. and H.M. Rodrigues, Bifurcation in the Duffing equation with independent parameters. I. Proc. Roy. Soc. Edinburgh 77A (1977) 57-65; II. Proc. Roy. Soc. Edinburgh 79A (1978) 317-326.

[89] Hale, J.K. and P.Z. Taboas, Interaction of damping and forcing in a second order equation. Nonlin. Anal., Th., Meth., Appl. 2 (1978) 77-84.

[90] Hale, J.K. and P.Z. Taboas, Bifurcation near degenerate families. Applicable Anal. 11 (1980) 21-37.

[91] Hale, J.K. and J. de Oliveira, Hopf bifurcation for functional equations. J. Math. Anal. Appl. 74 (1980) 41-59.

[92] Hamermesh, M., Group theory and its applications to physical problems. Addison-Wesley, Reading, Mass., 1962.

[93] Henry, D., Geometric theory of semilinear parabolic equations. Lecture Notes in Math. 840, Springer-Verlag, Berlin, 1981.

[94] Hildebrandt, T.H. and L.M. Graves, Implicit functions and their differentials in general analysis. Trans. Amer. Math. Soc. 29 (1927) 127-153.

[95] Hirsch, M. and S. Smale, Differential equations, dynamical systems and linear algebra. Academic Press, N.Y., 1974.

[96] Holmes, P.J. and D.A. Rand, The bifurcation of Duffing's equation : an application of catastrophe theory. J. Sound Vibr. 44 (1976) 237-253.

[97] Holmes, P.J. and D.A. Rand, Bifurcations of the forced van der Pol oscillator. Quart. Appl. Math. 35 (1978) 495-509.

[98] Hopf, E., Abzweigung einer periodischer Losung von einer stationären Losung eines Differentialsystems. Ber. Math. Phys. Sachsische Akad. der Wissenschaften, Leipzig 94 (1942) 1-22.

[99] Husain, T., Introduction to topological groups. W.B. Saunders, Philadelphia, 1966.

[100] Iooss, G., Existence et stabilité de la solution périodique secondaire intervenant dans les problèmes d'évolution du type Navier-Stokes. Arch. Rat. Mech. Anal. 47 (1972) 301-329.

[101] Iooss, G., Bifurcation et stabilité. Lecture Notes, Université de Paris XI, 1973.

[102] Iudovich, V.I., The onset of auto-oscillations in a fluid. J. Appl. Math. Mech. (P.M.M.) 35 (1971) 638-655.

[103] Ize, J., Bifurcation theory for Fredholm operators. Memoirs of the A.M.S. 174, 1976.

[104] Ize, J., Periodic solutions of nonlinear parabolic equations. Comm. Partial Diff. Eqns. 4 (1979) 1299-1387.

[105] Joseph, D. and D.H. Sattinger, Bifurcating time periodic solutions and their stability. Arch. Rat. Mech. Anal. 45 (1972) 75-109.

[106] Jost, R. and E. Zehnder, A generalization of the Hopf bifurcation theorem. Helv. Phys. Acta 45 (1972) 258-276.

[107] von Kármán, Th., Festigkeitsproblem in Machinenbau. Enz. d. Math. Wiss. IV-4 (1910) 348-352.

[108] Kato, T., Perturbation theory for linear operators. Springer-Verlag, Berlin, 1966.

[109] Keener, J.P., Buckling imperfection sensitivity of columns and spherical caps. Quart. Appl. Math. 32 (1974) 173-188.

[110] Keener, J.P. and H.B. Keller, Perturbed bifurcation and buckling of circular plates. Lecture Notes in Math. 280, Springer-Verlag, Berlin, 1972, p. 286-293.

[111] Keener, J.P. and H.B. Keller, Perturbed bifurcation theory. Arch. Rat. Mech. Anal. 50 (1973), 159-175.

[112] Keller, H.B., Nonlinear bifurcation. J. Diff. Eqns. 7 (1970) 417-434.

[113] Keller, H.B., J.B. Keller and E.L. Reiss, Buckled states of circular plates. Quart. Appl. Math. 20 (1962) 55-65.

[114] Keller, J.B. and S. Antman, Bifurcation theory and nonlinear eigenvalue problems. Benjamin, N.Y., 1969.

[115] Kellog, O.D., Foundations of potential theory. Springer-Verlag, Berlin, 1967.

[116] Kielhöfer, H., Stability and semilinear evolution equations in Hilbert space. Arch. Rat. Mech. Anal. 57 (1974) 150-165.

[117] Kielhöfer, H., Hopf bifurcation at multiple eigenvalues. Arch. Rat. Mech. Anal. 69 (1979) 53-83.

[118] Kielhöfer, H., Generalized Hopf bifurcation in Hilbert space. Math. Methods Appl. Sci. 1 (1979) 498-513.

[119] Kielhöfer, H., Degenerate bifurcation at simple eigenvalues and stability of bifurcating solutions. J. Funct. Anal. 38 (1980) 416-441.

[120] Kirchgässner, K. and H. Kielhöfer, Stability and bifurcation in fluid dynamics. Rocky Mountain J. of Math. 3 (1973) 275-318.

[121] Knightly, G.H., An existence theorem for the von Kårmån equations. Arch. Rat. Mech. Anal. 27 (1967) 233-242.

[122] Knightly, G.H., Some mathematical problems from plate and shell theory. In [33], p. 245-268.

[123] Knightly, G.H. and D. Sather, On nonuniqueness of solutions of the von Kårmån equations. Arch. Rat. Mech. Anal. 36 (1970) 65-78.

[124] Knightly, G.H. and D. Sather, Nonlinear buckled states of rectangular plates. Arch. Rat. Mech. Anal. 54 (1974) 356-372.

[125] Knightly, G.H. and D. Sather, Nonlinear axisymmetric buckled states of shallow spherical caps. SIAM J. Math. Anal. 6 (1975) 913-924.

[126] Knightly, G.H. and D. Sather, Existence and stability of axisymmetric buckled states of spherical shells. Arch. Rat. Mech. Anal. 63 (1977) 305-319.

[127] Knightly, G.H. and D. Sather, Nonlinear buckling of cylindrical panels. SIAM J. Math. Anal. 10 (1979) 389-403.

[128] Knobloch, H.W., Hopf bifurcation via integral manifolds. VII Int. Konf. über Nichtlineare Schwingungen, Band I.1, Abh. AdW der DDR, 3, 1977, 413-429.

[129] Knobloch, H.W. and F. Kappel, Gewöhnliche Differentialgleichungen. B.G. Teubner, Stuttgart, 1974.

[130] Knops, R.J. (Ed.), Nonlinear analsyis and mechanics : Heriot-Watt Symposium. Vol. 1. Pitman, London, 1977.

[131] Koiter, W.T., On the nonlinear theory of thin elastic shells. I. Proc. Kon. Ned. Akad. Wet. 69 (1966) 1-54.

[132] Kopell, N. and L.N. Howard, Bifurcations under nongeneric conditions. Adv. in Math. 13 (1974) 274-283.

[133] Krasnosel'skii, M.A., Positive solutions of operator equations. Noordhoff, Groningen (The Netherlands), 1964.

[134] Krasnosel'skii, M.A., Topological methods in the theory of nonlinear integral equations. Pergamon Press, Oxford - London, 1964.

[135] Krein, S.G., Linear differential equations in Banach space. A.M.S. Translations of Math. Monographs, Vol. 29, Providence, 1971.

[136] Krein, M.F. and M.A. Rutman, Linear operators leaving invariant a cone in a Banach space. A.M.S. Transl. Ser. 1, 10 (1962) 1-128.

[137] Ladas, G.E. and V. Lakshmikantham, Differential equations in abstract spaces. Academic Press, N.Y., 1972.

[138] Ladyzhenskaya, O.A. and N.N. Ural'tseva, Linear and quasi-linear elliptic equations. Academic Press, N.Y., 1968.

[139] Landau, L.D. and M.L. Lifschitz, Theory of elasticity. Pergamon Press, London, 1959.

[140] Lang, S., Real analysis. Addison-Wesley, Reading, Mass., 1969.

[141] Lang, S., Differential manifolds. Addison-Wesley, Reading, Mass.,1972.

[142] Leipholz, H., Theory of elasticity. Noordhoff, 1974.

[143] Liapunov, A.M., Sur les figures d'équilibre peu différentes des ellipsoids d'une masse liquide homogène douée d'un mouvement de rotation. Zap. Akad. Nauk. St. Petersburg 1 (1906) 1-225.

[144] Lima, P., Hopf bifurcation in equations with infinite delays. Ph.D. Thesis, Brown University, 1977.

[145] Lions, J.L. and E. Magenes, Problèmes aux limites non homogènes et applications. Vol. 1 & 2, Dunod, Paris, 1968.

[146] Lions, J.L., Quelques méthodes de résolution des problèmes aux limites non linéaires. Dunod, Paris, 1969.

[147] List, S., Generic bifurcation with application to the von Kármán equations. Ph.D. Thesis, Brown University, 1976. J. Diff. Eqns. 30 (1978) 89-118.

[148] Loginov, B.V. and V.A. Trenogin, The use of group properties to determine multi-parameter families of solutions of nonlinear equations. Math. USSR Sbornik 14 (1971) 438-452.

[149] Loginov, B.V. and V.A. Trenogin, On the application of continuous groups in the theory of branching. Soviet Math. Dokl. 12 (1971) 404-407.

[150] Lu, Y.-C., Singularity theory and an introduction to catastrophe theory. Springer-Verlag, Berlin, 1976.

[151] Magnus, R.J., On the local structure of the zero set of a Banach space valued mapping. J. Funct. Anal. 22 (1976) 58-72.

[152] Magnus, R.J. and T. Poston, On the full unfolding of the von Kármán equations at a double eigenvalue. Math. Report 109, Battelle,

Geneva, 1977.

[153] Mallet-Paret, J., Buckling of cylindrical shells with small curvature. Quart. Appl. Math. 35 (1977) 383-400.

[154] Marsden, J., The Hopf bifurcation for nonlinear semigroups. Bull. A.M.S. 79 (1973) 537-541.

[155] Marsden, J.E., Qualitative methods in bifurcation theory. Bull. A.M.S. 84 (1978) 1125-1148.

[156] Marsden, J.E. and M.F. McCracken, The Hopf bifurcation and its applications. Springer-Verlag, N.Y., 1976.

[157] Martin, R.H., Nonlinear operators and differential equations in Banach spaces. Wiley-Interscience, 1976.

[158] Matkowsky, B.J. and L.J. Putnick, Multiple buckled states of rectangular plates. Int. J. Nonlin. Mech. 9 (1974) 89-103.

[159] Matkowsky, B.J. and E.L. Reiss, Singular perturbations of bifurcations. SIAM J. Appl. Math. 33 (1977) 230-255.

[160] Mawhin, J., Degré topologique et solutions périodiques des systèmes différentielles non linéaires. Bull. Soc. Roy. Sci. Liège 38 (1969) 308-398.

[161] Mawhin, J., Nonlinear perturbations of Fredholm mappings in normed spaces and applications to differential equations. Lecture Notes Univ. de Brasilia, 1974.

[162] McLeod, J.B. and D.H. Sattinger, Loss of stability and bifurcation at a double eigenvalue. J. Funct. Anal. 14 (1973) 62-84.

[163] Meire, R. and A. Vanderbauwhede, A useful result for certain linear periodic ordinary differential equations. J. Comp. Appl. Math. 5 (1979) 59-61.

[164] Michor, P., The division theorem on Banach spaces. Preprint 1978.

[165] Mikhlin, S.G., Mathematical physics, an advanced course. North-Holland, Amsterdam, 1970.

[166] Miller, W., Symmetry groups and their representations. Academic Press, N.Y., 1972.

[167] Miranda, C., Partial differential equations of elliptic type. Springer-Verlag, Berlin, 1970.

[168] Muller, C., Spherical harmonics. Lecture Notes in Math. 17, Springer-Verlag, 1966.

[169] Nirenberg, L., Topics in nonlinear functional analysis. N.Y.U. Lecture Notes, 1973-74.

[170] de Oliveira, J., Hopf bifurcation for functional differential equations. Nonlin. Anal. 4 (1980), 217-229.

[171] Palmer, K.J., Topological equivalence and the Hopf bifurcation. J. Math. Anal. Appl. 66 (1978) 586-598.

[172] Pazy, A., Semi-groups of linear operators and applications to partial differential equations. Lecture Notes, Dept. of Math., Univ. of Maryland, 1974.

[173] Pimbley, G.H., Eigenfunction branches of nonlinear operators and their bifurcation. Lecture Notes in Math. 104, Springer-Verlag, Berlin, 1969.

[174] Poenaru,V., Singularities C^{∞} en présence de symétrie. Lectures Notes in Math. 510, Springer-Verlag, Berlin, 1976.

[175] Poore, A.B., On the theory and application of the Hopf-Friedrichs bifurcation theory. Arch. Rat. Mech. Anal. 60 (1976) 371-393.

[176] Poston, T. and I. Stewart, Catastrophe theory and its applications. Pitman, London, 1978.

[177] Rabinowitz, P.H., Some global results for nonlinear eigenvalue problems. J. Funct. Anal. 7 (1971) 487-513.

[178] Rabinowitz, P.H., A global theorem for nonlinear eigenvalue problems and applications. In [239], p. 11-36.

[179] Rabinowitz, P.H., Some aspects of nonlinear eigenvalue problems. Rocky Mountian J. Math. 3 (1973) 161-202.

[180] Rabinowitz, P.H., A survey of bifurcation theory. In [32], p. 83-96.

[181] Reiss, E.L., Bifurcation buckling of spherical caps. Comm. Pure Appl. Math. 18 (1965) 65-82.

[182] Rodrigues, H.M. and A.L. Vanderbauwhede, Symmetric perturbations of nonlinear equations : symmetry of small solutions. Nonlin. Anal., Th., Meth. and Appl. 2 (1978) 27-46.

[183] Rouche, N. and J. Mawhin, Equations différentielles ordinaires. Tome 1 : théorie générale. Masson, Paris, 1973.

[184] Rudin, W., Functional analysis. McGraw-Hill, N.Y., 1973.

[185] Ruelle, D., Bifurcations in the presence of a symmetry group. Arch. Rat. Mech. Anal. 51 (1973) 136-152.

[186] Ruelle, D. and F. Takens, On the nature of turbulence. Comm. Math. Phys. 20 (1971) 167-192.

[187] Sather, D., Branching of solutions of an equation in Hilbert space. Arch. Rat. Mech. Anal. 36 (1970) 47-64.

[188] Sather, D., Nonlinear gradient operators and the method of Liapunov-Schmidt. Arch. Rat. Mech. Anal. 43 (1971) 222-244.

[189] Sather, D., Branching of solutions of nonlinear equations. Rocky Mountain J. Math. 3 (1973) 203-250.

[190] Sather, D., Branching and stability for nonlinear shells. Lecture Notes in Math. 503, Springer-Verlag, Berlin, 1976, p.462-473.

[191] Sather, D., Bifurcation and stability for a class of shells. Arch. Rat. Mech. Anal. 63 (1977) 295-304.

[192] Sattinger, D.H., Bifurcation of periodic solutions of the Navier-Stokes equations. Arch. Rat. Mech. Anal. 41 (1971) 66-80.

[193] Sattinger, D.H., Topics in stability and bifurcation theory. Lecture Notes in Math. 309, Springer-Verlag, Berlin, 1973.

[194] Sattinger, D.H., Transformation groups and bifurcation at multiple eigenvalues. Bull. A.M.S. 79 (1973) 709-711.

[195] Sattinger, D.H., Group representation theory and branch points of nonlinear functional equations. SIAM J. Math. Anal. 8 (1977) 179-201.

[196] Sattinger, D.H., Group representation theory, bifurcation theory and pattern formation. J. Functional Anal. 28 (1978) 58-101.

[197] Sattinger, D.H., Bifurcation from rotationally invariant states. J. Math. Phys. 19 (1978) 1720-1732.

[198] Schechter, M., Principles of functional analysis. Acad. Press, N.Y., 1971.

[199] Schechter, M., Modern methods in partial differential equations. An introduction. McGraw-Hill, N.Y., 1977.

[200] Schmitt, B.V. and S. Mazzanti, Solutions périodiques symétriques de l'équation de Duffing sans dissipation. J. Diff. Eqns. 42 (1981) 199-214.

[201] Schmidt, D.S., Hopf's bifurcation theorem and the center theorem of Liapunov with resonance cases. J. Math. Anal. Appl. 63 (1978) 354-370.

[202] Schmidt, E., Zur Theorie der linearen und nichlinearen Integralgleichungen. Math. Ann. 65 (1908) 370-399.

[203] Schwartz, J.T., Nonlinear functional analysis. Gordon & Breach, N.Y., 1969.

[204] Séminaire équations aux dérivées partielles non linéaires. Série d'exposés sur : Bifurcation et applications Fredholm. Publ. Math. d'Orsay, no 77-76, 1976.

[205] Shearer, M., Small solutions of a non-linear equation in Banach space for a degenerate case. Proc. Roy. Soc. Edinburgh 79A (1977) 35-49.

[206] Shearer, M., Bifurcation of axisymmetric buckled states of a thin spherical shell. Nonlin. Anal., Th. Meth. Appl. 4 (1980) 699-713.

[207] Showalter, R.E., Hilbert space methods for partial differential equations. Pitman London, 1977.

[208] Sotomayor, J., Generic one-parameter families of vector fields on two-dimensional manifolds. Publ. Math. I.H.E.S. 43 (1974) 5-46.

[209] Stakgold, I., Branching of solutions of nonlinear equations. SIAM Rev. 13 (1971) 289-332.

[210] Stokes, A., Invariant subspaces, symmetry and alternative problems. Bull. Inst. Math. Acad. Sinica 3 (1975) 7-14.

[211] Stuart, C.A., Three fundamental theorems on bifurcation. Lecture Notes Ecole Polytechnique Fédérale de Lausanne, 1977.

[212] Takens, F., Singularities of functions and vector fields. Nieuw Arch. Wisk. 20 (1972) 107-130.

[213] Takens, F., Unfolding of certain singularities of vector fields : generalized Hopf bifurcations. J. Diff. Eqns. 14 (1973) 476-493.

[214] Takens, F., Singularities of vector fields. Publ. Math. I.H.E.S. 43 (1974) 47-100.

[215] Taylor, A.E., Introduction to functional analysis. Wiley, N.Y., 1958.

[216] Teman, R. (Ed.), Turbulence and Navier-Stokes equation (Orsay 1975). Lecture Notes in Math. 565, Springer-Verlag, Berlin, 1976.

[217] Teman, R., Navier-Stokes equations. North-Holland, Amsterdam, 1977.

[218] Temme, N.M. (Ed.), Nonlinear analysis, I and II. Mathematisch Centrum Amsterdam, 1976.

[219] Thom, R., Topological methods in biology. Topology 8 (1968) 313-335.

[220] Thom, R., Structural stability and morphogenesis. Benjamin, Reading, Mass., 1975.

[221] Thompson, J.M.T. and G.W. Hunt, A general theory of elastic stability. Wiley, London, 1973.

[222] Turner, R.E.L., Transversality in nonlinear eigenvalue problems. In [239], p. 37-68.

[223] Vainberg, M.M. and V.A. Trenogin, The methods of Liapunov and Schmidt in the theory of nonlinear equations and their further development. Russian Math. Surveys 17 (1962) 1-60.

[224] Vainberg, M.M. and V.A. Trenogin, Theory of branching of solutions of nonlinear equations. Noordhoff, 1974.

[225] Vanderbauwhede, A., Alternative problems and symmetry. J. Math. Anal. Appl. 62 (1978) 483-494.

[226] Vanderbauwhede, A., Alternative problems and invariant subspaces. J. Math. Anal. Appl. 63 (1978) 1-8.

[227] Vanderbauwhede, A., Some remarks on symmetry and bifurcation. Simon Stevin 51 (1978) 173-194.

[228] Vanderbauwhede, A., Generic and nongeneric bifurcation for the von Kårmån equations. J. Math. Anal. Appl. 66 (1978) 550-573.

[229] Vanderbauwhede, A., Generic bifurcation and symmetry, with an application to the von Kårmån equations. Proc. Roy. Soc. Edinburgh 81A (1978) 211-235.

[230] Vanderbauwhede, A., Bifurcation near multi-parameter families of solutions. Proc. Equadiff 78 Conference, Firenze, 1978, p. 9-23.

[231] Vanderbauwhede, A., Symmetry and bifurcation near families of solutions. J. Diff. Eqns. 36 (1980) 173-187.

[232] Vanderbauwhede, A., An abstract setting for the study of the Hopf bifurcation. Nonlin. Anal., Th. Meth. Appl. 4 (1980) 547-566.

[233] Wasserman, G., Stability of unfoldings. Lecture Notes in Math. 393, Springer-Verlag, Berlin, 1974.

[234] Weinberger, H.F., Variational methods for eigenvalue approximation. Reg. Conf. Ser. in Appl. Math., vol. 15, SIAM, Philadelphia, 1974.

[235] Westreich, D., Banach space bifurcation theory. Trans. Amer. Math. Soc. 171 (1972) 135-156.

[236] Wolkowisky, J.H., Existence of buckled states of circular plates. Comm. Pure Appl. Math. 20 (1967) 549-560.

[237] Yosida, K., Functional analysis. Springer-Verlag, Berlin, 1965.

[238] Zamansky, M., Linear algebra and analysis. Van Nostrand, London, 1969.

[239] Zarantonello, E.H. (Ed.), Contributions to nonlinear functional analysis. Academic Press, N.Y., 1971.

[240] Zeidler, E., Vorlesungen über nichtlineare Funktionalanalyse, I,II. Teubner, Leipzig, 1976.

[241] Amann, H., N. Bazley and K. Kirchgässner (Eds.), Applications of Nonlinear Analysis in the Physical Sciences, Surveys and Reference Works in Mathematics, Vol. 6, Pitman (London), 1981

[242] Andronov, A.A., Leontovich, E.A., Gordon, I.I. and Mayer, A.G., Theory of bifurcations of dynamics systems on a plane. Israel Program for Scientific Translations, Jeruzalem, 1971.

[243] Bourbaki, N., Vol. XXXII, Variétés différentielles et analytiques. Ed. Hermann, Paris, 1967.

[244] Bredon, G., Introduction to compact transformation groups. Academic Press, New York, 1972.

[245] Chow, S.N. and Hale, J., Methods of bifurcation theory. Grundlehren der mathematischen Wissenschaften, Band 251, Springer-Verlag, Berlin, 1982.

[246] Ciarlet, P.G. and Rabier, P., Les équations de von Kármán. Lecture Notes in Mathematics, Vol. 826, Springer-Verlag, Berlin, 1980.

[247] Dancer, E.N., An implicit function theorem with symmetries and its application to nonlinear eigenvalue problems. Bull. Austral. Math. Soc. 21 (1980) 81-91.

[248] Dancer, E.N., The G-invariant implicit function theorem in infinite dimensions. Preprint 1981.

[249] Dieudonné, J., Foundations of modern analysis. Academic Press, New York, 1960.

[250] Golubitsky, M. and Langford, W.F., Classification and unfoldings of degenerate Hopf bifurcations. J. Diff. Eqns. 41 (1981) 375-415.

[251] Golubitsky, M. and Schaeffer, D., Bifurcation with O(3)-symmetry including applications to the Bénard problem. Preprint 1980.

[252] Hale, J.K., Topics in dynamic bifurcation theory. CBMS Regional conference series in mathematics, Vol. 47, Am. Math. Soc., Providence, 1981.

[253] Hassard, B.D., Kazarinoff, N.D. and Wan, Y.-H., Theory and applications of Hopf bifurcation. London Math. Soc. Lecture Note Series, Vol. 41, Cambridge Univ. Press, Cambridge, 1981.

[254] Iooss, G., Bifurcation of maps and applications. Mathematics Studies, Vol. 36, North-Holland, Amsterdam, 1979.

[255] Iooss, G. and Joseph, D.D., Elementary stability and bifurcation theory. Springer-Verlag, Berlin, 1980.

[256] Kirchgraber, U., A note on Liapunov's center theorem. J. Math. Anal. Appl. 73 (1980) 568-570.

[257] Köthe, G., Topological vector spaces, I, Springer-Verlag, Berlin, 1969.

[258] Lewis, D.C., Autosynartetic solutions of differential equations. Amer. J. Math. 83 (1961) 1-32.

[259] Lewis, D.C. Some general methods for detecting the existence of periodic solutions. Studies in Appl. Math., Vol. 5, 156-158.

[260] Loud, W.S. and Vanderbauwhede, A., Functional dependence and boundary value problems with families of solutions. Preprint 1981.

[261] Martinet, J., Singularités des fonctions et applications différen-

tiables. Monografias de matematica da PUC, Vol. 1, Rio de Janeiro, 1977.

[262] Nashed, Z. (Ed.), Generalized inverses and applications. Academic Press, New York, 1976.

[263] Palmer, K.J., Linearization of reversible systems. J. Math. Anal. Appl. 60 (1977) 794-808.

[264] Rabinowitz, P.H. (Ed.), Applications of bifurcation theory. Academic Press, New York, 1977.

[265] Sattinger, D.H., Group theoretic methods in bifurcation theory. Lecture Notes in Math., Vol. 762, Springer-Verlag, Berlin, 1979.

[266] Sattinger, D.H., Bifurcation and symmetry breaking in applied mathematics. Bull. A.M.S. 3 (1980) 779-819.

[267] Sattinger, D.H., Spontaneous symmetry breaking : mathematical methods, applications and problems. In [241], p. 3-23.

[268] Vanderbauwhede, A., Hopf bifurcation for abstract nonlinear equations. In : R.J. Knops (Ed.), Trends in Applications of Pure Mathematics to Mechanics, Vol. III, Pitman, London, 1981, 218-234.

[269] Vanderbauwhede, A., Symmetry and bifurcation from multiple eigenvalues. In : W.N. Everitt & B.D. Sleeman (Eds.), Ordinary and Partial Differential Equations. Lecture Notes in Math., Vol. 846, 1981, 356-365.

[270] Vanderbauwhede, A., Bifurcation at double eigenvalues for a class of symmetric Dirichlet problems. Bull. Soc. Mat. Belgique. To appear.

[271] Vanderbauwhede, A., Families of periodic solutions for autonomous systems. Proceedings of the International Symposium on Dynamical Systems, Gainesville, 1981. To appear.

[272] Vanderbauwhede, A., Bifurcation for symmetric nonlinear boundary value problems. Proceedings of the Conference on Differential Equations and Applications, Graz, 1981. To appear.

[273] Vanderbauwhede, A., Note on symmetry and bifurcation near families of solutions. J. Diff. Eqns. To appear.

Index